T0257447

Crop Production: Technology and Methodology

Crop Production: Technology and Methodology

Edited by **Corey Aiken**

New York

Published by Callisto Reference,
106 Park Avenue, Suite 200,
New York, NY 10016, USA
www.callistoreference.com

Crop Production: Technology and Methodology
Edited by Corey Aiken

International Standard Book Number: 978-1-63239-134-6 (Hardback)

Contents

Preface

This book presents a comprehensive study of Crop Production, elucidating its technology as well as methodology. Crop production depends on the proper implementation of soil, water and nutrient management technologies. Food production must be increased by around 50 percent by 2020 to satisfy the needs of the then population of nearly 8 billion. Much of the increase would have to come from amplification of agricultural production. Significance of wise consumption of water, nutrient regulation and tillage in the agricultural sector for sustainable agricultural growth and regulating environmental degradation calls for immediate attention of researchers, planners, and policy makers. Crop models enable researchers to swiftly investigate on the long-term consequences of changes in agricultural practices. This book presents an interdisciplinary approach and contributes to this new vision. Leading authors have analyzed topics related to crop production technologies in this book. Emphasis has been laid on general descriptions and principles of each topic, technical details, original research work, and modeling aspects. This diverse subject has been presented using different topics for convenience of the readers.

Significant researches are present in this book. Intensive efforts have been employed by authors to make this book an outstanding discourse. This book contains the enlightening chapters which have been written on the basis of significant researches done by the experts.

Finally, I would also like to thank all the members involved in this book for being a team and meeting all the deadlines for the submission of their respective works. I would also like to thank my friends and family for being supportive in my efforts.

Editor

Part 1

Crop Modeling

Crop Models as Decision Support Systems in Crop Production

Simone Graeff, Johanna Link, Jochen Binder and Wilhelm Claupein
University Hohenheim, Crop Science (340a)
Germany

1. Introduction

The current challenges crop production faces in the context of required yield increases while reducing fertilizer, water and pesticide inputs have created an increasing demand for agronomic knowledge and enhanced decision support guidelines, which are difficult to obtain on spatial scales appropriate for use in a multitude of global cropping systems.

Nowadays crop models are increasingly being used to improve cropping techniques and cropping systems (Uehera and Tsuji, 1993; Penning de Vries and Teng, 1993; Boote et al., 1996). This trend results from a combination of mechanistic models designed by crop physiologists, soil scientists and meteorologists, and a growing awareness of the inadequacies of field experiments for responding to challenges like climate change. A general management decision to be made underlies the principle that a crop response to a certain input factor can only be expected if there is a physiological requirement and if other essential plant growth factors are in an optimum state. Hence, the challenge for a farmer is to determine how to use information with respect to the management decisions he has to make, in other words he has to find an efficient, relevant and accurate way how to evaluate data for specific management decisions. Crop models enable researchers to speculate on the long-term consequences of changes in agricultural practices and cropping systems on the level of an agro-ecosystem. Finally, models make it possible to identify very rapidly the adaptations required to enable cropping systems to respond to changes in the economic or regulatory context (Rossing et al., 1997).

The following chapter gives an overview on the current knowledge and use of crop models and addresses the problems associated with these methods. In a second part the use of crop growth models for decision support in terms of yield variability, fertilizer and irrigation strategies will be discussed in the context of two global case studies, one in China and the other one in Germany. The discussion focuses on the currently available modeling techniques and addresses the necessary future research areas in this context.

2. Decisions and uncertainty

Scientists have realized that farmers nowadays have multiple technologies like sensors, satellites etc. available to gather a myriad of data on weather, soil parameters, crop development and growth. The collected data is attributed to find the right management decision in the context of e.g. amount and timing of inputs, sowing and harvest date etc.

However, the collected data has opened a "Pandora's box" of uncertainty. The type of information gained, increases the level of uncertainty, because it emphasizes the apparent disorder, or in a physical sense the entropy, as is does not deliver the management decision itself, but rather provides information. The relationship between the three related attributes of uncertainty, information and entropy can be illustrated by different levels of uncertainty. In the first level of uncertainty, e.g. a single soil texture map represents minimal information but low entropy. In a second level representing e.g. multiple layers of information from a given field like yield, soil water and soil nitrogen (N), more information is presented, but it is of high entropy. The uncertainty of decision making is now realized and high. Finally, the third level reflects a decision support system, where the information content is high, so entropy is low leading to a reduced level of uncertainty in the taken management decision.

The current situation of cropping systems is represented in level two, where a lot of information is available, but it cannot be put in the right place yet. The challenge facing scientists now is to help farmers to understand this information. The farmer has to be enabled to use the information to manage his cropping system in a way that matches the underlying limitations. Thus, the optimum order will have minimum entropy and maximum information and a low level of uncertainty associated with the management decision. To capitalize on these features, however, the decision maker must have a more accurate assessment of the likely outcomes of action, thereby improving the likelihood of benefit to a degree, which justifies the additional effort required (Adams et al., 2000). Failure to do so has the consequence that uncertainty of outcomes remains to high, to provide reasonable benefit.

However, much of the uncertainty in management decisions cannot be removed in an easy way, because the outcome like crop yield is often timely separated from the decision like timing and amount of nitrogen fertilizer. Because of this, the outcome or consequence of a given action has to be predicted. Crop models can play a major role in trying to minimize the uncertainty associated with certain management actions as they integrate and consider multiple factors for the decision making process. The following section will focus on the different types of models available and their appropriateness for crop management.

3. Crop models

Model types are divided arbitrarily into the categories defined by Cook (1997) as 'conventional', 'intuitive', 'expert systems', 'deterministic' and 'stochastic'. The characteristics of the model types and their previous use in crop management have been discussed in Cook (1997) and Cook and Adams (1998).

Crop growth models have been used since the 1970s (Hoogenboom, 2003). The first crop growth models were based on approaches of simulating industrial processes (Forrester, 1961). Brouwer and De Wit (1968) and De Wit et al. (1970) developed some of the early crop growth models in a program called BACROS. The main aim of their modeling activities was to understand the underlying processes at the plant scale (Van Ittersum et al., 2003). While all models have achieved various degrees of success in application, they all have their weakness and fail under certain circumstances, wherefore authors of models should clarify the limitations of their models and ranges of applications (Ma and Schaffer, 2001).

In the past decade the dynamics of crop growth models has made substantial progress (Gerdes, 1993) and many crop models are available on the market. Many models exist for predicting how crops respond to climate, nutrients, water, light, and other conditions. One of the most widely used modeling systems across the world is the DSSAT model (Decision Support System for Agrotechnology Transfer). It was initially developed under the auspices of the International Benchmark Sites Network for Agrotechnology Transfer (Hoogenboom, 2003). Currently, the DSSAT shell is able to incorporate models of 27 different crops, including several cereal grains, grain legumes, and root crops (Hoogenboom, 2003). The models are process-oriented and are designed to work independent of location, season, crop cultivar, and management system. The models simulate the effects of weather, soil water, genotype, and soil and crop N dynamics on crop growth and yield (Jones et al., 2003). The models predict daily plant growth based on daily weather data and soil, management and genetic information. Growth is computed based on light interception and the daily photosynthesis, which can be reduced by temperature, water and N stress. Carbohydrate fixed by photosynthesis is then partitioned to plant components based on crop growth stage, and stress. Thus, the model is able to integrate daily effects of temporal stress on growth and yield. Information that is entered into the system, to be combined with yield potential, will include known variables such as soil types, soil depths, N leaching properties, soil nutrient status, information on pests, disease and weed populations, etc. Information on weather records is a further valuable source of information. Political factors can be taken into account, such as available subsidies or current legislation governing maximum rates of N application. Consideration can also be given to environmental issues, and the farmer's local knowledge and farm records may also be an essential input into the system. The farmer can test his preferred strategy, which could either be to increase the input (aiming at maximum yield) or to reduce inputs (aiming at reducing environmental pollution) to optimize gross margins.

With the aid of a crop model, a more detailed analysis of the management decisions and the possible effects on final yield can be undertaken. It also should be acknowledged that uncertainty exists in the final yield estimate as a result of uncertainty in the input data and errors in the models. Chen et al. (1997) used first order uncertainty analysis to examine the effect of uncertainty in input data on the model outcome of a mechanistic decision support system. They reported large uncertainty, which was contributed mostly to the given variability in specific model parameters. Therefore, the model output has to be critically assessed, which is mostly done based on regressions. The most commonly used criterion of model performance is the coefficient of determination (R^2), which is the ratio of the variance explained by the model to the total variance in the data. A number of authors have concluded that R^2 is not a good means of comparison between models representing yield response (Cerrato and Blackmer, 1990). A primary consideration for the unsuitability of R^2 is the fact that it gives no indication of how well a model performs when applied to data that were not used to create the model, as it does only provide information about the trend and does not provide information about the deviation of measured and simulated values. Over-fitting of calibrations leading to poor performance of the model on test data is often the result. Based on these considerations, another used measure of model accuracy is the root mean square error (RMSE). A major advantage of using RMSE over R^2 for model evaluation is that RMSE provides information about both on the calibration data and on new data not used in developing the model to estimate the true predictive ability of the model

(Drummond et al., 2003). Cross-validation is another more robust, reliable method of measuring prediction accuracy (Stone, 1973) of crop models.

Based upon the total information entered into the model, it will offer an agronomic recommendation on how to vary the inputs, change the management or optimize the overall gross margin. A crop growth model could therefore be used as a decision aid for determining different yields based on factors such as varying plant populations or nitrogen rates, which could help a farmer decide when to plant or replant areas within a field based on plant population data and risk factors for various soil types or how to manage nitrogen application rate and dates. It would give the farmer the analytical ability to identify relationships between different variables within the field and to find a best fit scenario on a high level of complexity.

Furthermore, crop models offer the possibility to aggregate knowledge on and over different scales. Linking the models with a GIS offers a mechanism to integrate many scales of data developed in and for agricultural research. Data access and final management decision can be expanded to a decision support system, which uses a mix of process-oriented models and biophysical data at different temporal and spatial scales (e.g. growing season, climate characteristics, soils). Thus, a need exists for an integrated GIS system which combines the different available information (e.g. soil map, yield, weather, management) to allow agricultural producers as well as policy makers to know the impact of differences between input and output spatially from one place or region to another to improve management, productivity and profitability.

3.1 Model applications

Crop yield and occurring yield gaps and are two important criteria for sustainable land management. Considering various agro-environments, several factors clearly account for crop yield and occurring yield gaps. Analysis of yield gaps in crop production is facilitated by using the concept of production ecology where different sets of eco-physiological variables affecting crop growth and development are distinguished (Penning de Vries et al., 1989). The approach recognizes three sets of factors affecting crop growth and development. Growth and yield determinants include mainly 1) crop genetics, 2) abiotic resources (water and nutrients, as well meteorological variables like temperature and solar radiation) which limit crop growth and development when their supply is suboptimal over different periods in the growing season, 3) biotic factors (pests, diseases, weeds). Hence, overall crop productivity is the result of growth and yield determining, limiting and reducing factors. Process-based crop models are increasingly being used in assessing these yield determining, limiting and reducing factors for a particular area or region with given agro-environmental conditions and are therefore a valuable tool to analyze occurring yield gaps.

Process-based crop growth models are a promising tool to help identify relationships between yield-limiting factors, management and environment. Crop models such as the DSSAT or the APOLLO (Batchelor et al., 2004b) model can be used to identify spatial yield-limiting factors (Batchelor et al., 2004b; Jones et al., 2003). Both models are based on the CROPGRO (Boote et al., 1998) and CERES (Ritchie et al., 1998) family of process-oriented crop models. Based on information about management (i.e. cultivar, planting, fertilization, plant protection, harvest) and environmental conditions (soil, weather), those process-oriented crop growth models compute the daily rate of plant growth, resulting in an estimation of final yield and plant biomass. Therefore those models simulate the daily interaction of plant growth, water, nitrogen, and pest stress on plant growth processes.

Future sustainable land management requires information on yield trends not only over time but also over space, to assess whether yields are stable or increasing, or whether they are decreasing and thereby signaling possible failure in the future. Spatial yield variability is a complex interaction of many factors including water stress, rooting depth, soil and drainage properties, weather, pests, fertility, and management. Spatial yield variability can also be related to spatial variability in soil fertility (Finke and Goense, 1993), soil organic matter content (Kravchenko and Bullock, 2000), and pest attacks (Plant et al., 1999). The challenge for farmers is to identify the factors that they can control and manage, and make appropriate management decisions to increase profits. Research advancements in data collection have given farmers the tools and capabilities to effectively map their fields, record yield histories and vary inputs and management strategies in response to variations in soil and environmental factors in the field. Recently, process-oriented crop growth models such as CROPGRO-Soybean (Hoogenboom et al., 1994), CERES-Maize (Jones and Kiniry, 1986) and CERES-Wheat (Godwin et al., 1989) have been used to study causes of spatial yield variability. The results have shown that the models can accurately simulate corn and soybean spatial yield variability, taking into account yield-limiting factors such as water stress in soybeans (Paz et al., 1998), soybean cyst nematodes (Paz et al., 2001), water stress in corn (Fraisse et al., 1998) and interaction of corn population and water stress (Paz et al., 1999). These studies also demonstrated that crop models can play an important role in understanding the causes of spatial yield variability. They can be used as a tool to explore hypotheses related to crop yield variability (Paz et al., 1998) and identify areas in the field, where problems due to varying growing conditions occur (Link et al., 2007). The spatial component of such a crop model is going to discretize a field studied in space and time into a finite number of regular cells and within those cells the model inputs are considered as uniform. With the appropriate computational hardware it is possible to discretize the modeled field into smaller and smaller grids in an effort to improve the quality of the model predictions. In theory, higher resolution modeling is expected to yield better predictions because of better resolved model inputs (e.g. soil texture, nutrients). The use of smaller grid sizes undeniably improves the appearance of simulation results, but the question raises how small can be to small to model as the potential benefits of higher resolution modeling have to be weighed against the increased demands on inputs. A separate issue might be the spatial coverage of data points and observations available to evaluate model performance. Observational data might not be sufficient to prove the benefits of higher resolution modeling and the model performance might get worse at smaller scales. It is therefore important to examine the relevant grid size for the model application on hand. The uncertainties in the underlying yield limiting parameters might not justify a high grid resolution. The aim of the following case study was to evaluate the reasons for spatial variability of corn yields using the APOLLO model to test its performance under German conditions, and to test it on various grid scales.

3.2 Case study Germany - Procedure to evaluate corn (*Zea mays* L.) yields in the upper Rhine valley using a crop growth model

Spatial yield variability is a result of complex interactions among different yield-limiting factors, such as soil properties, nutrient and water availability, rooting depth, pests and management. In order to manage spatial yield variability within a field, yield-limiting factors must be identified and understood. Initial efforts to study yield variability have focused on taking static measurements of soil, management, or plant properties and regressing these values against grid level yields (Sudduth et al., 1996). Classical statistics based on ordinary least squares have frequently been used to explore functional relationships between crop productivity and controlling factors (Long, 1998). Tomer and Anderson (1995) used linear regression to predict spatial patterns in yield based on soil fertility. However, it is difficult to represent the temporal effects of time dependent interactive stresses (i.e. water stress) on crop growth and yield using classical statistical techniques.

Characterization of yield variability requires the analysis of both spatial and temporal behavior of soil, weather, management and environmental factors. Thus, extending the use of crop models to examine within-field spatial yield variability is an intriguing challenge. In few studies, the APOLLO model was used to analyze causes of spatial yield variability in soybean (Batchelor et al., 2004b). The APOLLO model is a precision agriculture decision support system designed to use the CROPGRO-Soybean and CERES-Maize models to analyze causes of yield variability and to estimate the economic and environmental consequences of prescriptions (Batchelor et al., 2004a). Techniques in APOLLO have never been tested outside of the United States. To date, the APOLLO model has only been used in large fields with grid sizes ranging from 0.055 – 0.2 ha.

The overall goal of this work was to use the APOLLO model to study the spatial yield variability of three fields in the upper Rhine Valley (Germany) and to determine if crop model calibration techniques developed in the United States could be transferred to small fields in Germany. The specific objectives of this study were (i) to develop and test different calibration strategies to minimize the error between simulated and measured spatial corn yield, and (ii) to evaluate the impact of grid size on the error of simulated spatial yield variability.

3.2.1 Site, treatments and yield monitoring

The study was conducted as an on-farm study from 1998 through the 2002 growing season on three fields (I1, I2, I3) in the Upper Rhine Valley near Weisweil (48° 19′ N, 7° 67′ E), northwest of Freiburg, Germany. The mean annual precipitation in this area is 910 mm, the mean temperature is about 9.5° C and the sum of the yearly solar radiation averages about 11390 kJ m^{-2}. The major soil type is a silty loam. The aggregated size of the three fields was approximately 5.5 ha in total. Corn was grown each year from April – October during the years 1998 – 2002 in all three fields, with exception of field I1, where wheat was grown in 1999. The corn cultivars varied for each field and year (Table 1).

Each field was managed uniformly using the producer's current management practices. At sowing, a starter fertilizer of Ø 31 kg N ha^{-1} was applied uniformly as KAS (13 % NH_4-N, 13 % NO_3-N) to all fields in all 5 years. In the years 2000-2002 around the 4th leaf stage soil samples at a depth of 0 – 30, 30 – 60 and 60 – 90 cm were taken at 30 data collection points, which were set up at a distance of 40 x 40 m, and analyzed for soil available nitrogen. Table 2 shows the values of soil available nitrogen (kg N ha^{-1}) in the upper 90 cm of the soil layer around the 4th leaf stage. Urea (46 % N) was applied uniformly to each field based on the results of soil available nitrogen around the 4th leaf stage. Rates varied for each field and year, and ranged from 44 – 120 kg N ha^{-1}, to give an average of 250 kg N ha^{-1} in each field. In 2001 swine manure was applied in field I2, which provided an additional 40 kg N ha^{-1} in this field. No other field received swine manure. Herbicides and pesticides were applied as needed to control pests. After harvest in September or October, the corn residue was left on the surface of each field.

Field		1998	1999	2000	2001	2002
I1	Cultivars	Helix	*	Marista	Benicia	Marista
	Maturity	K220		K400	K250	K400
I2	Cultivars	Marista	Helix	Marista	Peso	Marista
	Maturity	K400	K220	K400	K290	K400
I3	Cultivars	Helix	Helix	Benicia	Benicia	DK514
	Maturity	K220	K220	K250	K250	K400

* in 1999 on field I1 wheat was grown

Table 1. Cultivars planted on field I1, I2, I3 during the 5-yr period (1998-2002). K indicates the maturity classification based on BSA (1998).

Geo-referenced corn grain yield data were collected over the 5-year period using a differentially corrected global positioning system and a yield monitor mounted on a combine harvester (Lexion, Claas, Harsewinkel, Germany). Corn grain yield and corn grain moisture content were measured every 5 seconds (10-m distance), resulting in about 200 yield monitor data points per hectare. Erroneous yield monitor data with missing values for yield or grain moisture content, or yield values greater than 15000 kg ha^{-1} were excluded from the yield monitoring dataset. In this paper, yield was calculated as corn grain yield at 0 % moisture content.

Yield monitor data were studied in four different scenarios (case A – D) to evaluate the impact of grid size on the accuracy of simulated spatial yield variability. A grid network was established using grid sizes defined for case A (grids of 10.5 x 10.5 m = 0.011 ha), case B (grids of 16.5 x 16.5 m = 0.027 ha) and case C (grids of 22.5 or 30.5 x 50.5 m = 0.114 or 0.154 ha). In case C two different grid sizes were used to better match the field boundaries. The smaller grids (22.5 x 50.5 m) were placed in the turning rows, and the larger grids (30.5 x 50.5 m) were placed in the middle of the field. The grids were overlaid onto yield maps and the average yield for each grid was computed using a software tool, developed and described by Thorp et al. (2004). Each grid contained at least three yield monitor points. Case D used the same grid configuration as in case C. In addition measured soil available nitrogen in the upper soil layers (0 – 30, 30 – 60 and 60 – 90 cm) around the 4th leaf stage in the years 2001 and 2002 (Table 2) was used to adjust model state variables on the measurement date during the simulation run.

Field		2000	2001	2002
I1	Mean	106	59	118
	Range		24-139	92-185
I2	Mean	45	98	176
	Range		25-359	119-230
I3	Mean	53	49	87
	Range		18-63	73-107

Table 2. Cumulative soil available nitrogen (kg N ha^{-1}) in the upper soil layer (0 – 90 cm) around 4th leaf stage; average for all data collection points in field I1, I2 and I3.

3.2.2 Apollo (application of precision agriculture for field management optimization)

APOLLO was developed to assist users in evaluating causes of spatial yield variability and to develop optimum prescriptions for nitrogen, water and seeding management (Batchelor et al., 2004a). It has modules to assist the user in 1) calibrating spatial soil inputs to minimize error between simulated and measured yield, 2) validating the calibrated model for independent seasons, and 3) developing management prescriptions. Management, soil, weather and cultivar information are required as input files to run the model. The management file (*.mzx) contains model inputs including weather file name, soil composition, initial soil water, nitrate, and ammonia content, planting date, row spacing, and residue amount. The soil profile characteristics for each grid are stored in the soil input file (*.sol). This file contains information such as bulk density, saturated hydraulic conductivity, upper and lower drained limit and root growth factor. Daily weather data, including daily maximum and minimum temperature, rainfall and solar radiation were stored in the weather file (*.wth). All weather data were obtained at the nearest German Weather Service station located at Emmendingen-Mundingen and Freiburg, which are about 16 and 25 km from the trial site, respectively. The cultivar file (*.cul) contains cultivar coefficient, which give information about the rate of development and the required growing degree days (GDDs) for each genotype. Yield data for each grid in the field were stored in a separate yield file over the 5 year period (*.mza).

Data about soil properties available are often mean values over the whole field (or even bigger areas) and thus do not take the existing variability into account. However, when trying to simulate spatial yield variability at small spatial scales, it is necessary to adjust soil properties over their expected range in order to more accurately reflect spatial soil properties within the field. APOLLO allows the user to adjust up to 10 soil parameters (Table 3) for each grid. The user can test if one or a combination of these soil parameters may help explain the spatial yield variability. When a user selects the parameters to adjust, APOLLO uses a simulated annealing optimization algorithm to estimate the parameter values that minimize the root mean square error (RMSE) between simulated and measured yield over selected years in each grid selected by the user. Calibration of the APOLLO model results in a unique set of soil properties for each grid.

		Parameters	Unit	Minimum	Maximum	Initial
1	CN	SCS curve number		40	90	70
2	DR	Drainage rate	fraction day^{-1}	0.1	0.5	0.4
3	ETDR	Effective tile drainage rate	1 day^{-1}	0.01	0.25	0.05
4	SHC	Saturated hydraulic conductivity of deep impermeable layer	cm day^{-1}	0.001	2	0.01
5	HPF	Hardpan factor	0.0 - 1.0	0.01	1.0	0.5
6	DHP	Depth to the hard pan	cm	5	150	30
7	RDRF	Root distribution reduction factor		-0.1	-0.001	-0.05
8	NMF	Nitrogen mineralization factor	0.0 - 1.0	0.1	1.0	0.8
9	SFF	Soil fertility factor	0.0 - 1.0	0.7	1.0	0.99
10	ASW	(adjust) Available soil water	%	-20	20	0

Table 3. Soil parameters available for calibration in the APOLLO model.

In this study several genetic coefficients were adjusted to set the maximum yield of different cultivars. In the next step APOLLO was used to compute soil inputs to minimize error between simulated and measured yield for each grid size scenario (case A, B and C) and for the scenario, where measured soil available nitrogen (kg N ha⁻¹) at 4th leaf stage was used to adjust simulated soil available nitrogen in the model database (case D).

Two calibration strategies were applied to the data set:

- soil parameters were calibrated one at a time to determine which parameter appeared to have the greatest power to explain spatial yield variability.
- combinations of soil parameters identified before were calibrated to determine if combinations of soil parameters improved the simulation of spatial yield variability.

These calibration strategies were applied to different scenarios, and the effects of grid resolution on model accuracy were examined. The accuracy of the model was evaluated by the correlation coefficient R between simulated and measured yields and RMSE.

3.2.3 Calibration strategies

The results of model calibration using single soil parameters showed that the adjustment of single soil parameters resulted in a good fit between simulated and measured yield. The calibration of the five soil properties (HPF + DHP, RDRF, NMF, SFF and ASW) reduced RMSE between simulated and measured yield compared to the default values, and thus, partially explained spatial yield variability (Table 4). However, the adjustment of soil parameters SCS CN, DR, ETDR + SHC (described in Table 3) did not significantly reduce

					Parameters					
		HPF + DHP		RDRF		NMF		SFF		ASW
Scale	R	RMSE (kg ha⁻¹)	R	RMSE (kg ha⁻¹)	R	RMSE (kg ha⁻¹)	R	RMSE (kg ha⁻¹)	R	RMSE (kg ha⁻¹)
Field I1										
Case A	0.47	1441	0.40	1852	0.42	1821	0.46	1483	0.43	1484
Case B	0.83	848	0.75	1328	0.81	1183	0.90	742	0.81	949
Case C	0.88	738	0.76	1304	0.81	1189	0.92	656	0.82	919
Case D	0.96	514	0.96	539	0.96	531	0.95	569	0.96	541
Field I2										
Case A	0.55	1308	0.30	1312	0.05	1514	0.21	1109	0.09	1275
Case B	0.77	1062	0.52	1177	0.14	1476	0.36	1062	0.21	1251
Case C	0.80	1020	0.50	1182	0.08	1516	0.34	1055	0.18	1258
Case D	0.74	993	0.54	930	-0.04	1009	0.33	1005	0.17	923
Field I3										
Case A	0.27	1058	0.18	963	-0.14	459	-0.05	1073	-0.22	781
Case B	0.70	884	0.54	823	-0.02	434	0.18	1070	-0.06	802
Case C	0.81	672	0.43	892	-0.33	421	-0.03	1056	-0.31	764
Case D	0.83	922	0.66	1198	0.51	841	0.31	1115	0.25	1033

Table 4. Correlation coefficient R and RMSE for simulated and measured yield after model calibration (2000 iterations) of field I1, I2 and I3 using multiple years of corn yield data and single soil parameters (5-10). Yield was calculated in dependency of the grids in case A (10.5 x 10.5 m grid size), case B (16.5 x 16.5 m grid size), case C (22.5 or 30.5 x 50.5 m grid size) and case D (22.5 or 30.5 x 50.5 m grid size). R >0.50 is written in bold letters.

error between simulated and measured yield and thus, did not explain spatial yield variability. Thus, not all soil parameters available for calibration by APOLLO contributed to explaining the given spatial yield variability. Table 4 shows the correlation coefficient R and RMSE for simulated and measured yields after model calibration using single soil parameters HPF + DHP, RDRF, NMF, SFF and ASW for the four scenarios (cases A – D). Based on these results HDF+DHP seem to have the greatest effect on within field variability in these fields.

In the next step combinations of parameters found in the previous calibration strategy that appeared to partially explain spatial yield variability. Based on the previous results of calibrating single soil parameters, the parameters HPF + DHP, RDRF, NMF, SFF and ASW were selected for further model calibration and applied to the different scenarios (case A – D). Table 5 shows the correlation coefficient R and RMSE for simulated and measured yield after model calibration of field I1, I2 and I3 using multiple soil parameters.

	Parameters							
	HPF + DHP + ASW		HPF + DHP + RDRF + NMF + SFF		HPF + DHP + RDRF + SFF + ASW		HPF + DHP + RDRF + NMF + SFF + ASW	
Scale	R	RMSE (kg ha^{-1})	R	RMSE (kg ha^{-1})	R	RMSE (kg ha^{-1})	R	RMSE (kg ha^{-1})
Field I1								
Case A	0.49	1318	0.59	1219	0.52	1330	0.62	1215
Case B	0.86	728	0.94	484	0.93	547	0.95	463
Case C	0.88	622	0.96	407	0.95	440	0.96	430
Case D	0.97	479	0.97	427	0.97	445	0.97	432
Field I2								
Case A	0.66	1154	0.60	1254	0.64	1142	0.66	1191
Case B	0.80	958	0.84	958	0.84	966	0.86	875
Case C	0.81	938	0.83	943	0.87	763	0.88	719
Case D	0.80	949	0.82	854	0.84	819	0.86	756
Field I3								
Case A	0.29	1035	0.32	978	0.34	951	0.35	957
Case B	0.82	672	0.80	704	0.82	674	0.75	718
Case C	0.82	667	0.83	667	0.85	590	0.82	657
Case D	0.86	859	0.86	757	0.86	809	0.88	696

Table 5. Correlation coefficient R and RMSE for simulated and measured yield after model calibration (2000 iterations) of field I1, I2 and I3 using multiple years of corn yield data and multiple soil parameters (5-10). Yield was calculated in dependency of the grids in case A (10.5 x 10.5 m grid size), case B (16.5 x 16.5 m grid size), case C (22.5 or 30.5 x 50.5 m grid size) and case D (22.5 or 30.5 x 50.5 m grid size). R >0.75 is written in bold letters.

Overall, the model explained the spatial yield variability in all grids over five years very well, when calibration was done using multiple soil parameters. In all three fields, slightly different combinations of soil parameters led to the best calibration of the model. The highest accuracy of the model was achieved in field I1 using yield values of case D and a combination of the soil parameters HPF + HPD + RDRF + NMF + SFF. These soil parameters explained about 94 % of the spatial yield variability in field I1 (Figure 1a). In field I2 a combination of the soil parameters HPF + HPD + RDRF + NMF + SFF + ASW explained about 77 % of the spatial yield variability (Figure 1b), when yield values of case C were used for the calibration process. However, in field I3 a combination of the soil parameters HPF + HPD + NMF + SFF + ASW explained about 77 % of the spatial yield variability, calculated by yield values of case D (Figure 1c). These results implied that the spatial yield variability was mostly influenced by six soil parameters. The soil parameters HPF + HPD + RDRF + NMF + SFF + ASW seem to count for at least 75 % of the spatial yield variability.

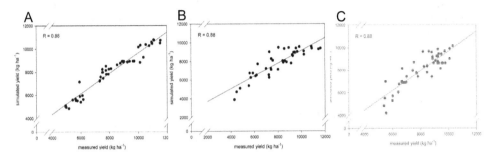

Fig. 1.a.-c. Simulated vs. measured corn yields (kg ha-1) in field I1 (A), I2 (B), I3 (C) in the years 1998 – 2002. The simulation is based on the calibration of multiple soil parameters leading to yield variability.

Although a strong influence of available soil parameters could be determined, the best simulation of yield was achieved for the soil parameters HPF + DHP. In all three fields good correlations between simulated and measured yields were determined, when these parameters were used for model calibration. These results implied that HPF + DHP were the parameters with the biggest impact in explaining spatial yield variability. Hardpan is described as a factor that is mainly induced by management practices. The effect of soil compaction after tillage is described in the literature (Lindstrom and Voorhees, 1994; Lipiec and Simonta, 1994). Due to continuous cultivation of corn at all three fields over the 5-year period, it is highly possible that the hardpan was strongly manifested in all fields. As a result of a hardpan in the field, root distribution could be affected and led to spatial yield variability, as also assumed in studies of Arvidsson and Håkansson (1996). RDRF explained much of the spatial yield variability especially in field I2 and I3. In model simulations where RDRF was considered, high correlation coefficients were achieved in all three fields, indicating a strong influence of RDRF factor on yield. In addition, soil fertility seemed to have an influence on the spatial yield variability, especially in field I1. ASW might be spatially different due to a probably inhomogeneous flint layers in the deeper soil, which affects the water supply in the field.

3.2.4 Effect of grid size
In this study grid size had a strong influence on the results of the model calibration. In general, smaller grids (case A) resulted in weak correlations between simulated and

measured yields (Table 4 and 5). In general, the model gave more accurate simulated yields for larger grids, suggesting that the applied calibration process may be more effective under large grid sizes. A slight improvement of the model accuracy was found when additional information on soil available nitrogen around the 4th leaf stage was imposed on the calibration process.

The larger grid sizes contained more yield monitor data points and thus, averaged over some of the spatial yield variation that occurred between two sequential yield monitor data points. Thus, the larger grid sizes averaged yield variability within the grid and thus were less acceptable for outlier, which was similar to results of Ping and Dobermann (2003). To work with spatial data sets in crop models, there appears to be a trade-off between maintaining spatial precision by selecting a small grid size and reducing noise in yield monitor data by selecting a larger grid size (Wong, 1995; Long, 1998). Combining area units into successively larger units, an agronomist will need to consider the scale at which the spatial variability of site-specific yield data has to be analyzed (Long, 1998). Considering the underlying soil factors, which had either a high range of variability or continuity, the model accuracy was improved by choosing larger grid sizes that captured the spatial variability and stability within different sites.

3.2.5 Evaluation of model performance
In general the APOLLO model preformed well in simulating yield of the three fields in the Upper Rhine Valley over the 5-year period. Among the yield limiting factors that were examined in this study, hardpan seemed to have a big impact on yield variability. However, one cannot discount the effect of other factors or interactions such as rooting depth, water availability etc. Nevertheless, the technique presented in this study demonstrates the value of using a crop growth model in quantifying individual as well as combined effects of factors leading to spatial yield variability. However, there is a need to further test and validate the model outputs by verifying the yield limiting factors through direct field measurements.

The case study has shown that yield variability may be explained by a combination of varying soil factors. The implemented crop models have proven to be useful tools to evaluate these complex interactions and to provide insight into causes of yield variability. However, the results of model calibration were affected by the grid size used for calibration. The model gave more accurate simulated yields for larger grids, suggesting that the applied calibration process may be more effective under large grid sizes.

Overall the ideal grid size for the calibration process seems to be determined by the underlying factors leading to spatial yield variability. Further research is needed to determine a suitable approach for the assessment of ideal grid sizes in model calibration and resulting grid resolution that captures enough information to represent spatial yield variability and temporal stability at a scale appropriate to finally optimize crop management and reduce yield gaps.

3.3 Case study China - model based analysis of a winter wheat - summer maize double cropping system in the North China Plain
Besides the prediction of crop yields and the analysis of yield gaps, there has been a substantial amount of crop model applications to improve crop management strategies like nitrogen (N) fertilization and irrigation. Especially in those crop production systems, where

water resources are limited and risk of groundwater contamination with nitrate is high, it is important to optimize irrigation and fertilization use efficiencies via use of sound water and nitrogen management practices. However, development and validation of guidelines for optimal timing and water and nitrogen requirements requires extensive and expensive field experiments. Since it is impossible to test all the interactions between the amount of water and nitrogen during the seasons, use of simulation models can greatly facilitate the evaluation of different production practices and/or environments and thereby streamline the decision-making process.

This case study outlines the use of the CERES-Wheat and CERES-Maize models, both implemented in DSSAT V.4.0 (Jones et al., 2003), to evaluate a double cropping system of winter wheat and summer maize under different fertilizer and irrigation input scenarios regarding water consumption, grain yield and gross margin. The models were calibrated and validated using data derived from a field experiment conducted in Dongbeiwang, near Beijing, China.

One of the most important regions of agricultural production in China is the North China Plain (NCP) (Kendy et al., 2003). Considering China's total grain yield, the NCP contributes approximately 41% of wheat and 25% of maize grain yield (Länderbericht China, 2000). The NCP, also known as the Huang-Huai-Hai Plain, is located in the north of the eastern part of China between 32° and 40° N latitude and 100° and 120° E longitude (Liu et al., 2001). Winter wheat (*Triticum aestivum* L.) and summer maize (*Zea mays* L.) are currently the two main crops combined in a single-year rotation also referred to as a double cropping system (Zhao et al., 2006). Winter wheat is sown at the beginning of October and harvested in mid June. Summer maize is sown immediately following winter wheat harvest and is harvested at beginning of October.

The climate in the NCP is warm-temperate with cold winter and hot summer. Precipitation shows a high spatial and temporal variability (Wu et al., 2006) and ranges from about 500 mm in the north to 800 mm in the south (Liu et al., 2001). About 50 to 75% of the total precipitation occurs from July to September during the summer monsoon. Depending on the seasonal precipitation situation, farmers usually irrigate winter wheat four to five times (Hu et al., 2006). In consequence, the irrigation water for wheat production comprises about 80% of the whole agricultural water consumption (Li, 1993).

Extensive use of fertilizer is also very common in the winter wheat-summer maize double cropping system. In Beijing area, for example, the average N application rates ranged around 309 kg N ha^{-1} for winter wheat and 256 kg N ha^{-1} for maize (Zhao et al., 1997). Besides the positive effects on yield, an increasing input of water and fertilizer is connected with increasing costs for the farmers and leads to environmental problems such as leaching or water scarcity. Therefore better management practices balancing both economic and environmental interests are required.

Numerous agricultural experiments were carried out to test the effects of different management strategies on grain yield and overall sustainability. However field experiments have their limitations as they are conducted at particular points in time and space. Besides, field experiments are time consuming, laborious and expensive (Jones et al., 2003) and therefore limited in extent and size. A viable alternative to these problems is to use crop models. Models can be applied as a valuable tool to propose better adapted crop management strategies and to test the hypothetical consequences of varying management practices (e.g. Saseendran et al., 2005). Furthermore, models can be used to optimize economic efficiency by finding best management strategies under given and future environmental conditions (Link et al., 2006). However, studies which use crop models to evaluate crop production systems in the NCP are rare (Yu et al., 2006). Hu et al. (2006) used the RZWQM model to assess N-management in a double cropping system (winter wheat and summer maize) at Luancheng in the NCP. Results of the study indicated, that both, N and water could be reduced by about half of the typical application rates without a strong reduction in yield. Yang et al. (2006) used the CERES-Wheat and CERES-Maize models to estimate agricultural water use and its impact on ground water depletion in the piedmont region of the NCP. The results showed a strong correlation between the agricultural water use and the ground water depletion. The authors concluded that there is still a sustainable water reduction possible if water-saving technologies are applied.

The aim of this study was to evaluate the double cropping of winter wheat and summer maize in the NCP under water shortage conditions. For this purpose the models CERES-Wheat and CERES-Maize were calibrated and validated. Based on a gross margin analysis, different irrigation and N-fertilizer treatments were evaluated. Originated from the most profitable treatment, four different irrigation scenarios for the dry growing season 2000/2001 and their effect on water consumption, grain yield and gross margin were simulated.

3.3.1 Study site and field experiment

The CERES-Maize and CERES-Wheat models were used with field study information of Böning-Zilkens (2004) for testing different irrigation scenarios in a double cropping system of winter wheat and summer maize in the NCP. The chosen field experiment was conducted from 1999-2002 (three vegetation periods, six harvests). The experimental site was Dongbeiwang located in the northwest of Beijing (40.0° N and 116.3° E). Soil type was a Calcaric Cambisol (FAO taxonomy) formed of silty loam. The winter wheat cultivar Jindong 8 was sown at the beginning of October with a row spacing of 15 cm and an aspired plant density of 480 plants m^{-2}. Wheat was harvested at the beginning of June each year. After winter wheat summer maize was sown directly without time lag and harvested at the beginning of October. The maize cultivar Jingkeng 114 was sown with a density of 6 plants m^{-2} and a row spacing of 70 cm. The experiment was designed as a three factorial split-split-plot design with four replications and included three different irrigation regimes, three N-fertilization rates as well as a treatment with and without straw. No major diseases were reported during the growing period. For the purpose of this study only the treatments without straw removal were used because this is the usual practice in the NCP.

Fertilization treatments varied between 0 and 600 kg N ha^{-1} a^{-1} (Table 6). In the traditional treatment N-fertilization was carried out according to farmers practice. The treatment reflects the standard production system in the NCP with the fertilizer inputs being very high. The fertilization of the optimized treatment was based on measured soil available nitrogen (Nmin-content) and target yield. No nitrogen fertilizer was applied in the control treatment.

Irrigation treatments varied between 195 and 354 mm (Table 6). The traditional irrigation was carried out according to farmers practice using border irrigation. The optimized irrigation was based on measurements of the volumetric soil water content aiming to keep the available field capacity between 45 and 80 % following Steiner et al. (1995). Available field capacity was defined as field capacity minus the amount of water which is retained through high soil water tension (pF < 4.2) and therefore not available to the plant. The volumetric soil water content was measured with time domain reflectometry probes (0-15, 15-30, 30-45, 45-60, 60-90, 90-120 cm depth) every four days. The suboptimal irrigation exemplified the control. Irrigation in this treatment was based on the amount of approximately two third of the optimized strategy. In the optimized and suboptimal irrigation treatments irrigation was carried out using sprinkler irrigation. Plot size for the factor irrigation was 50 x 70 m. Two-jet complete circle sprinkler with a height of 1 m were used, adjusted in a grid of 12 x 18 m. Summer maize was not irrigated in any of the experimental years (Böning-Zilkens, 2004), as precipitation is usually sufficient (Lohmar et al., 2003).

Treatment	Management		N-fertilization (kg N ha^{-1} a^{-1})			Irrigation (mm)	
	irrigation	fertilization	wheat	maize	total	wheat	maize
1	suboptimal	suboptimal	0	0	0	195	0
2	suboptimal	traditional	300	300	600	195	0
3	suboptimal	optimized	50	53	103	195	0
4	traditional	suboptimal	0	0	0	354	0
5	traditional	traditional	300	300	600	354	0
6	traditional	optimized	87	59	146	354	0
7	optimized	suboptimal	0	0	0	293	0
8	optimized	traditional	300	300	600	293	0
9	optimized	optimized	72	65	137	293	0

Table 6. Average amounts of nitrogen fertilizer (kg ha^{-1} a^{-1}) and irrigation amounts (mm) for winter wheat and summer maize over the three vegetation periods (1999/2000-2001/2002) of field experiments.

Different growth stages of winter wheat (emergence, hibernation, jointing, full flowering, medium milk, ripening) and maize (emergence, beginning of stem elongation, heading, end of flowering, medium milk) were recorded according to Zadoks et al. (1974). Four time harvests were done by hand in the growth stages mentioned before. Final harvest was done manually by cutting three times 3 m^2 of winter wheat and one time 39.2 m^2 of summer maize. Winter wheat ears and maize plants were counted. After harvest, plants were separated and grains and kernels were threshed and dried to constant weight at 105 °C to obtain total dry matter. Thousand kernel mass, grains per ear, kernels per row, kernels per cob and rows per cob were determined.

Weather data were collected from a local weather station at Dongbeiwang and including daily solar radiation, maximum and minimum air temperatures and precipitation. Mean annual temperature for 2000-2002 ranged between 11.5-11.7 °C and did not differ from the long-term average of 11.5 °C. The annual precipitation amounted 448, 366 and 520 mm. In comparison to the long-term average (556 mm), all three years were drier.

Statistical analyses of the data were performed with Sigma Stat 3.5 (Jandel Scientific Corp, San Rafae, CA). Differences between experimental groups were tested by using one factor analysis of variance (ANOVA). Tukey tests were carried out for comparison of means.

3.3.2 Model description

The CERES-Maize and CERES-Wheat models were calibrated and validated using data from Böning-Zilkens (2004). The growing seasons 1999/2000 and 2001/2002 were used for model calibration. Phenology and growth data such as biomass at different growth stages and the dates of phenological events were used to determine the cultivar coefficients for wheat and maize (Table 7). The calibration of these parameters followed a sequence of variables suggested by the DSSAT manual (Boote, 1999). As irrigation was carried out by border irrigation, irrigation efficiency was set to 0.45, according to Xinhua News (2001). For sprinkler irrigation the irrigation efficiency was set to 0.55 as strong winds during the experiment diminished the water distribution.

Winter wheat cv. Jindong 8		
Parameter	Description	Value
P1V	Sensitivity to vernalisation	35
P1D	Sensitivity to photoperiod	50
P5	Grain filling duration	500
G1	Kernel number per unit weight at anthesis	20
G2	Kernel weight under optimum conditions	36
G3	stem + spike dry weight at maturity	1.8
PHINT	Phyllochron interval	95
Summer maize cv. Jingkeng 114		
P1	Growing degree days from emergence to end of juvenile phase	180
P2	Photoperiod sensitivity	0.3
P5	Cumulative growing degree days from silking to maturity	685
G2	Potential kernel number	730
G3	Potential kernel growth rate	8.0
PHINT	Phyllochron interval	44

Table 7. Cultivar coefficients of winter wheat cv. Jindong 8 and summer maize cv. Jingkeng 114 used for model calibration and validation.

Model validation was carried out with an independent dataset from the dry growing season 2000/20001 using the cultivar coefficients obtained by calibration. The model validation involved the use of the model with the calibrated values without making any further adjustments of the constants (Gungula et al., 2003). The Root Mean Square Error (RMSE), between simulated and measured values was used to evaluate simulation results. For graphical representations the 1:1 line of measured vs. simulated values was used. Correlation coefficients were also reported to express the scatter of the simulated values compared with the measured data.

3.3.3 Economic analysis
Gross margins in Chinese Yuan (¥) (RMB, 1 ¥ ~ 0.12 US$) have been developed with information provided from literature to compare the different treatments form the field experiments used for model calibration and validation as well as to compare the simulated scenarios. The analysis was carried out according to the following equation:

$$gross\ margin = revenue - variable\ costs \qquad (1)$$

where gross margin is ¥ ha^{-1} a^{-1}.
The revenue parameter is the result of equation 2:

$$\text{revenue} = \text{YWt} \times \text{PW} + \text{YMt} \times \text{PM} \tag{2}$$

where YWt is the grain yield (kg ha^{-1}) of winter wheat W per year t, PW is the price of winter wheat W (¥ kg^{-1}), YMt is the grain yield (kg ha^{-1}) of summer maize M per year t, and PM is the price of summer maize M (¥ kg^{-1}).

The variable cost parameter is the result of equation 3:

$$\text{variable costs} = (\text{NWt} + \text{NMt}) \times \text{PN} + \text{WWt} \times \text{PW} \tag{3}$$

where NWt is the N-fertilizer rate (kg ha^{-1}) applied to winter wheat W in year t, NMt is the N-fertilizer rate (kg ha^{-1}) applied to summer maize M in year t, PN is the price of N fertilizer (¥ kg^{-1} N), WWt is the amount of irrigation water (m^3 ha^{-1}) used for winter wheat W in year t, and PW is the price for the irrigation water (¥ m^{-3}).

The prices for nitrogen fertilizer (4.0 ¥ kg N^{-1}), winter wheat (1.10 ¥ kg^{-1}) and summer maize (1.00 ¥ kg^{-1}) were obtained from Chen et al. (2004). In the study of Chen et al. (2004) prices for wheat and maize were based on grain moisture of 15 %. As the model input requires a grain moisture of 0% the price for maize was set to 1.18 ¥ kg^{-1} dry matter (DM) and to 1.29 ¥ kg^{-1} DM for winter wheat. The price for irrigation water was set to 0.60 ¥ m^{-3} based on information from Li et al. (2005). Note, that all the prices represent average prices for the NCP and will vary from region to region and year to year.

3.3.4 Simulated irrigation scenarios

To achieve high yields farmers in the NCP tend to overirrigate winter wheat (Zhang et al., 2004). The simulations show the potential saving of irrigation water and its effect on grain yield and gross margin under the conditions at the Dongbeiwang site. Based on the treatment with the highest gross margin of the field experiment, five different irrigation scenarios for the dry growing season 2000-2001 were simulated.

Treatment	Grain yield (kg DM ha^{-1} a^{-1})						Gross margin (¥ ha^{-1} a^{-1})*					
	wheat		maize		total		wheat		maize		total	
1	3280	c	5435	a	8715	d	3061	ab	6413	a	9475	abc
2	3493	c	5786	a	9279	cd	2136	b	5327	a	7763	d
3	3647	c	5834	a	9480	bcd	3334	a	6672	a	10006	ab
4	4077	abc	5397	a	9473	bcd	3135	ab	6368	a	9503	abc
5	4940	a	5590	a	10530	ab	3049	ab	5397	a	8445	cd
6	4947	a	5707	a	10653	a	3909	a	6498	a	10407	a
7	3730	bc	5364	a	9094	d	3054	ab	6330	a	9384	bc
8	4657	ab	5600	a	10257	abc	3049	ab	5408	a	8457	cd
9	4780	a	5780	a	10560	a	4120	a	6560	a	10680	a

*Currency Chinese Yuan (¥) (RMB, 1 ~ 0.12 US$)

Table 8. Average grain yield (kg DM ha^{-1}) and gross margin (¥ ha^{-1}) over the three vegetation periods (1999/2001-2001/2002) for winter wheat, summer maize and the double cropping of both cultivars regarding different irrigation and nitrogen fertilizer treatments in the field experiment. Different letters indicate significant differences at α=0.05.

From the calibration and validation results, the CERES-Maize and CERES-Wheat models were found to simulate yields well. The average RMSE between simulated and measured yield for winter wheat was 432 kg ha⁻¹ (calibration) and 342 kg ha⁻¹ (validation). Similar results were obtained for summer maize with an average RMSE of 253 kg ha⁻¹ (calibration) and 414 kg ha⁻¹ (validation). The results of the study showed that the differences between simulated and measured grain yields were within the range of differences reported in the literature.

Maximizing yield and gross margin as a function of inputs and production costs is one of the main goals when making management decisions such as fertilizer and irrigation applications (Bannayan et al., 2003). The effects of different management strategies on grain yield and gross margin are represented in Table 8.

The average winter wheat grain yields over the three vegetation periods indicated that, independent of irrigation regimes, the highest grain yields were reached by the treatments "optimized fertilization" (3, 6, 9), followed by the treatments "traditional" (2, 5, 8) and "suboptimal fertilization" (1, 4, 7). However between the treatments "optimized" and "traditional fertilization" no significant difference existed, whereas grain yields in the "suboptimal" treatments decreased significantly. Considering the different irrigation treatments the highest winter wheat grain yields were reached by the "traditional treatments" (4-6), followed by the "optimized" (7-9) and "suboptimal treatments" (1-3). The differences between the "traditional" and "optimized irrigation" treatments were not significant whereas grain yield of the "suboptimal irrigation" treatments was significantly reduced. Summer maize was not irrigated and the different irrigation treatments applied to winter wheat showed only small, or no significant effects on the grain yield of succeeding maize. The same was true for the fertilization treatments. The highest grain yield was reached by the treatments "optimized fertilization" followed by the treatments "traditional" and "suboptimal fertilization".

Overall, the results showed, that the differences in grain yield between "traditional" and "optimized fertilization" were not significant. One possible reason could be that crop N demand of wheat and summer maize is much lower than the applied rates of N in the NCP (Gao et al., 1999). This corresponds with results of Jia et al. (2001) and Chen et al. (2004) who showed that without any risk of yield decrease N fertilizer rates for winter wheat in the NCP could be reduced to <180 kg N ha⁻¹ when soil NO_3 testing or a yield response curve method was used. Similar results were found for irrigation management. According to Zhang et al. (2005) winter wheat could attain its maximum yield with less than full application of the current irrigation practice.

Beside the effects on grain yield, Table 8 indicates also the effects of different irrigation and N-fertilizer amounts on gross margin. The highest gross margin regarding the different fertilizer amounts was reached within the "optimized" treatments (3, 6, 9) followed by the "suboptimal" (1, 4, 7) and "traditional fertilizer" (2, 5, 8) treatments. Even if grain yields of the "suboptimal" treatments were significantly lower than grain yields of the "traditional" treatments they reached the same level of gross margin because the "suboptimal fertilizer" treatments went along without any costs for nitrogen fertilization. The gross margin of the different irrigation treatments increased by the order "suboptimal", "optimized" and "traditional irrigation". However, the differences between the "traditional" and "optimized" treatments were small. The highest total gross margin in the double cropping of winter wheat and summer maize was observed with treatment number 9 ("optimized irrigation" and "optimized N-fertilization"). The analysis of the different irrigation and N-fertilizer treatments of the field experiment showed that with a higher input of nitrogen fertilizer and irrigation water grain yields did not equally rise as the costs for the input factors. Therefore, the highest gross margin was not reached with the highest input of nitrogen and irrigation. This result corresponds with the report of Zhu and Chen (2002) who had reviewed the nitrogen fertilizer use in China. They found that the maximum yield, demonstrated by yield versus N application rate curves, is usually higher than the yield of the maximum economic efficiency.

In a next step, five different irrigation scenarios for winter wheat and their effects on grain yield, water consumption and gross margin were simulated for the dry season 2000/2001. As a starting point (= scenario 1) for the simulation of different irrigation scenarios treatment 9 ("optimized irrigation" and "optimized fertilization") was used because this treatment reached the highest gross margin (Table 8). The nitrogen fertilization amount for winter wheat in treatment 9 was 65 kg N ha-1. The costs for the N-fertilizer amounted to 260 ¥ ha-1 (4.00 ¥ kg N-1) and were considered in all succeeding scenarios. Table 9 gives an overview on the changes in irrigation amount, total water supply, grain yield and gross margin.

One of the most important aspects of water-saving in irrigated agriculture is the irrigation scheduling because the water sensitivity varies among different growth stages (Zhang et al., 1999). The irrigation in scenario one took place at six different growth stages. However the results of scenario three, four and five demonstrated that a reduction in irrigation frequency may not always lead to a decrease in grain yield. Commonly grain crops are more sensitive to water stress during flowering and early seed formation than during vegetative or grain filling phases (Doorenbos and Kassam, 1979). Results of Zhang et al. (2003) showed that irrigation before the over-wintering period could be omitted due to its loss to soil evaporation and its effects on increasing the non-effective tillers in spring. According to Zhang et al. (1999) wheat is particularly sensitive to water stress in the growth stages from jointing to heading and from heading to milk stage. Similar results were found in our study. The results showed that irrigation during seedling development (before winter) and at milk stage was not essential for the formation of grain yield. However, an additional irrigation during the regreening stage (scenario five) might lead to a further increase in grain yield.

	Irrigation		Total water supply**		Grain yield		Gross margin		
Scenario	mm	%	mm	%	kg ha-1 a-1	%	wheat ¥* ha-1 a-1	maize ¥ ha-1 a-1	total ¥ ha-1 a-1
1	310	100	452	100	4436	100	3602	7051	10654
2	0	-100	262	-42	1089	-75.5	1144	7244	8388
3	155	-50	370	-18.1	3721	-16.1	3610	6982	10592
4	200	-35.5	394	-12.8	4445	0.2	4274	6994	11268
5	290	-6.5	442	-2.2	5243	18.2	4763	7134	11897

*Currency Chinese Yuan (¥) (RMB, 1 ~ 0.12 US$)
** Total water supply = effective irrigation+ precipitation + depletion of the initial soil water content

Table 9. Irrigation amount (mm), total water supply (mm), grain yield (kg ha-1 a-1) and gross margin (¥ ha-1 a-1) of winter wheat.

Besides the irrigation frequency also the irrigation amount affected grain yield. With a complete renouncement of irrigation like in scenario 2 grain yield dropped to 1089 kg per ha[-1] leading to a decrease of 68.2 % in gross margin, as the yield reduction could not be compensated by the saved irrigation costs (Table 9). Hence, a supplemental irrigation is required in wheat to maintain high yields and to ensure an adequate gross margin for the farmers. Besides the necessity of irrigation for wheat, scenario 3 demonstrates the enormous potential for water saving without a financial deterioration for the farmers. The simulation indicated that a reduction in the irrigation frequency from six to four times and a reduction in the irrigation amount of up to 50 % are possible without any decrease in gross margin. As the Chinese government aims to achieve the goal of self-sufficiency (Huang 1998), scenario 4 tested the considerable potential for reducing the irrigation amount without any decrease in actual yield level. The simulation indicated that a reduction of the amount of irrigation water of up to one third is possible without any yield losses. The saved irrigation costs would lead to an increase in gross margin of up to 672 ¥ ha[-1] (18.6 %). A maximum gross margin of 4763 ¥ ha[-1] was reached in scenario 5. The maximum gross margin was connected with the highest grain yield of 5243 kg ha[-1]. Changes in the irrigation amounts and frequencies to winter wheat showed only small effect on yields of the following crop summer maize because there was enough precipitation. The total gross margin for the different scenarios of double cropping winter wheat and summer maize is given in Table 9 and indicated that scenario 4 and 5 would lead to the highest gross margin in the double cropping system while saving up to 35 % of irrigation water and having no yield loss.

The results of the simulated scenarios showed that there is a considerable potential for saving irrigation water even under dry conditions like in the growing season 2000/2001. For the purpose of improvement of gross margin, models can help to determine an optimum value for total water consumption where grain yield and gross margin were all relatively high. However this value largely depends on the cost of irrigation water. Crop models such as DSSAT offer the possibility to estimate crop water use and can help to develop appropriate irrigation strategies. It can be concluded that in areas with similar conditions as in the simulations, the common irrigation amount to wheat could be reduced by about one third without any yield losses. Furthermore, a reduction of about 50 % may be possible without a decrease in the initial gross margin. However, without irrigation gross margin would be very low, because the saved water costs could not balance the losses in grain yield. Therefore, a supplemental irrigation at critical growth stages seems to be essential to maintain high yields and to ensure an adequate gross margin for the farmers.

4. Conclusion

Crop models represent a means for agricultural scientists to provide farmers with information like possible crop yields for different levels of input factors and procedures for decision making by integrating the most relevant parameters affecting crop yield. The evolving area of crop model research represents an attempt of scientists to intervene and improve farmers' management and to fill this existing knowledge gap. However, the attempts to intervene and influence management decisions by delivering appropriate decision support tools have been harder than first thought. But crop models are considered to be an important tool for gaining a theoretical understanding of a crop production system. Finally, it has to be constituted and recalled that crop models do not offer a panacea for problem solving. They are limited in their ability to simulate various parts of a biological system and address complex systems in an often simplified manner. In summary, it can be adhered that crop models are a first step into the direction of decision making process. Besides all existing constrains to date, due to their highly innovative nature and potential for improved quality of production and crop management, models have the credentials to be a key candidate for a well-focused future research area.

5. Acknowledgement

We thank the ITADA-Projekt 1.1.1 "Nutzbarmachung von Verfahren der Präzisionslandwirtschaft am Oberrhein – Analyse und Interpretation der Variabilität von Ackerflächen in der Rheinebene", bearbeitet von Dr. Ivika Rühling (IfuL) und Didier Lasserre (ICTF) mit Kofinanzierung durch EU-Gemeinschaftsinitiative INTERREG II Oberrhein Mitte-Süd for providing some of the datasets.
This work was funded by the German Research Foundation (GRK 768 "Strategies to Reduce the Emission of Greenhouse Gases and Environmental Toxic Agents from Agriculture and Land Use" and GRK 107, "IRTG Sustainable Resource Use").

6. References

Adams, M.L., Cook, S.E., Corner, R. 2000. Managing Uncertainty in Site-Specific Management: What is the Best Model? Precision Agriculture 2: 39-54.

Arvidsson, J., Håkansson, I. 1996. Do effects of soil compaction persist after ploughing? Results from 21 long-term field experiments in Sweden. Soil Tillage Res. 39, 175-197.

Bannayan, M., Crout, N.M.J., Hoogenboom, G., 2003. Application of the CERES-Wheat model for within-season prediction of winter wheat yield in the United Kingdom. Agron. J. 95, 114-125.

Batchelor, W.D., Link, E.J., Thorp, K.R., Graeff, S., Paz, J.O. 2004b. The role of crop growth models in precision farming. Workshop Precision Farming, May 13, Renningen, Germany. pp 11-16.

Batchelor, W.D., Paz, J.O., Thorp, K.R. 2004a. Development and evaluation of a decision support system for precision farming. Proceedings of the 7th International Conference on Precision Agriculture. July 25-26, 2004, Minneapolis, MN. ASA, CSSA, SSSA Inc., Madison, WI, USA.

Böning-Zilkens, M.I., 2004. Comparative appraisal of different agronomic strategies in a winter wheat - summer maize double cropping system in the North China Plain with regard to their contribution to sustainability. Diss., Institute of Crop Production and Grassland Research, University of Hohenheim.

Boote K.J., Jones J.W., Pickering N.B. 1996. Potential uses and limitations of crop models, Agronomy Journal 88: 704–716.

Boote, K.J., 1999. Concepts of calibrating crop growth models. In: Hoogenboom, G., Wilkens, P.W., Tsuji, G.Y. (Eds.), DSSAT Version 3. A decision support system for agrotechnology transfer, pp. 179-200, Vol. 4, Univ. of Hawaii, Honolulu.

Boote, K.J., Jones, J.W., Hoogenboom, G.J, Pickering, N.B. 1998. The GROPGRO for grain legumes. In: Tsuji, G.Y., Hoogenboom, G.J., Thornton, P.K. (Eds.), Understanding Options for Agricultural Production. Kluwer Academic Publishers, Dordrecht, The Netherlands. pp 99-128.

Brouwer, R., De Wit, C. 1968. A simulation model of plant growth. In: Proc. Easter School in Agricultural Science, University of Nottingham, UK. Butterworths, Londres, RU.

Cerrato, M.E., Blackmer, A.M. 1990. Comparison of models for describing corn yield response to nitrogen fertilizer. Agronomy Journal 82(1): 138-143.

Chen, G., Yost, R.S., Li, Z.C., Wang, X., Cox, F.R. 1997. Uncertainty analysis for knowledgebased decision aids: application to phosphorus decision support system (PDSS). Agricultural Systems 55: 461-472.

Chen, X., Zhou, J., Wang, X., Blackmer, A.M., Zhang, F., 2004. Optimal rates of nitrogen fertilization for a winter wheat-corn cropping system in Northern China. Communications in Soil Sci. Plant Anal. 35, 583-597.

Cook, S.E. 1997. Data interpretation and risk analysis for precision agriculture. In: Precision agriculture: What can it offer the Australian Sugar Industry? Bramley, R.G.V., Cook, S.E., McMahon, G.G., (eds.) CSIRO Land and Water, Proceedings of a Workshop held in Townsville, pp. 77-85.

Cook, S.E., Adams, M.L. 1998. In: Precision Weed Management in Crop and Pastures. Medd, R.W., Tratley, J.E. (eds.) Richarson, Melbourne, Australia, p.101.

De Wit, C.T., Brouwer, R., Penning de Vries, F.W.T. 1970. The simulation of photosynthetic systems. In:, Prediction and measurement of photosynthetic productivity. Setlik, I. (ed.) Proceeding IBP/PP Technical Meeting Trebon 1969. Pudoc, Wageningen, The Netherlands, pp. 47-70.

Doorenbos, J., Kassam, A.H., 1979. Yield response to water. Irrig. and Drain. Pap. 33. FAO, Rome.

Drummond, S.T., Sudduth, K.A., Joshi, A., Birrell, S.J., Kitchen, N.R. 2003. Statistical and neural methods for site-specific yield prediction. Transactions of the ASAE 46(1): 5-14.

Finke, P.A., Goense, D. 1993. Differences in barley grain yields as a result of soil variability. Journal of Agricultural Science 120: 171–180.

Forrester, J.W. 1961. Industrial dynamics. Cambridge, Mass. U.S.A., Massachusetts Institute of Technology Press.

Fraisse, C.W., Sudduth, K.A., Kitchen, N.R. 1998. Evaluation of crop models to simulate site-specific crop development and yield. In: Robert, P.C, Rust, R.H., Larson, W.E. (Eds.),, Proceedings of the 4th International Conference on Precision Agriculture. ASA, CSSA, SSSA, Inc., Madison, WI, USA. pp 1297-1308.

Gao, W., Huang, J., Wu, D., Li, X., 1999. Investigation on nitrate pollution in ground water at intensive agricultural region in Huanghe-huaihe-haihe Plain. Eco-agric. Res. 7, 41-43 (in Chinese).

Gerdes, G. 1993. Calibration of the soybean growth simulation model SOYGRO for Central Portugal. Verlag Ulrich E. Grauer, Wendlingen.

Godwin, D.C., Ritchie, J.T., Singh, U., Hunt, L., 1989. A user´s guide to CERES Wheat-V2.10. International Fertilizer Development Center, Muscle Shoals, AL.

Gungula, D.T., Kling, J.G., Togun, A.O., 2003. CERES-Maize predictions of maize phenology under nitrogen-stressed conditions in Nigeria. Agron. J. 95, 892-899.

Hoogenboom, G. 2003. Crop growth and development. In: Handbook of Processes and Modeling in the Soil-Plant System. Bendi D.K., Nieder R. (eds.) The Haworth Press, Binghamton, New York, pp. 655-691.

Hoogenboom, G., Jones, J.W., Wilkens, P.W., Batchelor, W.D., Bowen, W.T., Hunt, L.A., Pickering, N.B., Singh, U., Godwin, D.C., Baer, B., Boote, K.J., Ritchie, J.T., White, J.W. 1994. Crop models. In: DSSAT Version 3, vol. 2. Tsuji, G.Y., Uehara, G., Balas, S. (eds.), University of Hawaii, Honolulu, HI, pp. 95-244.

Hu, C., Saseendran, A., Green, T.R., Ma, L., Li, X., Ahuja, L.R., 2006. Evaluation nitrogen and water management in a double-cropping system using RZWQM. Vadose Zone J. 5, 493-505.

Huang, J.K., 1998. Agricultural policy, development and food security in China. In: Tso, T.C., Tuan, F., Faust, M. (Eds.), Agriculture in China 1949-2030, pp. 209-257, IDEALS, Inc., Beltsville, Maryland.

Jia, L., Chen, X., Zhang, F., Liu, B., Wu, J., 2001. The study of optimum N supplying rate for winter wheat in Beijing area. J. Chinese Agric. Univ. 6, 67-73 (in Chinese).

Jones, C.A., Kiniry, N., 1986. CERES-Maize, a simulation model of maize growth and development. Texas A&M University Press, College Station, TX.

Jones, J.W., Hoogenboom, G., Porter, C.H., Boote, K.J., Batchelor, W.D., Hunt, L.A., Wilkens, P.W., Singh, U., Gijsman, A.J., Ritchie, J.T. 2003. DSSAT Cropping System Model. European Journal of Agronomy 18: 235-265.

Jones, J.W., Ritchie, J.T. 1990. Crop growth models. In: Management of farm irrigation systems, Hoffman, G.J., Howell, T.A., Solomon, K.H. (eds.) ASAE:St. Joseph, Michigan, pp. 63-89.

Kendy, E., Gerhard-Marchant, P., Walter, M.T., Zhang, Y.Q., Liu, C.M., Steenhuis, T.S., 2003. A soil-water-balance approach to quantify groundwater recharge from irrigated cropland in the North China Plain. Hydrol. Proces. 17, 2011-2031.

Kravchenko, A.N., Bullock, D.G. 2000. Correlation of grain yield with topography and soil properties. Agronomy Journal 92: 75-83.

Länderbericht China, 2000. Politik, Wirtschaft und Gesellschaft im chinesischen Kulturraum. Herrmann-Pillath, C., Lackner, M. (Eds.), Bundeszentrale für politische Bildung, Schriftreihe Bd. 351, Bonn.

Li, J., Eneji, A.E., Duan, L., Inanaga, S., Li, Z., 2005. Saving irrigation water for winter wheat with phosphorus application in the North China Plain. J. Plant Nutrition 28, 2001-2010.

Li, P.C., 1993. The development of water-saving type agriculture. Agri. Res. Arid Areas 11, 57-63.

Lindstrom, M.J., Voorhees, W.B. 1994. Response of temperates crops in North America to soil compaction. In: Soane, B.D., van Ouverkerk, C. (Eds.), Soil Compaction in Crop Production, Elsevier, Amsterdam. pp. 265-286.

Link, J., Graeff, S., Batchelor, W.D., Claupein, W., 2006. Evaluating the economic and environmental impact of environmental compensation payment policy under uniform and variable-rate nitrogen management. Agri. Syst. 91, 135-153.

Link, J., Graeff, S., Batchelor, W.D., Claupein, W., 2007. Identification of problem grids within a wheat and corn field by the implementation of a process-oriented precision farming crop growth model. Proceedings of GIL, 27th annual conference, p. 131-134.

Lipiec, J., Simota, C. 1994. Role of soil and climated factors in Influencing crop response to soil compaction in Central and Eastern Europe. In: Soane, B.D., van Ouverkerk, C. (Eds.), Soil Compaction in Crop Production, Elsevier, Amsterdam. pp. 365-390.

Liu, C., Yu, J., Kendy, E., 2001. Groundwater exploitation and its impact on the environment in the North China Plain. Water Intern. 26, 265-272.

Lohmar, B., Wang, J., Rozelle, S., Huang, J., Dawe, D., 2003. China´s agricultural water policy reforms: increasing investment, resolving conflicts, and revising incentives. USDA Economic Res. Service, AIB 782.

Long, D.S. 1998. Spatial autoregression modeling of site-specific wheat yield. Geoderma. 85, 181-197.

Ma. L., Schaffer, M.J. 2001. A Review for carbon and nitrogen processes in nine U.S. soil nitrogen dynamics models. In: Modeling Carbon and Nitrogen Dynamics for Soil Management. Schaffer, M.J., Ma, L., Hansen, S. (eds.) Lewis Publishers, pp. 651.

Paz, J.O., Batchelor, W.D., Babcock, B.A., Colvin, T.S., Logsdon, S.D. Kaspar, T.C., Karlen, D.L., 1999. Model-based technique to determine variable-rate nitrogen for corn. Agric. Syst. 61, 69-75.

Paz, J.O., Batchelor, W.D., Colvin, T.S., Logsdon, S.D., Kaspar, T.C., Karlen, D.L. 1998. Analysis of water stress effects causing spatial yield variability in soybeans. Agric. Systems 61, 69-75.

Paz, J.O., Batchelor, W.D., Tylka, G.L. 2001. Method to use crop growth models to estimated potential return for variable-rate management in soybeans. Trans. ASAE. 44, 1335-1341.

Penning de Vries, F.W.T. & Teng, P.S. 1993. Systems approach to agricultural development. Kluwer Academic Publishers, the Netherlands.

Penning de Vries, F.W.T., D.M. Jansen, H.F.M. ten Berge and A. Bakema, 1989. Simulation of Ecological Processes of Growth in Several Annual Crops, Pudoc, Wageningen and IRRI, Los Baños, 271 pp.

Ping, J.L., Dobermann, A. 2003. Creating spatially continuous yield classes for site-specific management. Agron. J. 95, 1121-1131.

Plant, R.E., Mermer, A., Pettygrove, G.S., Vayssieres, M.P., Young, J.A., Miller, R.O., Jackson, L.F., Denison, R.F., Phelps, K. 1999. Factors underlying grain yield spatial variability in three irrigated wheat fields. Transactions of the ASAE 42: 1187–1202.

Ritchie, J.T., Singh, U., Godwin, D.C., Bowen, W.T. 1998. Cereal growth, development and yield. In: Tsuji, G.Y., Hoogenboom, G.J., Thornton, P.K. (Eds.), Understanding Options for Agricultural Production. Kluwer Academic Publishers, Dordrecht, The Netherlands. pp 79-88.

Rossing W.A.H., Meynard J.M., van Ittersum M.K. 1997. Model-based explorations to support development of sustainable farming systems: case studies from France and the Netherlands. European Journal of Agronomy 7:271–283.

Saseendran, S.A., Ma, L., Nielsen, D.C., Vigil, M.F., Ahuja, L.R., 2005. Simulating planting date effects on corn production using RZWQM and CERES-Maize models. Agron. J. 97, 58-71.

Steiner, J.L., Smith, R.C.G., Meyer, W.S., Adeney, J.A., 1985. Water use, foliage temperature and yield of irrigated wheat in south eastern Australia. Australia J. Agric. Res. 36, 1-11.

Stone, M. 1973. Cross-validatory choice and assessment of statistical predictions. Journal of the Royal Statistical Society B 36: 111-147.

Sudduth, K.A., Drummond, S.T., Birrell, S.J., Kitchen, N.R. 1996. Analysis of spatial factors influencing crop yield. In:. Robert, P.C, Rust, R.H., Larson, W.E. (Eds.), Proceedings of the 3rd International Conference on Precision Agriculture. ASA, CSSA, SSSA, Inc., Madison, WI, USA. pp 129-140.

Thorp, K.R., Batchelor, W.D., Paz, J.O. and Steward, B.L. 2004. Estimating Yield and Environmental Risks Associated With Variable Rate Nitrogen Management for Corn Using Apollo. Proceedings of the 7th International Conference on Precision Agriculture. July 25-26, 2004, Minneapolis, MN. ASA, CSSA, SSSA Inc., Madison, WI, USA.

Tomer, M.D., Anderson, J.L. 1995. Field evaluation of a soil water-capacitance probe in a fine sand. Soil Sci. 159, 90-98.

Tsuji, G.Y., Hoogenboom, G., Thorton, P.K. (Eds.), 1998. Understanding options for agricultural production. Kluwer Academic Publishers, Dordrecht, The Netherlands.

Uehera, G., Tsuji, G.Y. 1993. The IBSNAT project. In: Systems approach to agricultural development. Penning de Vries, F.W.T., Teng, P.S. (eds.) Kluwer Academic Publishers, The Netherlands. p. 505-514.

Van Ittersum, M.K., Leffelaar, P.A., van Keulen, H., Kropff, M.J., Bastiaans, L., Goudriaan, J. 2003. On approaches and applications of the Wageningen crop models. European Journal of Agronomy 18: 201-234.

Wong, D.W.S. 1995. Aggregation effects in geo-referenced data. In: Arlinghaus, S.L., Griffith, D.A. (Eds.), Practical Handbook of Spatial Statistics. CRC Press, Boca Raton, FL. pp 83-106.

Wu, D., Yu, Q., Hengsdijk, H., 2006. Quantification of production potentials of winter wheat in the North China Plain. Eur. J. Agron. 24, 226-235.

Xinhua News Agency, 2001. China Plans to Promote Efficient Irrigation. http://service.china.org.cn/link/wcm/Show_Text?info_id=16509&pqry=efficient%20and%20irrigation.

Yang, Y., Watanabe, M., Zhang, X., Hao, X., Zhang, J., 2006. Estimation of groundwater use by crop production simulated by DSSAT-wheat and DSSAT-maize models in the piedmont region of the North China Plain. Hydrol. Process. 20, 2787-2802.

Yu, Q., Saseendran, S.A., Ma, L., Flerchinger, G.N., Green, T.R., Ahuja, L.R., 2006. Modeling a winter wheat-corn double cropping system using RZWQM and the RZWQM-CERES Hybrid. Agric. Syst. 89, 457-477.

Zadoks, J.C., Chang, T.T. Konzak, C.F., 1974. A decimal code for the growth stages of cereals. Weed Res. 14, 415-421.

Zhang, H., Wang, X., You, M., Liu, C., 1999. Water-yield relations and water-use efficiency of winter wheat in the North China Plain. Irrig. Sci. 19, 37-45.

Zhang, X., Chen, S., Liu, M., Pei, D., Sun, H., 2005. Improved water use efficiency associated with cultivars and agronomic management in the North China Plain. Agron. J. 97, 783-790.

Zhang, X.Y., Pei, D., Hu, C.S., 2003. Conserving groundwater for irrigation in the North China Plain. Irrig. Sci. 21, 159-166.

Zhang, Y., Kendy, E., Qiang, Y., Changming, L., Yanjun, S., Hongyong, S., 2004. Effect of soil water deficit on evapotranspiration, crop yield, and water use efficiency in the North China Plain. Agric. Water Manage. 64, 107-122.

Zhao, J.R., Guo, Q., Guo, J.R., Wie, D.M., Wang, C.W., Liu, Y., Lin, K., 1997. Investigation and analysis on current status of chemical fertilizer inputs and crop yields in agricultural field of Beijing Subburb. J. Beijing Agric. Sci. 15, 36-38 (in Chinese).

Zhao, R., Chen, X., Zhang, F., Schroder, H., Römheld, V., 2006. Fertilization and Nitrogen Balance in a Wheat-Maize Rotation System in North China. Agron. J. 98, 938-945.

Zhu, Z.L., Chen, D.L., 2002. Nitrogen fertilizer use in China - Contributions to food production, impacts on the environment and best management strategies. Nutr. Cycling Agroecosyst. 63, 117-127.

Part 2

Water Management and Crop Production

Crop Water Requirements in Cameroon's Savanna Zones Under Climate Change Scenarios and Adaptation Needs

Genesis T. Yengoh[1], Sara Brogaard[2] and Lennart Olsson[2]
[1]Department of Earth and Ecosystem Sciences Division of Physical Geography and Ecosystem Analysis Lund University Sölvegatan 12, SE-223 62 Lund,
[2]Lund University Centre for Sustainability Studies Geocentrum 1, Sölvegatan 10, Lund
Sweden

1. Introduction

Rain-fed agriculture is practiced on approximately 80 percent of global agricultural land area (Wani et al. 2009). It accounts for about 70 percent of the global staple foods production (Cooper et al. 2009). This is the main mode of production favored by poor farmers in the developing world and other economically deprived societies (Wani et al. 2009). The contribution to global food supply from rain-fed agriculture is forecasted to decline from 65 percent at present to 48 percent in 2030 (Bruinsma 2003). The decline of precipitation forecast for some regions of the African savanna may affect agricultural production in different ways. During the already dry months, the decline of precipitation is likely to reduce the resilience of some plants (Vanacker et al. 2005). This is especially true for many ecosystems in Sub-Saharan Africa, particularly grass and shrub savannahs, which are shown to be highly sensitive to short-term availability of water due to climate variability (Vanacker et al. 2005). While the ranges of species may shift as a result of changing climate, this shift may probably not be in cohesive and intact units and are likely to become more fragmented (Channell and Lomolino 2000). This stimulates interest in understanding the effects of climate change on the potential for rain-fed agriculture for particular regions where this practice has important economic and social implications.

While observed and measured data increasingly support predictions of a warmer world in the next 50 - 100 years, the impact of rising temperatures on rainfall distribution patterns in the semi-arid tropics of Africa remain far less certain (Cooper et al. 2009). African countries are particularly vulnerable to climate change because of their dependence on rain-fed agriculture, low economic power, low levels of human and physical capital, and poor infrastructure (Nelson 2009). The negative effects of climate change on agricultural production are especially pronounced in Sub-Saharan Africa because of the significant contribution of the agricultural sector to the GDP, export earnings, and employment (Fan et al. 2009).

Cameroon, like many countries in sub-Saharan Africa has a high share of total poverty in the rural sector and a high share of GDP growth originating in agriculture (De Janvry 2009). Less than 1 percent of total agricultural land area in Cameroon is equipped for irrigation

(FAOSTAT 2010). Small-scale food crop production in Cameroon's savanna zones is heavily dependent on timing and length of the rainy season. The timing and length of the rainy season is strongly dependent on and directly influenced by the movement of the Inter-tropical Tropical Convergence Front. This subjects the region to a pronounced seasonality with four to six months of rainy season in which most of the rain-fed food crop production takes place and six to eight months of dry season with no opportunities for non-irrigated food production (Yengoh et al. 2011). Food production among small-scale farmers is therefore very dependent on the reliability of the onset of rains and the distribution of rainfall during the rainy season (Yengoh et al. 2010) and rain-fed agriculture is one of the most vulnerable livelihood and economic sectors to climate change in the these regions. Since 1960, four major droughts and two floods have affected different parts of Cameroon's savanna zones with considerable effects on food production and human well-being (CRED 2011). The frequency of weather-related crop failures in Cameroon's savanna zones points to the vulnerability of food production in this region to future climate change. On a larger context, the vulnerability of this region mirrors the situation of the country's socio-economic life to forces of climate because of the importance of this region as a major production zone for cereals, pulses and livestock (Yengoh et al. 2011).

While installed irrigation capacity is generally low in Sub-Saharan Africa compared to that of many other regions (FAOSTAT 2010), climate change may further worsen the irrigation water supply reliability in areas with installed irrigation infrastructure (Nelson 2009). An appreciation of the expected impact of climate change to agricultural communities dependent on rain-fed production can guide policy-makers in designing strategies for mitigation and adaptation (Yengoh et al. 2010).

In this study, the CROPWAT model (Clarke et al. 1998) is used to compute crop water requirements and yield reduction due to soil moisture stress for the Guinea- and Sudan Savanna agro-ecological zones of Cameroon for baseline climate conditions (1961-1990) as well as three scenarios from the Special Report on Emissions Scenarios (SRES). Tools of geographical information systems are used to estimate the impact of climate change on the potential area for rain-fed agriculture and crop water requirements of five main food crops in this region. The goal of the analysis is to assess the impact of climate change at finer scales and explore avenues for adaptation. Besides assessing the impact of climate change on yield reduction, we estimate the crop area that will be suitable for the cultivation of major food crops of the region under different water requirement situations. The results are used to discuss the implications for agricultural planning at national, regional and farm level, as well as implications on the vulnerability of small-scale farmers. Based on such implications, a climate change adaptation portfolio for small-scale farming systems for savanna zones of Sub-Saharan Africa is proposed. This study complements others that assess the impact of climate change at country and regional level (Parry et al. 2004, Molua and Lambi 2006, Iglesias et al. 2007, Morton 2007).

2. Description of study area

The study area is located between latitude 6°N and 13°N (Figure 1). This region covers an area of about 163 513 Km² and makes up more than 35 percent of the total area of the country, 475 500 Km². It comprises the Guinea- and Sudan-Savanna agro-ecological zones of the country and is the seat of most of the country's food crops and livestock production (Yengoh et al. 2011). This zone is interesting as a case study because its agro-ecological characteristics, farming systems and the economic situation of farmers here can be seen in large swaths of the continent, from Guinea in the west to Ethiopia in the East. Knowledge of production constraints as well as issues of mitigation and adaptation in this study area can therefore be applicable to similar regions in Sub-Saharan Africa.

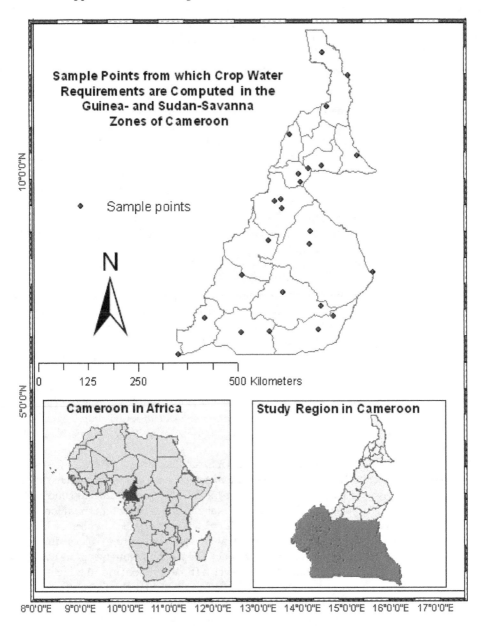

Fig. 1. Location of the study area.

Rainfall varies between 1 200 mm and 2 300 mm annually around latitude 6°N to about 850 mm annually in the neighborhood of latitude 13°N. Population density is low in the southern parts of the region, at 13 and 20 persons per km² in the Adamawa and North administrative regions respectively and 88 persons per km² in the Far North administrative region (above latitude 10°N). The national average population density is approximately 38 persons per km². Agriculture is the main economic activity in Cameroon, practiced by more than 80 percent of the working population and contributing to >45 percent of GDP (Laux et al. 2010).

3. Sources of data and analysis methods

We use monthly climate data from the National Meteorological Service of Cameroon in Douala. These data which become inputs into the CROPWAT model for the computation of crop water requirements include temperature, humidity, wind speed and sunshine. For nine out of the twenty-five stations where such station data is absent, New LocClim is used to derive values for these areas by interpolation (see appendix for list and location of stations). New LocClim is a tool, capable of producing average monthly climate values (up to eight variables) taken from the agro-climatic database of the Agromet Group of the Food and Agriculture Organization of the United Nations (Grieser et al. 2006). Besides allowing for an extensive investigation of interpolation errors, and the influence of different settings on the results, New LocClim also allows for the optimization of the interpolation with respect to the data analyzed. Average monthly temperature and rainfall for three SRES emission scenarios for 2050s are derived from the HadCM3 model of the Hadley Centre for Climate Prediction and Research, Meteorological Office, United Kingdom.

The data is used to calculate crop water requirements using CROPWAT 8.0 for Windows. CROPWAT is a decision support system developed by the Land and Water Development Division of the Food and Agriculture Organization of the United Nations Organization to calculate potential evapotranspiration, crop water requirements and crop irrigation requirements (Clarke et al. 1998). Procedures for calculating variables for crop water requirements used in CROPWAT are based on FAO guidelines extensively presented in the publication No. 56 of the Irrigation and Drainage Series of FAO "Crop Evapotranspiration - Guidelines for computing crop water requirements" (Allen et al. 1998). CROPWAT has been used extensively for computing crop water requirements for different crops, case studies and at different scales (Molua and Lambi 2006, Iglesias et al. 2007, Traore et al. 2007, Stancalie et al. 2010).

We begin by calculating crop water requirements for 27 stations using climate data for baseline conditions for 1961-1990. Then baseline rainfall and temperature data is replaced with data from the HadCM3 model output. This allows for the calculation of simulated conditions of evapotranspiration, and effective rainfall based on rainfall and temperature conditions of the 2050s. While we assume constant relative humidity, wind speed and number of hours of sunshine, radiation changes with the introduction of new temperature values during the calculation of evapotranspiration using modeled data. Crop water requirements are calculated for three SRES story-lines: the moderate, mid-level A1B carbon scenario, the higher, more extreme A2 carbon scenario, and the more optimistic B1 carbon scenario. These story-lines are alternative images of how the future might unfold (IPCC 2007).

Crop water requirements for baseline, A2, A1B, and B1 scenarios are then interpolated on the study area using inverse distance weighting (IDW) with ArcEditor 9.3. The choice of

IDW was made because this interpolation method provides a simple way of guessing the values of a field at locations where no measurement is available. It estimates unknown measurements as weighted averages over the known measurements at nearby points, giving the greatest weight to the nearest points. IDW is therefore ideal in obtaining a smooth surface whose value at any point is more like the values at nearby points than the values at distant points (Longley 2005). In calculating the potential area for rain-fed crop production potential, land occupied by a number of land land-use types is eliminated. These include land occupied by urban and rural settlements, protected areas, swamps, water bodies and land occupied by irrigated agriculture.

Yield reduction due to soil moisture stress is expressed as a percentage of the maximum production achievable in the area under optimal soil moisture conditions. It is computed with reference to the whole growing season based on (Allen et al. 1998, Clarke et al. 1998):

$$(1 - Ya/Ymax) = Ky \ (1 - ETc \ adj \ / \ ETc)$$

Ya = Yield achievable under actual conditions
$Ymax$ = Maximum crop yield achievable in case of full satisfaction of crop water needs
Ky = Yield response factor
$ETc \ adj$ = Crop evapotranspiration under non-standard conditions
Etc = Crop evapotranspiration under standard conditions

A choice of five crops, (all of which are important for farmers in this region) is used to assess the effects of climate change on the potential for rain-fed agriculture in Cameroon's savanna zones.

Beans are the main source of non-animal protein among resource poor farmers, have a short growing cycle and are cultivated widely among small-holder farmers. Surpluses are sold in local markets and end up in markets in the south of the country.

Groundnuts are an important source of cooking oil in the savanna zones where oil palms do not grow. Together with beans, they have a short growing cycle, are widely cultivated, and help boost soil nitrogen levels. Besides being widely used on traditional foods, groundnuts are also a very important export crop. Producers in the Far North Region sell to buyers from Nigeria, Chad and the Central African Republic.

Maize has, over the last decade been replacing millet as the stable food in some traditional diets of the region. Besides, its growing importance as a food crop, the relatively high yields of maize, relative to the millet it is replacing and the steady demand fueled by consumption both within and out of the country is giving maize and increasing importance in the region.

Potatoes are seen as local luxury food stuff with high demand from both local buyers and merchants from neighbouring countries. Recently introduced into the region, potato cultivation is growing rapidly.

Sorghum remains one of the main food crops among most communities in this part of the country. It continues to command both local and foreign demand and so is widely grown.

4. Results and discussion

4.1 Yield reduction due to soil moisture stress

Mean yield reduction due to soil moisture stress is greater for all crops in the A1B and A2 scenarios relative to the baseline (Figure 2). Potato shows the greatest mean yield reductions of 4.1 percent and 7.7 percent for the A1B and A2 scenarios respectively and 1.5 percent for the B1 scenario. The mean yield reduction for potato in the baselines scenario is 2.3 percent. The relatively drier conditions of the Sudan-savanna zone in the far north of the study area severely restricts the cultivation of potato and accounts for this high mean reduction in yields. In these parts of the country, maximum yield reductions of up to 35.2 and 50.4 percent are obtained in the A1B and A2 scenarios respectively (Figure 2). This is followed by bean, with mean yield reductions of 3.4 percent for the A1B scenario 7.2 percent for A2 and 0.6 percent for the B1 scenario. The largest maximum yield reductions are observed in bean, 36.7 and 55.1 percent for the A1B and A2 scenarios respectively. While maize and groundnuts reveal lower mean and maximum yield reductions compared to potatoes and beans, sorghum has the least reductions in yield, 0.6 and 1.9 percent in the A1B and A2 scenarios respectively. In the baseline and the B1 scenario, the yield reduction is negligible (Figure 2). All crops serve the dual purpose of household consumption and income generation from surpluses. However, some crops are more important for income generation, such as potato, groundnuts and beans, and are more heavily impacted by yield reduction that those required for basic household food supply, sorghum and maize.

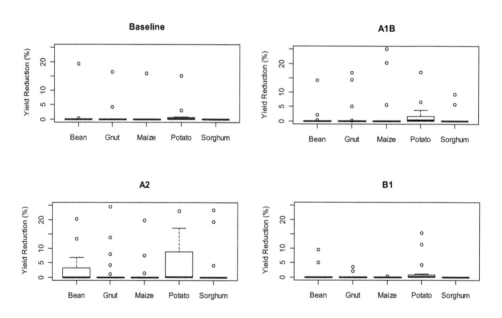

Fig. 2. Yield reduction due to soil moisture stress for all crops and scenarios. Notice the large variations resulting from differences in agro-climatic resources between the Guinea- and Sudan-savanna agro-ecological zones. Also see Appendix 1 for a table of the descriptive statistics.

While the study area is collectively described as the savanna zone of Cameroon, some parts of this zone are marginal in their agricultural production potential. In the Sudan-Savanna agro-ecological zone (above latitude 12oN), the short rainy season of only about three months, loose sandy soils, and high temperatures impose problems of agricultural water availability which reduce crop production potential here. Crop production potential in this agro-ecological zone is therefore in a precarious balance based on these water-related constraints to the length of the growing season, the soil's water-holding capacity, rates of crop evapotranspiration, and others. Small changes in temperature, as well as the amount or pattern of precipitation may result in these zones tipping over to areas of limited crop production potential. The high maximum reduction in yields resulting from soil moisture stress in the A1B and the A2 scenarios is attributed to the steep decline of crop productivity in the Sudan-Savanna agro-ecological zone due to such changes in temperature and precipitation changes forecasted for these scenarios. For maize and sorghum, much of the area which was formerly favorable for rain-fed production becomes unfavorable (requiring irrigation for optimal yields without water stress).

Other studies have established relationships between climate change and yields at different levels and different time scales. Ex-post statistical analysis of relationships between growing season temperature, precipitation and global average yield for six major crops, estimated that warming since 1981 has resulted in annual combined losses of 40 million tons or US$5 billion (Lobell and Field 2007). Using two scenarios of future temperature increases (1.5°C from MAGICC/SCENGEN model and 3.6°C rise from Global Circulation Model results), previous studies have predicted a reduction in wheat grain and other field crops yield of up to 30 percent and increase in its water needs of about 3 percent (El-Shaer et al. 1997, Eid and El-Mowelhi 1998) in Egypt by the year of 2050. Studies using crop simulation models such as CropSyst also forecasts yield reductions in Egyptian grain wheat by 2038 (Khalil et al. 2009). Sub-Saharan Africa is increasingly being corroborated as being the region with the most effects of climate change on yield reduction. By using an integrated approach to analyze the correlation between historical crop yield and meteorological drought, Africa emerges with the highest drought risk index (from 95.77 at present to 205.46 in 2100), with yield reductions increasing accordingly by >50 percent in 2050 and almost 90 percent in 2100 (Li et al. 2009). Average rice, wheat and maize yields are forecast to decline by up to 14 percent, 22 percent, and 5 percent, respectively, in Sub-Saharan Africa as a result of climate change (Nelson 2009). It is estimated that yield reductions of up to 50 percent may be experienced for some crops by 2020 (Boko 2007).

5. Fall in area with potential for rain-fed agriculture

Under baseline conditions, sorghum and maize offer the greatest opportunities for rain-fed cultivation. The potential is low for bean and groundnuts and inexistent for potatoes. The amount of area with potential for rain-fed agriculture falls considerably for sorghum and maize, especially for the baseline scenario relative to the A1B and A2 scenarios (Figure 3). This emphasizes the idea of certain agro-ecological zones being in a precarious balance in relation to crop water requirements discussed previously. The precarious balance of agricultural water conditions for the cultivation of rain-fed maize and sorghum is offset by precipitation and temperature changes resulting from climate change by the middle of the century. The outcome is a significant fall in the amount of area with rain-fed potential for the cultivation of sorghum and maize.

The area with rain-fed potential for bean increases in the B1 scenario relative to the baseline but also fall for the A1B and A2 scenarios. A fall in the potential area for the rain-fed cultivation or its absence does not however mean that these crops are not being cultivated anyways. It indicates that based on the water requirements of these crops, maximum obtainable yields cannot be achieved without irrigation (Clarke et al. 1998). The amount of area with the potential for rain-fed cultivation of different crops in the baseline scenario relative to the total cultivable area is sorghum (79.47 percent), potato (0 percent), maize (39.48 percent) groundnuts (2.3 percent), and bean (4.28 percent). This points water availability being an important constraint to food crop production in this region even at the present.

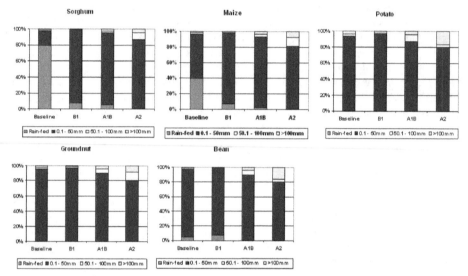

Fig. 3. Amount of land (in percent of total arable land) for cultivation under non-water stressed conditions of analyzed crops in the baseline conditions (1960-1990) and three climate scenarios for 2050 average conditions. The amount of land under different water requirements is visualized in three different intervals: rain-fed, ≤50mm, 50.1 - ≤100mm, ≥100mm.

The low potential for rain-fed cultivation of potato, groundnut and beans even in the baseline scenario (Figure 3) also highlights the importance of water as an important element determining the yield of food crops in this region. This is especially important for food security at household level for two reasons.

1. Small-scale farmers who make up more than 90 percent of farmers in this region are heavily dependent on rain-fed food production.
2. The few economic resources of these small-scale farmers reduce their ability to invest in water-provision technologies to improve yields.

The above reasons may likely be reinforced by the effects of climate change. The average optimum temperature range for sorghum is 21°C to 35°C for seed germination, 26°C to 34°C for vegetative growth and development, and 25°C to 28°C for reproductive growth (Maiti 1996). Mean annual temperatures in Cameroon's savanna regions are 22.02°C, 27.95°C, and 28.34°C for Ngoundere, Garoua and Maroua respectively with diurnal ranges of up to 13°C. Given the strong seasonality, growing season temperatures ranges are between 7°C and 10°C. Mean annual temperature is projected to increase by about 1.0°C to 2.9°C by the 2060s with 'hot' days occurring in 20-51 percent of days at this time (McSweeney et al. 2010). Such changes could have significant negative effects on the productivity of grain crops, including sorghum (Prasad et al. 2008). This may explain the significant fall in the area with rain-fed potential for sorghum and maize (Figure 3).

6. Requirements for irrigated agriculture

A majority of the area in Cameroon's savanna regions will be suitable for the cultivation of the five main crops being investigated if crop water requirements of up to 50mm are met (Figure 3). With the exception of sorghum in which most of the land area is assessed to be suitable for rain-fed agriculture in the baseline, all other crops are suitable for cultivation on >80 percent of the land given irrigation of up to 50mm (Figure 3). The potential area for the cultivation of crops under non-water stressed conditions given crop water satisfaction of both 50mm – 100mm and >100mm is low (Figure 3). This area increases however for all crops under the A1B and A2 scenarios relative to the baseline (Figure 4). With the exception of potato and beans in the A2 and A1B scenarios, less that 10 percent of the total area of Cameroons savanna will require irrigation of > 50 mm for the cultivation of any of the five crops.

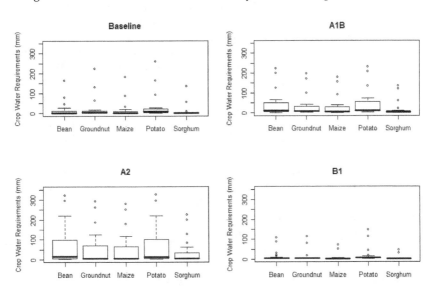

Fig. 4. Crop water requirements for all crops and scenarios.

Crop water requirements under different scenarios will however be different for different crops. Water requirements increase for all crops in the A1B scenario with the exception of sorghum (Figure 4). In the A2 scenario on the other hand, water requirements are appreciably higher for bean and potato, but generally high for all crops (including sorghum). Such requirements fall in the B1 scenario relative to the baseline for all crops (Figure 4).

7. Implications for agricultural planning and the drive towards sustainable agriculture

Concerns over human-induced land degradation, including a large-scale decline in soil fertility by 'soil mining' especially in sub-Saharan Africa have been raised (Mokwunye et al. 1996). The impact of climate change on the potential area for rain-fed agriculture will likely compound these challenges. Population growth, increase of urbanization and associated changes in dietary preferences contribute to increasing the production pressure on African agriculture. These challenges contribute to imposing substantial pressure for the transformation of agriculture in Sub-Saharan Africa. The impetus for such transformation will likely be the pressure of shrinking agricultural land in a continent whose demography and economics is arguably poised to transform significantly in the next few decades (Sanchez et al. 2009, Sánchez 2010). Evidence of such Boserupian drivers of change has been reported in many cases of agricultural transformations (Boserup and Kaldor 2005, Demont et al. 2007). Studies already point to the possibility of tripling yields per hectare with investments in the appropriate combination of governance and material resources within a framework of a strong commitment to sustainable organic agriculture (Sanchez et al. 2009, Sánchez 2010).

Priorities at national and regional level

The adoption of sustainable systems of agricultural intensification may therefore be necessary to meet the numerous food production challenges associated with this transformation (Davies and Chaves 2010). While some of such technologies like crop residue management, improved fallow, agro-forestry and others have proven successful in many parts of the world, their success in Africa will be seriously influenced by many of the peculiar socio-economic and political realities of this region (Yengoh et al. 2009).

The drive towards sustainable intensification is urged as an approach which can permit the more sustainable use of resources in agriculture, giving rise to higher water use efficiencies, more sustainable nutrient management, optimal use of available agricultural land and others (Baulcombe et al. 2009, Williams 2009). The attainment of sustainable implementation as a means of adapting to the need to increase food production while meeting challenges imposed by global environmental changes, changes in human demographics and culinary preferences also demand addressing non-biophysical challenges of small-scale farmers. The attainment of sustainable intensification calls for the dissemination and support for tested practices of sustainable agriculture (Table 1) which respond to the social, economic and cultural realities of would-be practitioners. For small-scale farmers in Sub-Saharan Africa, it also entails addressing problems associated with the economic and social context within which agriculture is practiced. Such measures should ideally address three key issues:

-    Make agriculture profitable enough for farmers to be able to afford alternative means of fertilization, farm clearing and the provision of forage for livestock.
-    Provide affordable access to the resources and information necessary for adapting more sustainable techniques in agriculture.
-    Develop agricultural governance structures at local level and ensure their effectiveness by eliminating setbacks like corruption and administrative red-tape in the implementation of sustainable agriculture projects.

Priorities at farm level

Besides an increase in atmospheric temperatures of between 1.4°C to 5.8°C, climate change is expected to bring greater variability in some climatic event that are important for food crop production (IPCC 2007). These include events such as the increased frequency of short episodes of extreme temperatures increases and associated stresses to crops (IPCC 2007). Farmers may have to revise their structure of decision-making to accommodate an increasingly variable climate. There will be a need for resilience in decision-making with regards to choice of crop-types to cultivate on a season-to-season basis, the timing of farm operations, the choice of systems to improve water availability and use, method of erosion control, and other production-related decisions. Short- and medium-term decisions on food production would have to be made on-the-fly based on weather and climate forecasts in a new order of increased climate variability.

Sustainable practice	Agro-ecological goals	Benefits for small-scale farming systems
Crop residue management	- Optimize benefits from multifunctional uses of crop residues - Improve soil organic matter	Naturally decaying residues improve soil organic matter structure and content (Moyin-Jesu 2007). Residues may serve other purposes like being used for animal feed or straw
Agro-forestry	- Introduce multifunctional trees into agro-ecosystems - Improve agro-ecosystem resilience	Agricultural risk is diversified as agro-ecosystems become more resilient (Wojtkowski 2008). The system offers other needed products besides food like firewood & forage. It is effective in restoring disturbed land (Peng et al. 2009)
Integrated water management	- Sustain plant water availability for longer periods - Improve on plant water use efficiency	Minimizes the problem of increasing scarcity of water for agricultural purposes and improves water productivity (Tenywa and Bekunda 2009)
Integrated pest management	- Control pests and diseases - Minimize/eliminate the use of pesticides	Increases agricultural biodiversity both in the soil and above ground. Reduces the risk of environmental pollution from pesticide use (Baulcombe et al. 2009)
Integrated nutrient management	- Generate nutrients within farm systems - Minimize/eliminate dependence on external nutrient sources	Plant residue can become inputs into composting systems for organic fertilizers (Moyin-Jesu 2007). Optimizes the flow of farm-scale biogeochemical processes
Conservation tillage	- Conserve soil structure - Improve on the flow of biogeochemicals	Naturally decaying residues improve soil organic matter structure and content (Moyin-Jesu 2007).
Livestock integration	- Generate nutrients within farm systems - Improve the system & efficiency of nutrient cycling & flows	Creates a system that is less dependent on outside sources of fertilizers & more resilient to food crop natural and price fluctuations
Integrated erosion management	- Conserve soil structure - Conserve the top soil & soil organic matter	Sustains soil fertility and productivity (Baulcombe et al. 2009)

Table 1. The contribution of sustainable agricultural practices in enhancing farmland productivity among small-scale farmers.

8. Implications for the vulnerability of small-scale farmers

Climate change is expected to increase global net crop irrigation requirements by 5-8 percent by 2070, with considerable regional variation (Döll 2002). The vulnerability of agriculture, like most other sectors in Sub-Saharan Africa is aggravated by the interaction of multiple stressors affecting the sector at different levels (Boko 2007). The estimated increase of 5 – 8 percent in the proportion of arid and semi-arid lands on the continent (Boko 2007) points to a future with reduced input of precipitation especially for the savanna regions which buffer the already dry African Sahel. The situation in Cameroon's savanna zone is a mirror of what can be found in many parts of the continent. Small-scale farmers continue to practice food production notwithstanding the reduction in yields imposed by water stress. While semi-arid conditions are already imposing challenging conditions for food production among many small-scale farmers, it is estimated that yield reductions of up to 50 percent may be experienced by 2020 (Boko 2007). Studies warn of impending global problems associated to food security and access to water if appropriate action is taken to improve water management and increase water use efficiency (Falkenmark and Rockström 2008, Ringler et al. 2010).

Stagnation in land and agricultural productivity in sub-Saharan Africa, and hence the slow drive towards attaining food security goals is attributed to two main factors: lack of expansion of area under irrigation and limited use of chemical fertilizers (De Janvry 2009, Jayne et al. 2010). While the donor community aspires to replicate the success of irrigation projects from other regions of the world to benefit agriculture in Sub-Saharan Africa, the scale of need involves overcoming some huge challenges. These include issues related to the choice of technology, competition for water between sectors, a relative lack of water management expertise, the cost of maintaining and replacing worn out equipment; uncertain markets and related barriers; and the uncertainties of Sub-Saharan African weather (Lankford 2009).

At the level of farmers, adaptation may take the form of either changing or increasing their use of farm inputs with which they are already familiar (factor substitution), or adopt new technologies and production methods through a learning process (Binswanger and Pingali 1988). The availability of new technology does not however mean instant adoption. The process of technology adoption among small-holder farmers in Sub-Saharan Africa is a complex one involving the interplay of many complex drivers (Yengoh et al. 2009). The effects of adaptation to climate change on yields are expected to be different for different regions. With adaptation in the low latitudes, the yields of maize, wheat and rice maintains at current levels given temperature changes of +1 to +2°C. At changes of +2 to +3°C, adaptation maintains yields of all crops above baseline. At this range, yields drop below baseline for all crops without adaptation. At changes of +3 to +5°C, yields of major crops like maize and wheat are reduced below baseline regardless of adaptation, but adaptation maintains rice yield at baseline levels (Easterling et al. 2007).

9. Equipping small-scale farming systems in savanna regions of Sub-Saharan Africa to face challenges of climate change and variability

The challenge of equipping small-scale, farming communities to face agricultural water shortages resulting from climate entails developing a mitigation portfolio which addresses key aspects of mitigation preparedness. These key components are access to appropriate resources, the presence of a clearly thought out strategy, at cooperation in information development, knowledge and skill sharing (Figure 5). These components are essential in addressing key social, economic and knowledge gaps of small-scale farming communities in the Sub-Saharan savanna.

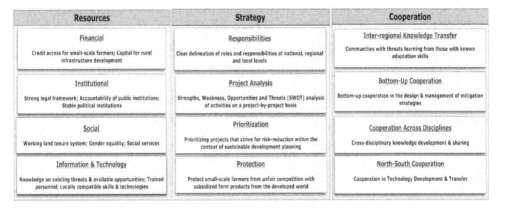

Resources	Strategy	Cooperation
Financial Credit access for small-scale farmers; Capital for rural infrastructure development	**Responsibilities** Clear delineation of roles and responsibilities at national, regional and local levels	**Inter-regional Knowledge Transfer** Communities with threats learning from those with known adaptation skills
Institutional Strong legal framework; Accountability of public institutions; Stable political institutions	**Project Analysis** Strengths, Weakness, Opportunities and Threats (SWOT) analysis of activities on a project-by-project basis	**Bottom-Up Cooperation** Bottom-up cooperation in the design & management of mitigation strategies
Social Working land tenure system; Gender equality; Social services	**Prioritization** Prioritizing projects that strive for risk-reduction within the context of sustainable development planning	**Cooperation Across Disciplines** Cross-disciplinary knowledge development & sharing
Information & Technology Knowledge on existing threats & available opportunities; Trained personnel; Locally compatible skills & technologies	**Protection** Protect small-scale farmers from unfair competition with subsidized farm products from the developed world	**North-South Cooperation** Cooperation in Technology Development & Transfer

Fig. 5. Components of climate change adaptation portfolio for small-scale farming systems in savanna zones of Sub-Saharan Africa.

In Cameroon and many countries in the regions, a number of obstacles challenge the development of each of the components in this climate change mitigation portfolio.

1. Resources:
   Access to financial, institutional, social and information resources are necessary to plan, implement and sustain adaptation efforts. Financial resources are inadequate, and investment in credit systems for small-scale farmers or development of rural infrastructure is limited. The absence of a strong legal system breeds and perpetuates the proliferation of corruption and lack of public accountability at different levels of governance. Where political instability prevails, resources cannot be marshaled for purposes of enhancing mitigation, since people strive to survive on a day to day basis. While many Cameroonians increasingly question the fairness and representation of the democratic in their country, Cameroon has remained an island of relative peace in a continent troubled by many different forms of political strife. The land tenure system remains largely based on traditional cultures with small-scale farmers having virtually no formalized rights to landed property their families have owned for several generations. Social safety nets are very limited and reduce risk-taking in production decisions among small-scale farmers.

2. Strategy:
   The importance, urgency and scale of threats posed by climate change on agriculture demands careful strategizing. Strategies should develop a clear road-map which addresses different scales and factors for achieving desired mitigation goals. They should also address the distribution of responsibilities and identify projects to meet desired short- and long-term objectives. There is need to prioritize projects and activities towards attaining the desired level of mitigation based on the availability of

human and technological resources. Agricultural policies in OECD countries are seen to damage agricultural production in developing countries. Small-scale farmers in developing countries are unlikely grow and attain different goals of sustainability if they continue to compete subsidized products from the developed world. Notwithstanding the insufficiencies of small-scale farming systems in Sub-Saharan Africa, world trade reform in some crops stand a good chance of benefiting many of the small-scale farmers of the region (Jayne et al. 2010). A level of protection for the small-scale farmers is needed in the face of subsidies of agricultural products from the developed world to create an environment in which small-scale farmers can expand their output, profits and viability.

3. Cooperation:

Cooperation between regions and cultures in sharing experiences and knowledge on different aspects of mitigation can be very useful (Yengoh et al. 2010). In order to meet the challenge of integrating different sectors that relate to and influence small-scale farming systems, there is need for cooperation across disciplines and socio-economic sectors. A bottom-up approach ensures that the outcomes of climate change mitigation efforts respond to the needs of those truly in need (the small-scale farmers at local level). The need for cooperation at international level in knowledge development and resources transfer has been raised as important in assisting systems of developing countries in their drive towards developing mitigation strategies and adapting to climate change.

10. Conclusion

Cameroon's savanna regions do not offer the full potential for rain-fed production for some of the crops studied in this area. In the baseline scenario, the potential is especially low for potatoes, groundnuts and beans, and relatively high for sorghum and maize. This potential area for rain-fed production falls considerably for sorghum and maize by 2050 in the A2, A1B and B1 SRES scenarios relative to the baseline (1961-1990). By meeting crop water requirements of up to 50 mm the greatest amount of area in which maximum obtainable yields (in the absence of water stress) can be expected for all crops studied is achieved in all SRES scenarios. Mean yield reduction due to soil moisture stress is greatest for potato, maize and bean, especially in the A2 and A1B scenarios. National and regional level planning should accelerate the process of introducing and up-scaling tested processes and practices of agricultural intensification. At the farm level, decision-making should become more resilient, capable of changing at short and medium scales to accommodate new climate related adaptive imperatives. The adaptation portfolio of small-scale farmers in Cameroon's savanna regions should comprise of three groups of components: the necessary resources for adaptation, a clear strategy of response with carefully shared responsibilities for meeting prioritized targets, and an indispensable cooperation strategy.

11. Supplementary material

Appendix 1. Descriptive statistical table of yield reduction due to soil moisture stress for the study region. It is expressed as a percentage of the maximum production achievable in the area under optimal conditions.

Baseline Scenario	Minimum	Maximum	Mean	Median	Standard Deviation	Standard Error
Bean	0.0	19.2	0.8	0.0	3.8	0.8
Groundnut	0.0	16.4	0.8	0.0	3.4	0.7
Maize	0.0	16.0	0.6	0.0	3.2	0.6
Potato	0.0	34.5	2.3	0.1	7.4	1.5
Sorghum	0.0	0.0	0.0	0.0	0.0	0.0
Scenario A1B						
Bean	0.0	36.7	3.4	0.0	9.8	2.0
Groundnut	0.0	16.7	1.4	0.0	4.4	0.9
Maize	0.0	25.1	2.0	0.0	6.3	1.3
Potato	0.0	35.4	4.1	0.2	9.6	1.9
Sorghum	0.0	9.1	0.6	0.0	2.1	0.4
Scenario A2						
Bean	0.0	55.1	7.2	0.0	15.6	3.1
Groundnut	0.0	27.2	3.2	0.0	7.5	1.5
Maize	0.0	45.8	4.6	0.0	12.3	2.5
Potato	0.0	50.4	7.7	0.1	14.8	3.0
Sorghum	0.0	23.4	1.9	0.0	5.9	1.2
Scenario B1						
Bean	0.0	9.5	0.6	0.0	2.1	0.4
Groundnut	0.0	3.6	0.2	0.0	0.8	0.2
Maize	0.0	0.2	0.0	0.0	0.0	0.0
Potato	0.0	15.3	1.5	0.1	3.7	0.7
Sorghum	0.0	0.0	0.0	0.0	0.0	0.0

Appendix 2. Combined maps of potential area for the cultivation of different crops given different crop water requirements (derived from interpolation).

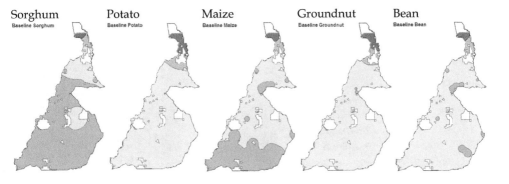

Sorghum   Potato   Maize   Groundnut   Bean

Baseline Sorghum   Baseline Potato   Baseline Maize   Baseline Groundnut   Baseline Bean

Sorghum   Potato   Maize   Groundnut   Bean

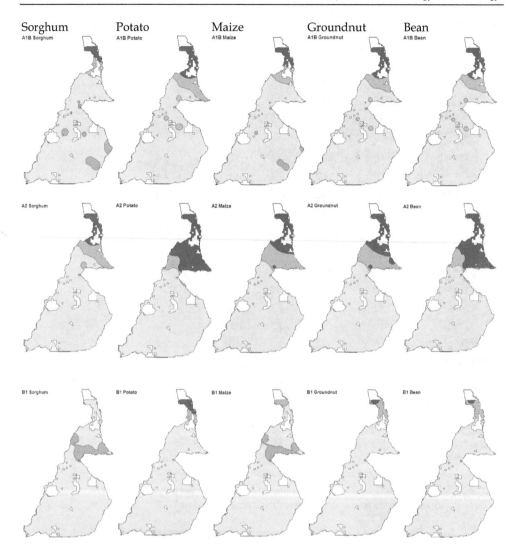

Legend

0mm Water Requirement

0.1mm - 50mm Water Requirement

50.1 - 100mm Water Requirement

>100mm Water Requirement

Area for Other Land Uses Clipped Out

Appendix 3. Data derived from modeling and used for interpolation for all crops and all stations.

Location	Crop	Crop Water Requirements (mm)				Yield Reduction (%)			
		Baseline	A1B	A2	B1	Baseline	A1B	A2	B1
Yagoua Lat: 10,33 Long: 15.23	Beans	0,7	50,4	131,9	4,8	0,0	0,2	13,3	0,0
	Groundnuts	17,7	30,8	104,0	5,4	0,0	0,0	4,3	0,0
	Maize	0,3	29,8	96,6	0,0	0,0	0,0	1,4	0,0
	Potato	21,3	55,6	134,4	10,1	0,7	3,7	17,1	1,0
	Sorghum	0,7	10,8	61,1	0,0	0,0	0,0	0,0	0,0
Figuil Lat: 9,75 Long: 13,97	Beans	0,0	9,2	44,5	1,3	0,0	0,0	0,0	0,0
	Groundnuts	0,0	3,5	24,2	1,9	0,0	0,0	0,0	0,0
	Maize	0,0	0,8	30,6	0,0	0,0	0,0	0,0	0,0
	Potato	2,4	7,1	43,3	3,6	0,6	1,2	1,7	0,7
	Sorghum	0,0	0,0	5,9	0,0	0,0	0,0	0,0	0,0
Guider Lat: 9,93 Long: 13,93	Beans	0,3	0,0	4,9	1,3	0,0	0,0	0,0	0,0
	Groundnuts	0,0	3,5	2,5	1,9	0,0	0,0	0,0	0,0
	Maize	0,0	0,0	4,5	0,0	0,0	0,0	0,0	0,0
	Potato	2,5	4,5	3,2	2,4	0,7	1,2	0,0	0,7
	Sorghum	0,0	0,0	4,5	0,0	0,0	0,0	0,0	0,0
Mora Lat: 10,05 Long: 14,15	Beans	0,0	64,6	152,4	6,3	0,0	2,1	20,3	0,0
	Groundnuts	14,2	42,5	123,2	7,1	0,0	0,1	8,0	0,0
	Maize	0,0	40,7	117,4	0,8	0,0	0,0	7,6	0,0
	Potato	19,9	72,9	156,8	14,1	0,5	6,5	23,0	1,0
	Sorghum	0,6	22,0	82,2	0,8	0,0	0,0	0,0	0,0
Mokolo Lat: 10,78 Long: 13,73	Beans	0,0	28,8	97,3	1,5	0,0	0,0	3,2	0,0
	Groundnuts	15,0	13,2	68,8	2,4	0,0	0,0	0,1	0,0
	Maize	0,0	14,6	64,8	0,0	0,0	0,0	0,0	0,0
	Potato	20,8	27,8	101,2	7,1	0,0	1,7	8,9	0,8
	Sorghum	0,7	1,9	34,2	0,0	0,0	0,0	0,0	0,0
Kaele Lat: 10,1 Long: 14,45	Beans	0,3	31,5	109,9	3,5	0,0	0,0	6,8	0,0
	Groundnuts	9,5	14,5	82,7	4,1	0,0	0,0	1,0	0,0
	Maize	0,0	14,0	75,3	0,0	0,0	0,0	0,0	0,0
	Potato	14,7	32,0	112,5	8,7	0,7	1,5	11,9	0,9
	Sorghum	0,0	0,7	41,1	0,0	0,0	0,0	0,0	0,0
Kousseri Lat: 12,08 Long: 15,03	Beans	79,0	199,2	293,6	86,7	0,3	32,0	50,1	5,0
	Groundnuts	129,6	172,5	261,5	81,0	4,2	14,3	24,5	2,0
	Maize	87,2	156,2	251,6	50,5	0,0	20,2	40,3	0,0
	Potato	162,8	206,9	295,7	113,6	15,1	31,9	46,6	11,3
	Sorghum	55,7	118,9	200,3	31,8	0,0	5,7	19,3	0,0

Makari Lat: 12,58 Long: 14,47	Beans	164,3	223,9	321,9	108,1	19,2	36,7	55,1	9,5
	Groundnuts	223,7	196,6	291,0	112,9	16,4	16,7	27,2	3,6
	Maize	183,2	179,9	278,9	72,7	16,0	25,1	45,8	0,2
	Potato	258,9	231,8	325,9	146,5	34,5	35,4	50,4	15,3
	Sorghum	134,9	135,5	224,2	47,2	0,0	9,1	23,4	0,0
Waza Lat: 11,4 Long: 14,57	Beans	46,3	125,2	218,8	30,3	0,0	14,2	31,1	0,0
	Groundnuts	62,3	99,2	185,1	17,2	0,0	5,0	13,8	0,0
	Maize	33,1	91,8	177,4	3,6	0,0	5,5	19,7	0,0
	Potato	92,6	134,8	220,8	44,8	3,0	17,0	31,0	4,2
	Sorghum	8,9	60,9	126,9	3,0	0,0	0,0	4,1	0,0
Garoua Lat: 9,33 Long: 13,38	Beans	12,2	33,6	18,1	7,4	0,0	0,0	0,0	0,0
	Groundnuts	4,4	18,5	3,7	7,4	0,0	0,0	0,0	0,0
	Maize	7,8	15,1	5,6	5,6	0,0	0,0	0,0	0,0
	Potato	14,2	37,6	16,8	9,3	0,1	0,3	0,1	0,1
	Sorghum	0,0	3,6	4,5	5,6	0,0	0,0	0,0	0,0
Belel Lat: 7,05 Long: 14,43	Beans	0,0	2,0	1,6	4,7	0,0	0,0	0,0	0,0
	Groundnuts	3,4	2,6	1,6	4,7	0,0	0,0	0,0	0,0
	Maize	0,0	0,0	3,2	3,5	0,0	0,0	0,0	0,0
	Potato	4,3	3,4	2,0	5,8	0,0	0,0	0,0	0,0
	Sorghum	0,0	0,0	3,2	3,5	0,0	0,0	0,0	0,0
Bebeni Lat: 9,17 Long: 13,54	Beans	12,0	58,8	44,2	20,5	0,0	0,0	0,0	0,0
	Groundnuts	4,3	41,5	22,9	7,9	0,0	0,0	0,0	0,0
	Maize	7,6	34,5	25,5	8,0	0,0	0,0	0,0	0,0
	Potato	14,1	61,9	47,1	18,7	0,1	0,7	0,2	0,1
	Sorghum	0,0	9,0	5,7	5,7	0,0	0,0	0,0	0,0
Djohong Lat: 6,83 Long: 14,7	Beans	0,0	2,1	1,6	4,5	0,0	0,0	0,0	0,0
	Groundnuts	3,3	2,7	1,6	4,5	0,0	0,0	0,0	0,0
	Maize	0,0	0,0	3,2	3,4	0,0	0,0	0,0	0,0
	Potato	4,2	3,4	2,0	5,7	0,0	0,0	0,0	0,0
	Sorghum	0,0	0,0	3,1	3,4	0,0	0,0	0,0	0,0
Magba Lat: 6 Long: 11,22	Beans	2,0	6,8	5,2	3,4	0,0	0,0	0,0	0,0
	Groundnuts	1,5	6,8	5,2	3,4	0,0	0,0	0,0	0,0
	Maize	0,0	5,1	3,9	3,5	0,0	0,0	0,0	0,0
	Potato	1,9	8,5	6,5	4,2	0,0	0,0	0,0	0,0
	Sorghum	0,0	5,1	3,9	3,5	0,0	0,0	0,0	0,0
Touboro Lat: 7,78 Long: 15,6	Beans	0,0	5,8	2,1	6,0	0,0	0,0	0,0	0,0
	Groundnuts	4,0	5,7	2,1	6,0	0,0	0,0	0,0	0,0
	Maize	0,0	0,0	4,0	4,5	0,0	0,0	0,0	0,0
	Potato	4,9	9,7	2,6	7,5	0,0	0,0	0,0	0,0

	Sorghum	0,0	0,0	3,9	4,5	0,0	0,0	0,0	0,0
Tchollire Lat: 8,4 Long: 14,17	Beans	27,3	27,1	15,7	8,9	0,0	0,0	0,0	0,0
	Groundnuts	12,3	12,6	3,2	7,6	0,0	0,0	0,0	0,0
	Maize	18,7	11,4	7,0	5,7	0,0	0,0	0,0	0,0
	Potato	26,8	25,0	8,1	9,5	0,1	0,2	0,1	0,0
	Sorghum	2,1	0,0	4,8	5,7	0,0	0,0	0,0	0,0
Rey Bouba Lat: 8,67 Long: 14,18	Beans	27,1	57,6	40,7	15,7	0,0	0,0	0,0	0,0
	Groundnuts	12,4	38,6	17,9	8,3	0,0	0,0	0,0	0,0
	Maize	18,5	30,4	20,0	5,8	0,0	0,0	0,0	0,0
	Potato	26,6	62,7	44,6	16,5	0,1	0,8	0,3	0,1
	Sorghum	2,0	7,8	5,3	5,8	0,0	0,0	0,0	0,0
Poli Lat: 8,48 Long: 13,25	Beans	0,0	11,9	4,3	7,0	0,0	0,0	0,0	0,0
	Groundnuts	4,1	5,5	2,4	7,0	0,0	0,0	0,0	0,0
	Maize	0,0	0,0	4,2	5,3	0,0	0,0	0,0	0,0
	Potato	5,1	11,3	3,0	8,8	0,1	0,1	0,0	0,0
	Sorghum	0,0	0,0	4,3	5,3	0,0	0,0	0,0	0,0
Pitoa Lat: 9,38 Long: 13,53	Beans	12,7	29,2	17,9	7,6	0,0	0,0	0,0	0,0
	Groundnuts	4,7	13,8	2,5	7,4	0,0	0,0	0,0	0,0
	Maize	8,2	12,3	6,4	5,6	0,0	0,0	0,0	0,0
	Potato	14,6	28,4	11,6	9,3	0,1	0,0	0,1	0,1
	Sorghum	0,0	0,0	4,5	5,6	0,0	0,0	0,0	0,0
Banyo Lat: 6,78 Long: 11,81	Beans	2,1	6,3	4,6	3,1	0,0	0,0	0,0	0,0
	Groundnuts	1,6	6,3	4,6	3,1	0,0	0,0	0,0	0,0
	Maize	0,0	4,7	3,5	3,3	0,0	0,0	0,0	0,0
	Potato	2,0	7,9	5,8	3,9	0,0	0,0	0,0	0,0
	Sorghum	0,0	4,7	3,5	3,3	0,0	0,0	0,0	0,0
Tibati Lat: 6,48 Long: 12,63	Beans	2,1	6,4	4,8	3,2	0,0	0,0	0,0	0,0
	Groundnuts	1,5	6,4	4,8	3,2	0,0	0,0	0,0	0,0
	Maize	0,0	4,8	3,6	3,4	0,0	0,0	0,0	0,0
	Potato	3,5	8,0	6,0	3,9	0,0	0,0	0,0	0,0
	Sorghum	0,0	4,8	3,6	3,3	0,0	0,0	0,0	0,0
Ngoundal Lat: 6,5 Long: 13,27	Beans	2,4	7,2	5,8	3,6	0,0	0,0	0,0	0,0
	Groundnuts	2,0	7,2	5,8	3,6	0,0	0,0	0,0	0,0
	Maize	0,0	5,4	4,3	3,7	0,0	0,0	0,0	0,0
	Potato	5,4	9,0	7,2	4,5	0,0	0,0	0,0	0,0
	Sorghum	0,0	5,4	4,3	3,6	0,0	0,0	0,0	0,0
Ngoundere Lat: 7,35 Long: 13,56	Beans	4,6	7,4	6,4	3,7	0,0	0,0	0,0	0,0
	Groundnuts	4,2	7,4	6,4	3,7	0,0	0,0	0,0	0,0
	Maize	0,0	5,6	4,4	3,7	0,0	0,0	0,0	0,0

	Potato	7,6	9,3	13,0	4,6	0,0	0,0	0,0	0,0
	Sorghum	0,0	5,6	4,4	3,7	0,0	0,0	0,0	0,0
	Beans	2,7	7,2	5,7	3,5	0,0	0,0	0,0	0,0
	Groundnuts	2,3	7,2	5,7	3,5	0,0	0,0	0,0	0,0
Meiganga Lat: 6,53 Long: 14,36	Maize	0,0	5,4	4,3	3,6	0,0	0,0	0,0	0,0
	Potato	5,7	9,0	7,1	4,4	0,0	0,0	0,0	0,0
	Sorghum	0,0	5,4	4,3	3,6	0,0	0,0	0,0	0,0
	Beans	2,5	7,6	6,0	3,8	0,0	0,0	0,0	0,0
	Groundnuts	2,0	7,6	6,0	3,8	0,0	0,0	0,0	0,0
Tignere Lat: 7,73 Long: 12,65	Maize	0,0	5,7	4,5	3,8	0,0	0,0	0,0	0,0
	Potato	5,5	9,5	11,6	4,7	0,0	0,0	0,0	0,0
	Sorghum	0,0	5,7	4,5	3,8	0,0	0,0	0,0	0,0

12. References

Allen, R., L. Pereira, D. Raes, and M. Smith. 1998. Crop evapotranspiration-Guidelines for computing crop water requirements-FAO Irrigation and drainage paper 56. FAO, Rome 300.

Baulcombe, D., I. Crute, B. Davies, J. Dunwell, M. Gale, J. Jones, J. Pretty, W. Sutherland, and C. Toulmin. 2009. Reaping the benefits: science and the sustainable intensification of global agriculture. London, UK: The Royal Society.

Binswanger, H. and P. Pingali. 1988. Technological priorities for farming in sub-Saharan Africa. The World Bank Research Observer 3:81.

Boko, M. N., A. Nyong, C. Vogel, A. Githeko, M. Medany, B. Osman-Elasha, R. Tabo and P. Yanda. 2007. Africa. Climate Change 2007: Impacts, Adaptation and Vulnerability. Contribution of Working Group II to the Fourth Assessment Report of the Intergovernmental Panel on Climate Change. Cambridge Univ Pr.

Boserup, E. and N. Kaldor. 2005. The conditions of agricultural growth: the economics of agrarian change under population pressure. Aldine De Gruyter.

Bruinsma, J. 2003. World agriculture: towards 2015/2030: an FAO perspective. Earthscan/James & James.

Channell, R. and M. Lomolino. 2000. Dynamic biogeography and conservation of endangered species. Nature 403:84-86.

Clarke, D., M. Smith, and K. El-Askari. 1998. CropWat for Windows: user guide. Food and Agriculture Organization:1-45.

Cooper, P., K. Rao, P. Singh, J. Dimes, P. Traore, K. Rao, P. Dixit, and S. Twomlow. 2009. Farming with current and future climate risk: advancing a 'Hypothesis of Hope'for rainfed agriculture in the semi-arid tropics.

CRED. 2011. The International Disaster Database. Centre for Research on the Epidemiology of Disasters, Université Catholique de Louvain. Brussels, Belgium.

Davies, B. and M. Chaves. 2010. Drought effects and water use efficiency: improving crop production in dry environments. Functional Plant Biology 37.

De Janvry, A. 2009. AERC Conference on Agriculture for Development in Sub-Saharan Africa: Introduction.

Demont, M., P. Jouve, J. Stessens, and E. Tollens. 2007. Boserup versus Malthus revisited: Evolution of farming systems in Northern Côte d'Ivoire. Agricultural Systems 93:215-228.

Döll, P. 2002. Impact of climate change and variability on irrigation requirements: a global perspective. Climatic Change 54:269-293.

Easterling, W., P. Aggarwal, P. Batima, K. Brander, L. Erda, S. Howden, A. Kirilenko, J. Morton, J. Soussana, and J. Schmidhuber. 2007. Food, fibre and forest products. Climate change 2007: impacts, adaptation and vulnerability. Contribution of Working Group II to the Fourth Assessment Report of the Intergovernmental Panel on Climate Change, Parry, pp273-313, ML et al (eds).

Eid, H. and N. El-Mowelhi. 1998. Impact of climate change on field crops and water needs in Egypt.

El-Shaer, H., C. Rosenzweig, A. Iglesias, M. Eid, and D. Hillel. 1997. Impact of climate change on possible scenarios for Egyptian agriculture in the future. Mitigation and Adaptation Strategies for Global Change 1:233-250.

Falkenmark, M. and J. Rockström. 2008. Building resilience to drought in desertification-prone savannas in Sub-Saharan Africa: The water perspective. Pages 93-102. John Wiley & Sons.

Fan, S., T. Mogues, and S. Benin. 2009. Setting priorities for public spending for agricultural and rural development in Africa. Policy briefs.

FAOSTAT, F. 2010. Statistical Databases. Food and Agriculture Organization of the United Nations, Washington, DC:128-130.

Grieser, J., R. Gommes, and M. Bernardi. 2006. New LocClim-the local climate estimator of FAO.

Iglesias, A., L. Garrote, F. Flores, and M. Moneo. 2007. Challenges to manage the risk of water scarcity and climate change in the Mediterranean. Water Resources Management 21:775-788.

IPCC. 2007. Climate Change 2007: Synthesis Report. IPCC Secretariat, 7 bis Avenue de la Paix C. P. 2300 Geneva 2 CH- 1211 Switzerland.

Jayne, T., D. Mather, and E. Mghenyi. 2010. Principal Challenges Confronting Smallholder Agriculture in Sub-Saharan Africa. World Development.

Khalil, F., H. Farag, G. El Afandi, and S. Ouda. 2009. Vulnerability and adaptation of wheat to climate change in Middle Egypt. Pages 12-15.

Lankford, B. 2009. Viewpoint-The right irrigation? Policy directions for agricultural water management in Sub-Saharan Africa. Water Alternatives 2:476-480.

Laux, P., G. Jäckel, R. Tingem, and H. Kunstmann. 2010. Impact of climate change on agricultural productivity under rainfed conditions in Cameroon--A method to improve attainable crop yields by planting date adaptations. Agricultural and Forest Meteorology.

Li, Y., W. Ye, M. Wang, and X. Yan. 2009. Climate change and drought: a risk assessment of crop-yield impacts. Clim. Res 39:31-46.

Lobell, D. and C. Field. 2007. Global scale climate-crop yield relationships and the impacts of recent warming. Environmental Research Letters 2:014002.

Longley, P. 2005. Geographic information systems and science. John Wiley & Sons Inc.

Maiti, R. K. 1996. Sorghum science. Science Publishers, Lebanon, NH.

McSweeney, C., G. Lizcano, M. New, and X. Lu. 2010. The UNDP Climate Change Country Profiles: Cameroon. Bulletin of the American Meteorological Society 91:157-166.

Mokwunye, A., A. de Jager, and E. Smaling. 1996. Restoring and maintaining the productivity of West African soils: key to sustainable development. International Fertilizer Development Center.

Molua, E. and C. Lambi. 2006. Assessing the impact of climate on crop water use and crop water productivity: The CROPWAT analysis of three districts in Cameroon. CEEPA Discussion paper.

Morton, J. 2007. The impact of climate change on smallholder and subsistence agriculture. Proceedings of the National Academy of Sciences 104:19680.

Moyin-Jesu, E. 2007. Use of plant residues for improving soil fertility, pod nutrients, root growth and pod weight of okra (Abelmoschus esculentum L). Bioresource technology 98:2057-2064.

Nelson, G. 2009. Climate Change: Impact on agriculture and costs of adaptation. Intl Food Policy Res Inst.

Parry, M., C. Rosenzweig, A. Iglesias, M. Livermore, and G. Fischer. 2004. Effects of climate change on global food production under SRES emissions and socio-economic scenarios. Global Environmental Change 14:53-67.

Peng, X., Y. Zhang, J. Cai, Z. Jiang, and S. Zhang. 2009. Photosynthesis, growth and yield of soybean and maize in a tree-based agroforestry intercropping system on the Loess Plateau. Agroforestry systems 76:569-577.

Prasad, P., S. Pisipati, R. Mutava, and M. Tuinstra. 2008. Sensitivity of grain sorghum to high temperature stress during reproductive development. Crop Sci 48:1911-1917.

Ringler, C., E. Bryan, A. Biswas, and S. Cline. 2010. Water and Food Security Under Global Change. Global Change: Impacts on Water and food Security:3-15.

Sánchez, P. 2010. Tripling crop yields in tropical Africa. Nature Geoscience 3:299-300.

Sanchez, P., G. Denning, and G. Nziguheba. 2009. The African green revolution moves forward. Food Security 1:37-44.

Stancalie, G., A. Marica, and L. Toulios. 2010. Using earth observation data and CROPWAT model to estimate the actual crop evapotranspiration. Physics and Chemistry of the Earth, Parts A/B/C 35:25-30.

Tenywa, M. and M. Bekunda. 2009. Managing soils in Sub-Saharan Africa: Challenges and opportunities for soil and water conservation. Journal of Soil and Water Conservation 64:44A.

Traore, S., Y. Wang, T. Kerh, and A. Ouedraogo. 2007. Application of CROPWAT simulation model for rainfed and irrigated agriculture water planning in Burkina Faso. Journal of International Cooperation 3:1-26.

Vanacker, V., M. Linderman, F. Lupo, S. Flasse, and E. Lambin. 2005. Impact of short term rainfall fluctuation on interannual land cover change in sub Saharan Africa. Global Ecology and Biogeography 14:123-135.

Wani, S., J. Rockström, and T. Oweis. 2009. Rainfed Agriculture: unlocking the potential. CABI Publishing.

Williams, N. 2009. Feeding the future world. Current Biology 19:R968-R969.

Wojtkowski, P. 2008. Ensuring Food Security. science 320:611.

Yengoh, G., A. Ato, and M. Svensson. 2009. Technology Adoption in Small-Scale Agriculture: The Case of Cameroon and Ghana. Science, Technology & Innovation Studies 5.

Yengoh, G., A. Tchuinte, F. Armah, and J. Odoi. 2010. Impact of prolonged rainy seasons on food crop production in Cameroon. Mitigation and Adaptation Strategies for Global Change:1-17.

Yengoh, G. T., T. Hickler, and A. Tchuinte. 2011. Agro-climatic Resources and Challenges to Food Production in Cameroon. Geocarto International:1-1.

3

Optimizing Water Consumption Using Crop Water Production Functions

Ali Reza Kiani[1] and Fariborz Abbasi[2]

*[1]Department of Agricultural Engineering, Agricultural and
Natural Resources Research Centre of Golestan Province, Gorgan
[2]Agricultural Engineering Research Institute, AERI, Karaj,
Iran*

1. Introduction

There are two options for increasing crop production: Increasing acreage and improving land utilization. Food supply in the early period of agriculture was obtained only by expanding the physical area of land. Gradually, improving land productivity using different agronomy and management techniques for food production has had a more prominent role than increase in physical land area. As today, the acreage increase does not play key role in the crop production. Historical samples are proof of this claim. Historical trends indicate that, although application of fertilizers, pesticides and plant breeding could improve yield per unit area, but gradually due to improper management of resources, pollution caused by chemical fertilizer and pesticides has led to agricultural sustainability is threatened. While in terms of plant breeding, in many cases the plants have reached their biological production limit. But irrigation management has still key role to increase land productivity and food production. Currently, 2.5 billion people in terms of employment, food and income are dependent on irrigated agriculture. It is estimated that within the next 30 years, 80 percent of the additional food needed for world population is rely on irrigation. Although the increase in irrigated agriculture can play a significant role in providing world food needs, but water resources for such development is limited. Therefore, more effective use of current resources in irrigation is emphasized. Groundwater extraction in the short term is effective in producing food, but its continuance would not be sustainable in the long run.

Agricultural is the largest consumer of water in the world. About 80 to 90 percent of exploited water resources are used in the agricultural section. In this section, it is possible to increase water productivity and use of saved water for the new land could increase crop production. Generally, traditional management to reduce risk from water shortages in the agricultural sector was demand management, mainly to develop water storage using dams or water captivates had been focused. Of course, it is led to various problems of environmental because of changing in water hydrologic cycle. Today, most countries have reached the peak utilization of their water resources, extracting the main accessible water. It is natural that in these circumstances water extraction costs more and more expensive every day, and in practical point of view, agricultural practices will be non-economic. In these conditions, the agricultural sector had to compete with other sectors to use of each unit of water. In other words, a serious question arises whether in the long term water extraction in large scale to crop production is economy? or use of water in other sectors like industry has a higher comparative advantage? Rapid population growth and the need for more production has caused the agricultural sector would have more demand for water than any other water consuming sectors (industry, domestic, drinking). So, the main challenge of the agricultural sector is to produce more crops from less water. In this context, increasing of water use efficiency is one of the important strategies for more production. One of the important solutions to reduce hunger, poverty and maintaining the previous level of production will be   increasing water productivity proportional to increase in demand for food. Increasing water use efficiency while maintaining stable production and reducing water consumption, a significant volume of water can store without a new structure to be constructed. There are two basic solutions for increasing water productivity: These are reducing water consumption with proven production (irrigation management) and increasing production with a fixed amount of water (agronomy management) that eventually will lead to more production with less water.

Reducing water consumption while maintaining the previous production, includes all activities that are leading to improve irrigation management. Ways to increase water productivity from the perspective of improving irrigation management can be summarized as follows:

- improving irrigation efficiency (improving conventional irrigation systems and change to modern irrigation systems)
- optimal allocation of water resources
- deficit or alternate irrigation
- optimize water use by determination of water – yield relations (production functions)
- the use of uncommon water
- reduce evaporation

Increasing production per unit water consumption is achievable by improving the production management. The production management refers to all agricultural activities that are leading to increase production. The following can be cited In this context:

- breading activities to make drought resistant cultivars
- reduce the period of plant growth
- increasing the depth and root density in soils
- change and reform cropping pattern on the basis of the highest plant productivity
- crop rotation
- application of conservation tillage techniques for more efficient use of available moisture in the soil, irrigation water and rain
- controlling pests and diseases
- improving soil fertility

Irrigation strategy management to increase production would not be the same in all areas. In areas where enough water is available, full irrigation strategy could be a suitable option, while in areas where water is limited, deficit irrigation is appropriate strategy and finally in areas where water resources are saline, use of management strategies for achieving sustainable production as well as economic yield is suitable management. However, any optional strategies are needed to specific management, and they are a function of time and space. But in all cases, knowledge of water-yield relation (production function) is necessary in order to achieve optimal amounts of irrigation water and sustainable production. The purpose of this section is to optimize water consumption using production functions in terms of saline and non-saline water application.

2. Crop production functions

To express the relationship between inputs and crop yield, production functions are used. Crop production functions are mathematical relationship between yield and inputs used in the production process. In other words, the crop production function identifies the conversion rate of input to output. The statistical data obtained from field observations or controlled experimental design can be used to estimate of production functions. Overall form of the crop production function can be written as follows:

$$Y = F(X_1, X_2, \dots X_n) \tag{1}$$

This equation shows the amount of production determined by different amounts of inputs (n). Production factors can be classified in different ways. Some factors are variables and some others are fixed. Some of the factors are very important and some others not significant. The crop production function is usually estimated based on a few variable factors under controlled. Using the estimated production function can be defined different scenarios based on user-defined. The amount of yield in different levels of inputs used to crop production function, marginal production, the final value of each of the factors of production and marginal rate of technical substitution factors could be calculated.

3. Optimizing water consumption

Optimizing water consumption is a type of management options that may establish relationship between land and water under limitation of water/land conditions. So that crop production is economically affordable and technically possible. Generally, Figure 1 shows relationship between the applied water (AW) as a function of gross income and cost (English, 1990). As shown, there is curvilinear relation between gross income and applied water and a linear relation with cost function. In fact, the gross income increases with water and reaches the maximum point, then decreases mainly due to increasing water cost and also decreasing in production. But cost function rises linearly with increasing water amount. Net income is equal to the difference between gross income and cost (distance between two curves in Figure 1).

Selection of appropriate strategy depends on the presence or absence of limiting factors previously mentioned. If land is limiting factor, using the amount of water maximizing net benefit per unit of land, is the best strategy (AW_L in Figure 1). Although with increasing amount of water, gross income could be increased so that maximum crop production is obtained with full irrigation (AW_F on the right AW_L in Figure 1), but net income resulting from application of AW_F is less than AW_L. Reducing the amount of water applied from AW_L and move to the left curve gradually decreases the net income, so that net income in AW_d per unit of land is equal to net income from full irrigation. It is clear that full irrigation needed to achieve maximum production, but to get the highest net profit not need to full irrigation. In practical point of view, in water scarcity areas economic benefit could be achieved using deficit irrigation scheduling. In areas where water is a limiting factor, the costs of supply, transmission and distribution of water are the most serious challenges. In these areas, due to water shortages a large part of the farms are not cultivated or remained as rainfed. According to presented analysis, in terms of water consumption there is a point that irrigation depth is less than the amount of crop water requirement and full irrigation depth, however its net profit per unit of land is equal to net income from full irrigation (AW$_d$). So with the saved water for irrigating new lands can produce more revenue achieved without paying additional cost for new water supply. To achieve optimal levels of water consumption and choosing appropriate management options, it is necessary to estimate variable amounts of derived functions such as AW$_L$, AW$_F$ and AW$_d$.

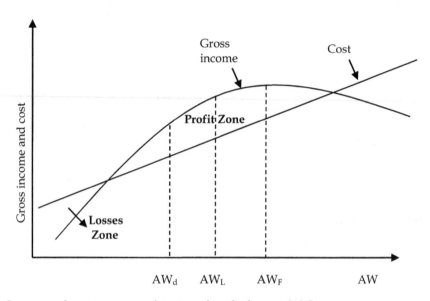

Fig. 1. Income and cost curves as a function of applied water (AW)

3.1 Determination of threshold values
For optimum irrigation scheduling, driving water-yield relation (crop water production function) is required for determination of optimum amount of water. Generally, the quadratic form of crop water production is used to describe relationship between irrigation water and crop yield:

$$Y = a_0 + a_1(I) + a_2(I^2) \qquad (2)$$

where, Y is crop yield (t ha^{-1}), I is irrigation water (mm) and a_0, a_1 and a_2 are constants.
The cost equation is used as:

$$C = b_0 + b_1(I) \qquad (3)$$

where, C is production cost per hectare and b_0 and b_1 are constants.
Constant coefficients in above equations can be derived using various levels of water against their corresponding yields as follows (English, 1990):

$$I_F = -a_1/2a_2 \qquad (4)$$

$$I_L = (b_1 - P_c a_1)/(2P_c a_2) \tag{5}$$

$$I_d = [(P_c \times a_0 - b_0)/(P_c \times a_2)]^{0.5} \tag{6}$$

where, I_F, I_L and I_d are full irrigation, optimum amount of irrigation where land is limiting and water is limiting, respectively (mm) and Pc is the crop price per kg.

Above equations, were modified by Sepaskha & Akbari (2005) for case having rainfall (P) as follows:

$$I_F = -a_1'/a_2' - P \tag{7}$$

$$I_L = (b_1 - P_c a_1')/(2P_c a_2') - P \tag{8}$$

$$I_d = \{[P_c(a_0'P + a_1'P^2 + a_2') - b_0]/P_c a_2'\}^{0.5} \tag{9}$$

where, , a_0', a_1' and a_2' are the similar constants as equation 2 when total applied water (AW) is replaced with I.

The following sections present how to determine the optimal amount of water consumption by using analysis of production functions in different conditions.

4. Wheat production function under supplementary irrigation

Supplementary irrigation plays a key role in crop production in many countries in the world. In most rainfed areas, the amount of rainfall and its distribution is not suitable throughout growing season of winter crops. Therefore, increasing crop production is necessary to timing irrigation with supplementary irrigation. At present, supplementary irrigation covers 80 percent of cultivated areas over the world, producing 60 percent of the global production (Harris, 1991). Wheat yield under rainfed conditions in arid and semiarid regions of the world, including West Asia and North Africa, varies between 0.6 to 1.5 t/ha. In these areas, supplemental irrigation significantly increases yield and water productivity. The experimental results on wheat show that water productivity is increased with deficit irrigation scheduling (Sun et al., 2006; Zhang et al., 2005; Zhang et al., 2006). Supplemental irrigation potentially increases wheat yield and water productivity. Irrigation water for achieving optimum water productivity is not the same in different areas of the world, mainly due to different climate. For example, the highest water productivity of wheat in northern Syria (Zhang and Oweis, 1999) was corresponded with 440-500 mm of water application (140-180 mm irrigation water), in northern China (Zhang et al., 1999), with 400 mm (120-160 mm irrigation water) and in Oregon America (English and Nakamora, 1989) with 750-850 mm of water applied (350-450 mm irrigation water).

A field study was conducted to compare various genotypes of wheat to water using line source sprinkler irrigation during growing season of wheat (2005-2006). The geographical location of the farm was 36° 54' N, 54° 25' E and 155 m above mean sea level. This area represents medium annual rainfall, mostly falls in winter (November–April). Seasonal rainfall during the wheat growth stages was about 250 mm. Four irrigation treatments were provided by the decline with distance from the line source, during the growing season to meet 100 %(W_1), 76 %(W_2), 52 %(W_3) and 39% (W_4) crop water requirement. Optimal irrigation water application was considered by generated crop water production functions for six cultivars of wheat under supplementary irrigation. The experiment used a strip plot design to examine the effect of the fixed irrigation rates on six cultivars treatments (C_1=TAJAN, C_2= N-80-6, C_3= N-80-7, C_4= N-80-19, C_5= N-81-18, C_6= Desconcido) with four replications. The soil texture of the experimental plots was silty clay loam in the surface layer (0-30 cm) and silty loam in the deeper layers (30-90 cm), having soil bulk density of 1.3 and 1.4 gr /cm³, respectively.

Wheat yield variation of different cultivars as a function of applied water (AW) illustrated in figure 1, separately. A quadratic form of production function was used to describe the relationship between applied water and yield. The general trend is similar to all the cultivars, so that with increasing applied water, yield increases and at a certain point that yield reaches the peak value, more increasing water will result in decreasing the yield. Estimated wheat production function for any cultivars showed that they have different constants as well as different response to water. As a result, appropriate cultivar could be selected due to provide different production from a certain amount of water. For example, the cultivars C_1 – C_6 produced 4.2, 4.8, 4.5, 4.4, 4.6 and 4.2 t/ha against 350 mm applied water, respectively (Fig. 2).

Crop water production and cost functions were used to describe optimum scheduling of irrigation water. Due to the presence of effective rainfall in the region, optimum irrigation water under full irrigation (I_f) and deficit irrigation (I_d) were estimated as function of seasonal rainfall (with the assumption that wheat price equal to 1700 Rials/kg and production cost without irrigation costs are 2.5 million Rials/ha, 10000Rials=US\$1). The results revealed that, optimum irrigation water depth will be decreased when seasonal rainfall increased. Required water under full irrigation strategy for maximizing production in cultivars C_1 to C_6 was 362, 370, 335, 342, 340 and 345 mm. Optimum irrigation water under deficit irrigation for all cultivars as a function of seasonal rainfall was presented in Table 1. It is observed that under deficit irrigation, whenever during the wheat growing season 250 mm of rainfall occurs with a similar distribution, all cultivars will not require to irrigate. If rain does not happen (in terms of deficit irrigation) C_2 cultivar will need more water than the others (331 mm). So, if assumes that the average value of 250 mm rainfall occurs during the wheat growing season, in case of full supplemental irrigation two or three irrigation event will be needed to all the cultivars. However, in case of deficit irrigation cultivars will not require irrigation water (Table 2).

P (mm)	Cultivars					
	C_1	C_2	C_3	C_4	C_5	C_6
0	325	331	311	315	315	315
10	314	320	300	305	305	305
20	303	310	290	293	293	293
30	291	297	279	282	282	282
40	280	286	269	270	270	270
50	268	274	257	260	260	260
70	245	250	235	238	238	236
100	208	214	200	202	203	200
150	140	146	139	139	141	135
200	30	43	57	50	59	35

Table 2. Optimum irrigation water as a function of seasonal rainfall in different cultivars of wheat

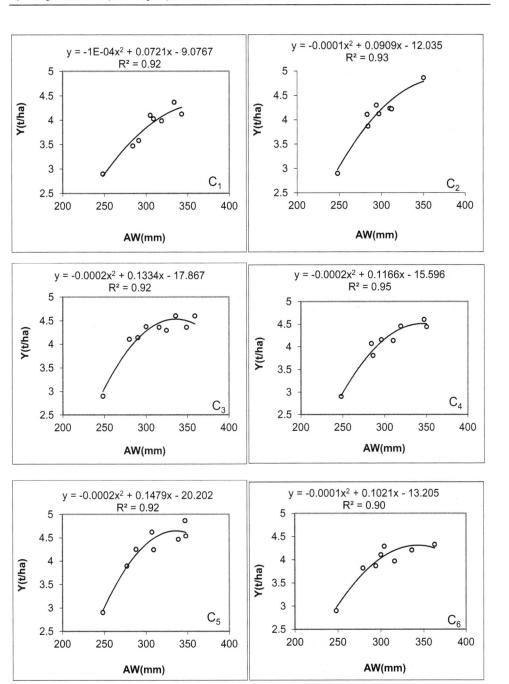

Fig. 2. Wheat grain yield as a function of applied water in different cultivars

4.1 Comparison of deficit and full irrigation strategies

Generally, when farmers face water scarcity, as water resources are not enough to follow the full irrigation, they have two options (Oweis and Hachum, 2003): in the first case may be, they apply the available water for irrigating a part of the farm and leave the rest as rainfed. In the second case may be, they apply less water to irrigate the whole farm (deficit irrigation). The quantitative comparison of deficit and full irrigation strategies are showed in Table 2 that deficit irrigation is more useful strategy for obtaining more production as compared to full irrigation. For example, in C_1 cultivar, if full irrigation (99 mm) is applied to irrigate one hectare and left a 1.54 hectare as a rainfed, totally it can be obtained 7.27 ton grain yield. But, if instead of full irrigation of one hectare, it is applied deficit irrigation (39 mm) for all the 2.54 hectare, totally 9.02 ton grain yield can be obtained. This trend with more appropriate result for other cultivars is presented in Table 2. It is observed that in W_4 treatment total yield in C_1 to C_6 cultivars were increased 24%, 40%, 47%, 40%, 48% and 32%, respectively, when compared to the W_1 treatment. To minimize risk with water stress on crop yield reduction, it is needed to know sensitivity of different growth stages of wheat to water stress.

C_i	IT^*	I (mm)	Y (t/ha)	Area (ha) Irrigated	Rainfed	Total	Y (t/ha) Irrigated	Rainfed	Total	Increased Yield (%)
C_1	W_1	99	4.19	1	1.54	2.54	4.19	3.1	7.27	24
	W_4	39	3.55	2.54	0	2.54	9.02	0	9.02	
C_2	W_1	109	4.74	1	2.11	3.11	4.74	4.22	8.95	40
	W_4	35	4.02	3.11	0	3.11	12.5	0	12.5	
C_3	W_1	111	4.15	1	2.26	3.26	4.51	4.54	9.03	47
	W_4	34	4.06	3.26	0	3.26	13.2	0	13.2	
C_4	W_1	100	4.26	1	1.7	2.7	4.26	3.4	7.66	40
	W_4	37	3.98	2.7	0	2.7	10.8	0	10.8	
C_5	W_1	101	4.52	1	2.37	3.37	4.52	4.74	9.26	48
	W_4	30	4.07	3.37	0	3.37	13.7	0	13.7	
C_6	W_1	100	4.5	1	1.7	2.7	4.5	3.4	7.9	32
	W_4	37	3.86	2.7	0	2.7	10.4	0	10.4	

* Irrigation Treatments

Table 2. Comparison of total production in different cultivars of wheat at full (W_1) and deficit (W_4) irrigation

5. Soybean production function

Soybean is one of the most important crops for oil and protein production in the world. Generally, soybean is planted in warm and semi-warm climate and relatively resistance to low and very high temperatures, but its growth rate is reduced at temperatures higher than 35 °C and less than 18 °C (FAOSTAT, 2001). Because soybean having a high concentration of protein (36%), oil (18%) and carbohydrate (20%) planted in almost all parts of the world for human consumption, livestock and plants (Boydak et al., 2002). A field study was conducted to consider water-yield relation in three genotypes of soybeans, optimum irrigation depth, comparison of deficit and full irrigation at Gorgan Research Station in two lasted growing seasons (2005 & 2006). In this experiment, four irrigation treatments (W_1, W_2, W_3 and W_4) were provided by the decline in irrigation with distance from a line source. The experiment was based on a strip plot design to examine the effect of the fixed irrigation rates on three cultivars treatments (SAHAR, G_3 & DPX) with four replications.

Generally, obtained highest soybean yield is consisted with treatment that has received the highest water (W_1). In W_4 treatment, grain yield of soybean were 47% and 40% of the W_1 treatment, for 2005 and 2006, respectively. The average values of 2-years irrigation water in W_2, W_3 and W_4 were 80, 49 and 24% of W_1. Irrigation water application in both years with distance from the line source was reduced for all three cultivars (Fig. 3). The largest amount of irrigation was W_1 treatment being 360 mm in 2005 and 342 mm in 2006. In the first year in terms of irrigation water, W_4 treatment in Sahar, G_3 and DPX has received 17, 19 and 24% as compared with W_1 treatment, respectively (the same trend also observed in the second year).

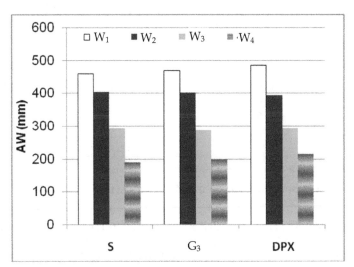

Fig. 3. Applied water (AP) in different treatments (2-year average)

The estimated coefficient using equation 2 and their statistical analysis in different cultivars were presented for 2005, 2006 and average 2-years in Table 3.

As shown in Table 4, the determination of coefficient (R^2) of the quadratic function for all cultivars are more than 84% and it is found that the effects of water levels on soybean grain yield is significant. This indicates that 84% of the yield variability of soybean is explained by the variations of applied water (AW). A quadratic form has been reported between yield and applied water by many researchers (Stewart and Hagan, 1983; Zhang and Oweis, 1999; Sepaskhah and Akbari, 2005). The grain yield of soybean cultivars as a function of AW in 2005, 2006 and 2005-2006 are illustrated in Fig. 4. Figure 4 states that the yield of soybean cultivars is increased as a decline slope with increasing water amount. In other words, for every certain amount of water increase, soybean yield will not increase proportionally. After a certain level of water, with increasing water, the yield starts to decrease (Fig. 4). The equation coefficients on three cultivars are different. It is found that the yields have a different response to given water. For example, the most productive use of water in the Sahar (2.5 t/ha), G_3 (3 t/ha) and DPX (3.7 t/ha) were reached with about 490, 490 and 510 mm of AW, respectively. For a certain amount of AW, the DPX cultivar produced more grain yield than the 2 other cultivars (Fig. 4).

Factors*	2005			2006		
	S	G_3	DPX	S	G_3	DPX
$a_0^{'}$	-0.487	0.0264	-1.204	-2.502	-3.261	-0.439
$a_1^{'}$	0.0085	0.0041	0.0152	0.022	0.0267	0.0098
$a_2^{'}$	-9.19E-6	-1.39E-6	-1.53E-5	-1.96E-5	-2.26E-5	-5.89E-7
R^2	0.84	0.91	0.91	0.85	0.97	0.98
SE	0.17	0.14	0.20	0.47	0.24	0.16
F	12	26	25	15	92	128
SigF	0.01	0.002	0.002	0.008	0.0001	0.0001
	2-years average					
	S		G_3		DPX	
$a_0^{'}$	-0.242		-1.669		-1.229	
$a_1^{'}$	0.0138		0.0166		0.0151	
$a_2^{'}$	-1.247E-5		-1.424E-5		-1.176E-5	
R^2	0.85		0.99		0.97	
SE	0.31		0.11		0.15	
F	14		184		82	
SigF	0.008		0.000		0.0002	

* R^2 = determination of coefficient; SE= Standard Error; F= Function statistic; SigF= Significant function

Table 3. Equation constants and their statistical analysis in three soybean cultivars

Based on derived equations, irrigation water under deficit irrigation conditions (I_d) as a function of seasonal rainfall are calculated and presented in Table 4. Using rainfall-irrigation water relation can be decided that for a given rainfall how much water is needed to achieve optimal production. Unit price of soybean yield equal to US$35/kg (2-years average) and total production cost for soybean (without irrigation cost) equal to US$350/ha including: land preparation (US$35), planting (US$50), total fertilizer, pesticide, herbicide and thinning (US$75), harvest (US$55) and land rent (US$135) were used to calculate the best amount of irrigation water. Table 4 shows that in deficit irrigation condition with increasing effective rainfall during the soybean growing season amount of optimum irrigation water decreases to achieve optimal production. In full irrigation strategy to obtain maximum grain yield the cultivars Sahar, G_3, and DPX are required to 550, 580 and 640 mm of AW, respectively. Water requirement of soybean is 450-600 mm to produce maximum depending on climatic conditions and variety (FAO, 2002). But, in deficit irrigation strategy, if no rainfall occurs during the growing season, Sahar, G_3 and DPX need 425, 435 and 435 mm of irrigation water, and if 200 mm effective rainfall occurs the cultivars will not need to irrigation water.

Under no irrigation water and with 200 mm effective rainfall the grain yield of Sahar, G_3 and DPX were obtained at 1, 1.2 and 1.5 t/ha, respectively. However, with increasing the amounts of AW from optimal level increasing yield is possible, but this strategy for areas that facing with water shortages are not appropriate.

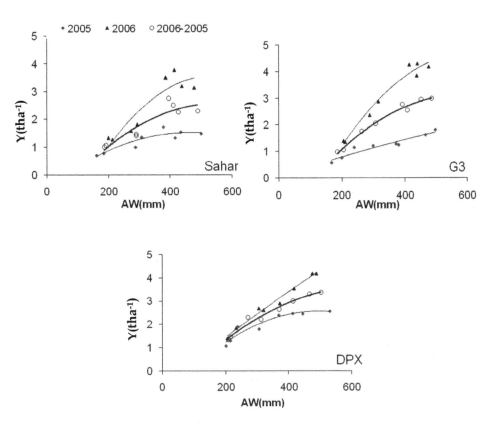

Fig. 4. Soybean grain yield as a function of applied water (AW)

| | | | | | P (mm) | | | |
Cultivars	0	30	60	90	120	150	180	200
S	424	384	342	297	248	191	115	0
G3	433	391	348	301	249	188	101	0
DPX	435	385	331	269	192	58	0	0

Table 4. Optimum irrigation water (I_d, mm) as a function of seasonal rainfall in three soybean cultivars

5.1 Comparison of deficit and full irrigation strategies

Using collected data, W_1 (full irrigation) and W_4 (deficit irrigation) treatments were compared and presented in Table 5. The quantitative comparison of two strategies showed that for all cultivars total production under deficit irrigation (W_4) was more than full irrigation (W_1) for a given levels of farm. For example, in Sahar cultivar, if full irrigation (345 mm) is applied to irrigate one hectare and left a 3.85 hectare as rainfed, totally 4.5 ton grain yield can be obtained. Instead of full irrigation of one hectare, if deficit irrigation (71mm) is applied for all the 4.85 hectare, totally 5.8 ton grain yield is achievable. This trend with more appropriate results for G_3 and DPX cultivars are presented in Table 5.

Cultivars	irrigation treatment	I (mm)	Y (t/ ha)	Areas (ha)			Y (t/ha)		
				irrigated	rainfall	total	irrigated	rainfall	total
S	W_1	345	2.585	1	3.85	4.85	2.585	1.923	4.508
	W_4	71	1.2	4.85	0	4.85	5.82	0	5.82
G_3	W_1	342	3.349	1	3.22	4.22	3.349	1.61	4.959
	W_4	81	1.389	4.22	0	4.22	5.86	0	5.86
DPX	W_1	367	3.984	1	3.08	4.08	3.984	1.54	5.524
	W_4	90	1.648	4.08	0	4.08	6.724	0	6.724

Table 5. Quantitative comparison of full (W_1) and deficit irrigation (W_4) in three soybean cultivars

Two-year average has shown that there was a tendency for more production in DPX cultivar than other cultivars, mainly due to appropriate response to water. As a result, using deficit irrigation and allocation of saved water to other lands where facing with water scarcity is the optimal strategy to maximize production as well as farmer's income. It is clear that water stress does not have similar effect on the whole crop growth. The advantage of deficit irrigation can be achieved by saving water during those periods when the crop is less sensitive to water stress.

6. Water–salinity production function

Water scarcity and salinity are major problems in reducing crop production in the arid and semi-arid regions of the word. Due to scarcity of fresh water, use of saline water is becoming inevitable to meet agricultural crop water requirement. In some water shortage areas of the world, farmers usually apply saline water especially to winter cereals in the pre-sowing and early stages of the crop growth. Farmers are faced with the challenges of optimal water applications for given saline water and economical effect of using saline water. Therefore, it is necessary to estimate crop yield in response to joint salinity and water stress conditions resulting from the use of any given set of water quantity and quality.

Several studies have explored the response of crops to salinity and water stress, i.e., use of saline water for irrigation purposes (Ayers and Westcot, 1985; Tanji, 1990; Rhoades et al., 1992). Blending non-saline and saline water, or cyclic use of these types of water, have been developed to use saline drainage water for irrigation. Water-salinity production functions are estimated by two different approaches. The first one uses mathematical models to describe the combined effects of the amount of applied water and its salinity on crop yields (Letey et al., 1985; Letey and Dinar, 1986). The second approach experimentally estimates production functions by statistical analysis from a wide range of water qualities and quantities (Dinar et al., 1985; Dinar and Knapp, 1986; Russo and Bakker, 1986; Datta et al., 1998; Datta and Dayal, 2000; Kiani & Abbasi, 2009). The first approach is valuable. Because it

allows exploration of different management strategies by using appropriate water–salinity production functions saving time and costs. But, due to the implicit assumptions, the practical application is restricted. However, several studies have been published where production functions have been verified by field data, or have been used to assess the dual effects of salinity and the amount of applied water on crop yields (Letey *et al.*, 1985; Shani and Dudley, 2001; Oster *et al.*, 2007). An experiment was conducted in the northern region of Golestan province (37°07' N, 54°07' E) in Iran, for two growing seasons (2001–2002 and 2002–2003). The mean annual rainfall in this region is 330 mm of which more than 80% occurs in autumn and winter (November to April). The average annual temperature and relative humidity are 17.8 °C and 75%, respectively. The experimental design was a randomised complete block design with split plot layout considering water quantity as the main plot and water quality as subplot with three replications. The treatments consisted of four levels of irrigation water providing 50(W_1), 75(W_2), 100(W_3) and 125(W_4) percent of the crop water requirement and four levels of irrigation water salinity of 1.5(S_1), 8.5(S_2), 11.5(S_3) and 14.2(S_4) dS/m. Saline waters for different irrigation salinity levels were obtained by mixing various ratios of well (1 dS/m) and drainage channel (10-30 dS/m) waters.
In the first year, a pre-sowing of 26 mm non-saline water was applied to all treatments using a sprinkler irrigation system, and three irrigations were subsequently applied using surface irrigation. In the second year, due to suitable rainfall at sowing date, water was applied in four irrigations after the vegetative stage. The amount of irrigation water during the growing season, for W_1, W_2, W_3 and W_4 treatments was equal to 118, 163, 209 and 246 mm in the first year and 104, 160, 212 and 264 mm in the second year, respectively. Effective rainfall in the growing season was 163 and 184 mm in 2001-02 and 2002 - 2003, respectively. Data obtained from the experimental plots were used to investigate crop response to both electrical conductivity of soil water (EC_{sw}) and soil water content (θ), assuming that the crop responded only to these two factors. Other possible factors affecting the yield were assumed to be constant according the following relationship:

$$Y = f(\theta, EC_{sw}, X) \tag{10}$$

Where, X is the constant vector for considering other factors affecting the yield.
The following various production functions were explored using the collected experimental data:
Linear:

$$Y = a_0 + a_1\theta + a_2(ECsw) \tag{11}$$

Cobb-Douglas:

$$Y = a_0\theta^{a_1}(ECsw)^{a_2} \Rightarrow Ln(y) = Ln(a_0) + a_1Ln(\theta) + a_2Ln(ECsw) \tag{12}$$

Quadratic:

$$Y = a_0 + a_1\theta + a_2\theta^2 + a_3(ECsw) + a_4(ECsw)^2 + a_5\theta(ECsw) \tag{13}$$

Transcendental:

$$Y = a_0\theta^{a_1}(Ecsw)^{a_2}Exp(a_3\theta + a_4(Ecsw)) \Rightarrow Ln(y) =$$
$$Ln(a_0) + a_1Ln(\theta) + a_2Ln(ECsw) + a_3\theta + a_4(ECsw) \tag{14}$$

The optimal production function was selected based upon statistical analysis. The coefficients of various production functions were estimated using the Ordinary Least Square (OLS) technique (SPSS, version 11.5). T-statistic of model's coefficients for determination of significantly different, F-value, R^2, Standard Error (SE) and Relative Error (RE) were estimated for comparison of various production functions. The RE was computed by:

$$RE = \frac{(Y_m - Y_p)}{Y_m} \qquad (15)$$

where, Y_m and Y_p are measured and predicted yield.

The marginal production of water content (MP_θ) and salinity (MP_{ECsw}), the marginal value product of θ (VMP_θ) and EC_{sw} (VMP_{ECsw}) and the marginal rate of technical substitution of θ and EC_{sw} ($MRTS_{\theta ECsw}$) were determined using the selected production function as:

$$MP_\theta = \frac{\partial Y}{\partial \theta} \qquad (16)$$

$$MP_{ECsw} = \frac{\partial Y}{\partial (ECsw)} \qquad (17)$$

$$VMP_\theta = P_y . MP_\theta \qquad (18)$$

$$VMP_{ECsw} = P_y . MP_{(ECsw)} \qquad (19)$$

$$MRTS_{\theta (ECsw)} = \frac{MP_\theta}{MP_{(ECsw)}} \qquad (20)$$

where, P_y is wheat price.

Table 6 shows the amount of matric potential (h), osmotic potential (h_o) and wheat grain yield in different treatments. Generally, a rather consistent decrease in yield with increasing salinity application for a given irrigation treatment and a rather constant increase in yield with increasing water application for a given saline water. In the W_1S_1 treatment, the total water potential was -6678 cm, which is somewhat lower than the total potential of –6,403 cm in the W_3S_3 treatment. The yield reduction for the W_1S_1 treatment compared to W_3S_1 was 17%, and for the W_3S_3 treatment it was 7%. For the W_1S_1 treatment, the matric potential contributed 38% of the total water potential as compared to 26% for the W_3S_3 treatment. The results suggest that the reduction of the yield due to joint salinity and water stress are not confirmed by the previously proposed simple linear additive and multiplicative concepts. For example, in the first year, yield reduction for the W_1S_1 and W_3S_4 treatments were, respectively, 0.17 and 0.12 compared to W_3S_1, while the yield reduction for the combined stress W_1S_4 was 0.26. The same results were obtained in the second year. Matric and osmotic potentials were additive in their effect on yield, but the amount of yield reduction due to unit increase of matric potential was more than that due to unit increase of osmotic potential.

Treatments		Matric Potential		Osmotic Potential		Yield	
		h (cm)		h_o (cm)		(kg ha^{-1})	
Irrigation	Salinity	2001-2002	2002-2003	2001-2002	2002-2003	2001-2002	2002-2003
	S1	-2,565	-1,528	-4,113	-3,615	3,213	3,790
W_1	S2	-2,421	-1,489	-4,303	-4,212	3,216	3,663
	S3	-2,595	-1,492	-5,267	-5,111	3,000	3,540
	S4	-3,391	-1,427	-5,922	-5,330	2,862	3,430
	S1	-2,584	-1,420	-3,673	-3,901	3,388	4,439
W_2	S2	-2,519	-1,306	-4,112	-4,478	3,358	4,330
	S3	-2,159	-1,290	-4,666	-4,621	3,325	3,951
	S4	-2,023	-1,227	-5,179	-5,012	3,125	3,938
	S1	-1,527	-1,056	-3,122	-3,804	3,845	4,777
W_3	S2	-1,324	-1,038	-4,227	-4,328	3,605	4,651
	S3	-1,680	-976	-4,723	-4,701	3,578	4,539
	S4	-1,535	-945	-4,732	-5,172	3,400	4,374
W_4	S1	-1,575	-877	-2,555	-2,904	3,825	4,839
	S2	-1,397	-853	-4,248	-3,833	3,783	4,690
	S3	-1,285	-825	-4,531	-4,880	3,773	4,405
	S4	-1,089	-807	-5,383	-5,402	3,467	4,331
LSD(0.05)						510	474

Table 6. Average seasonal matric and osmotic potential and grain yield per treatment combination

The estimated coefficients and the statistical analysis of various production functions are shown in Table 8. The results indicate that transcendental and quadratic production functions provide a better fit to the data. The determination coefficient (R^2) of the transcendental function is about 0.94 (two years average) and larger than the other functional forms. The R^2 values in the transcendental and quadratic production functions suggest that 94% of the yield variability is explained by the variations of EC_{sw} and θ. It is also found that the transcendental model provides a higher F-value and lower standard error (SE) compared to the quadratic model (Table 7). Furthermore, RE of the various production functions suggests that the transcendental model is the best. In addition, the transcendental form of the production function is found useful in describing input-output data including all three stages of the production curve with increasing positive, decreasing positive and negative marginal products (Sankhayan, 1988). As a result, in this study the transcendental model is found as a suitable wheat production function under salinity–water stress conditions.

The isoquant curves of wheat relative yield obtained using the transcendental production function are shown in Figure 5. These curves provide the different combinations of θ and EC_{sw} that result in the same yield. For a given θ, increase of EC_{sw} results in the decrease of relative yield and for a given EC_{sw} relative yield increases with increasing θ. In general, the relative yield is strongly affected by both θ and EC_{sw}. Isoquant curves (Figure 5) indicate that each one of the two factors (EC_{sw} and θ) can be substituted by the other one for a wide range in order to achieve equal amount of yield. The higher level of EC_{sw} can be used with increasing soil water content, without yield reduction. For a given relative yield, results

Variables	Linear		Cobb-Douglas		Quadratic		Transcendental	
	2001-02	2002-03	2001-02	2002-03	2001-02	2002-03	2001-02	2002-03
Constant	618 (1.12)[1]	-2422 (-1.6)	5.01 (8.1)	2.5 (2)	1047 (0.29)	-130925 (-3.4)	-14.3 (-.96)	-123.9 (-2.7)
Θ	147** (7.1) (20)[2]	272** (5.7) (47)	-	-	-188ns (-0.94) (200)	9255** (3.3) (2767)	-0.29ns (-1.2) (0.24)	-1.7* (-2.8) (0.64)
EC_{sw}	-104** (-8.1) (12)	-119** (-3.7) (32)	-	-	630ns (1.5) (429)	-715ns (-0.84) (851)	-0.09** (-4.1) (0.02)	-0.14* (-2.2) (0.06)
$Ln\theta$	-	-	1.16** (6.3) (0.19)	1.93** (5.3) (0.35)	-	-	9ns (1.4) (6.4)	54.1* (2.9) (18.8)
$Ln\ EC_{sw}$	-	-	-0.3** (-6.8) (0.04)	-0.28** (-3.1) (0.09)	-	-	0.829* (2.4) (0.34)	1.2ns (1.8) (0.67)
Θ^2	-	-	-	-	10.2ns (2.2) (4.7)	-158** (-3.2) (48)	-	-
EC_{sw}^2	-	-	-	-	-10.6ns (-2.1) (5.1)	-9.9ns (-0.65) (15.3)	-	-
$\Theta\ EC_{sw}$	-	-	-	-	-18.9ns (-1.5) (13)	-18.9ns (1.3) (27)	-	-
F value	58	24	44	21	33	24	55	27
R^2	0.90	0.78	0.87	0.76	0.94	0.92	0.95	0.92
SE	105	224	0.04	0.06	90	153	0.02	0.04
RE	3.2	4.2	0.33	3.9	1.7	2.3	0.28	0.21

1,2 Values in the parenthesis show t statistic values and standard error (SE), respectively. **significant at level $P < 0.01$ by LSD range. *significant at level $P < 0.05$ by LSD range. ns= not significant

Table 7. Estimated coefficients for each of the examined wheat water-salinity production functions.

show that the optimal combination of EC_{sw} and θ is located where the slope of the isoquant lines is parallel to the θ axis. From this point forward, an increase of θ will not result in increasing the yield. For example, a relative yield of 85% can be obtained with water content of 25% and EC_{sw} of 8 dS/m. If EC_{sw} increases to 12.2 dS/m, previous relative yield (85%) can be achieved, but the water content must be increased to 27%. The direction of the isoquant curves changes for EC_{sw} less than 6.5 dS/m. The slope of the isoquant curves becomes parallel to EC_{sw} axis. This means that by increasing EC_{sw} from 0 to 6.5 dS/m the relative yield will increase slightly with low water contents, but sharply with relatively high water contents.

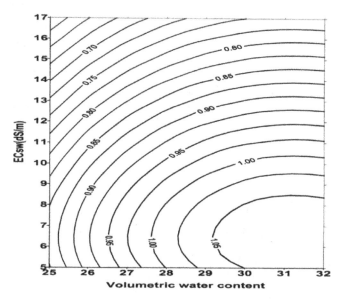

Fig. 5. Calculated isoquant curves of wheat relative yield (Yr) as a function of the 2001-2002 seasonal average θ and ECsw

Maas and Hoffman (1977) did not consider soil water contents. They obtained and reported yield as a function of the ECe. In fact, soil water content was not a limiting factor in their study. Consequently, their function applies only to well watered conditions and inherently cannot be construed to have implications to situations where soil water content is limiting. Letey et al. (1985) assumed that the relative yield becomes maximal at EC_{sw} equal to zero. However, it is clear that the relative yield does not become maximal at EC_{sw} equal to zero because $EC_{sw} = 0$ implies that there is not any mass (i.e., fertilizer) in the soil, which it can provide minimum relative yield. Of course, Letey et al. (1985) did not intend an extrapolation to a situation where all nutrients are removed from the soil.

The $MRTS_{\theta EC_{sw}}$ for the two growing seasons were 0.96 and 0.67, respectively. This means that one unit increase in EC_{sw} requires the soil water content to be increased by a factor of 0.96 and 0.67 to prevent the yield reduction in two studied years. The marginal production of θ (MP_{θ}) and EC_{sw} ($MP_{EC_{sw}}$) were 125 and -121 kg/ha, respectively in the first year, and 232 and -158 kg/ha, respectively in the second year. It means that the marginal value of income per hectare due to one unit increase in water content is US$50 (two years average) and the marginal value of damage per hectare due to an increase in EC_{sw} is US$38 (Py = US$0.28). Using the Cobb-Douglas model, Datta and Dayal (2000) reported that the marginal value product of water quantity of irrigation water and the marginal value of damage of water quality were US$1.2/ha and US$4.2/ha, respectively.

The iso-($\partial Y/\partial \theta$) and iso-($\partial Y/\partial ECsw$) lines determined using the transcendental model are shown in Figs. 6a and 6b, respectively. In the dry soil (for θ less than 21%), the rate of change of the yield with θ ($\partial Y/\partial \theta$) increase as θ increase and EC_{sw} decrease. On the contrary, in the wet soil (for θ more than 22%) $\partial Y/\partial \theta$ decreases as both θ and EC_{sw} increase (Figure 6a). The rate of change of yield with θ becomes negative at water content of over 32% (data was not showed in Figure 6a). Figure 5a revealed that $\partial Y/\partial \theta$ is influenced by θ changes, but slightly changes with EC_{sw} variations. Findings of Letey et al. (1985) for a given EC_e showed that $\partial Y/\partial W$ was affected by applied water. The decreasing yield per unit increase in soil water content is consistent with the findings of Letey et al. (1985). The $\partial Y/\partial ECsw$ increases (with negative slope) as both θ and EC_{sw} increase (Figure 6b). In the dry soil, the EC_{sw} variation has a little effect on $\partial Y/\partial ECsw$ and θ variation has a more effect on $\partial Y/\partial ECsw$. In contrary, in the wet soil, the EC_{sw} variation has a more effect on $\partial Y/\partial ECsw$ and θ variation has a little effect on $\partial Y/\partial ECsw$.

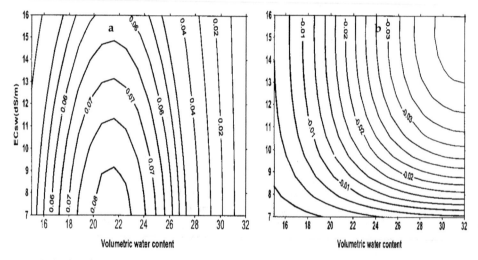

Fig. 6. Calculated isoquant curves representing the variation of $\delta Y/\delta \theta$ (a) and $\delta Y/\delta EC_{sw}$ as a function of the 2001-02 seasonal average θ and EC_{sw}

7. Conclusion

Knowledge of water-yield relation (production function) is necessary in order to achieve optimal amounts of irrigation water and sustainable production. Optimizing water use is a type of management options that may establish relationship between land and water under limitation of water/ land conditions so that the crop production is economically affordable and technically also be possible. Optimal irrigation water application was considered by generated crop water production functions for some cultivars of wheat and soybean in different amount of water under non–saline and also combined salinity and water stress conditions. Estimated water production function for any cultivars showed that they have different constants as well as different response to water, resulting the appropriate varieties for higher productivity of water can be selected. Quantitative comparison of two strategies of full irrigation and deficit irrigation has shown that if producers instead of full irrigation the part of farm, use deficit irrigation to increase irrigated area, total production and total revenue will increase. To minimize risk with water stress on crop yield reduction, it is needed to know sensitivity of different growth stages of crop to water stress. Considering the trend of crop yield under different amounts of irrigation water (steep slope in the deficit irrigation and less slope in full irrigation) priority allocation water resource to little water areas is a very effective strategy for optimal use of water resources, increasing production and farmers income. The reduction in yield due to joint salinity and water stress was not confirmed by the simple linear additive and multiplicative concepts. It was found that transcendental and quadratic production functions reasonably well predicted the yield under salinity and water stress conditions. The results revealed that yield reduction due to a unit increase of matric potential (i.e., water content) was not the same as that due to a unit increase of osmotic potential (i.e., salinity). The effect of soil water content on the rate of change of wheat yield was more pronounced than the effect of salinity. Iso-quant curves indicated that soil water content and salinity can be substituted by one another for a wide range to achieve equal amount of yield. Crop-water productions derived with the experimental approaches are site- and year–specific. Therefore, generalization of the production functions would not be so easy.

8. References

Ayers RS, Westcot DW. 1985. Water quality for agriculture. FAO Irrigation and Drainage Paper, No. 29, Rev.1.

Boydak, E., Alpaslan, M., Hayta, M., Gercek, S. and Simsek, M. 2002. Seed composition of soybeans grown in the Harran region of Turkey as affected by row spacing and irrigation. J. Agric. Food Chem. 50: 4718–4720.

Datta KK, Dayal B. 2000. Irrigation with poor quality water: An empirical study of input use economic loss and coping strategies. Ind. J. Agric. Econ. 55: 26-37.

Datta KK, Sharma VP, Sharma DP. 1998. Estimation of a production functions for wheat under saline conditions. Agric. Water Manag. 36: 85-94. Agric. Econ. 11 (1): 58-66.

Dinar R, Letey J, Vaux HJJr. 1985. Optimal rates of saline and non-saline irrigation waters for crop production. Soil Sci. Soc. Am. J. 50: 440-443.

Dinar A, Knapp KC. 1986. A dynamic analysis of optimal water use under saline conditions. Western J. Soil Sci. Soc. Am. J., 49: 1005-1009.

English, M.J. 1990. Deficit irrigation-I: analytical framework. Journal of Irrigation and Drainage Engineering, 116: 399-412.

English, M.J., and Nakamura, B.C. 1989. Effects of deficit irrigation and irrigation frequency on wheat yields. Journal of Irrigation and Drainage Engineering, 115(2), 172-184.

FAOSTAT 2001. FAOSTAT statistical database. Food and Agriculture Organization http://www.fao.org/

Harris, H.C. 1991. Implications of climate variability. In: Harris, H.C., Cooper, P.J.M. and Pala, M. (eds) Soil and Crop Management for Improved Water Use Efficiency in rain-fed areas. Proceedings of an international workshop 1989, Ankara, Turkey. ICARDA, Alepo, Syria, P. 352.

Kiani, A.R. & Abbasi, F. 2009. Assessment of the water-salinity crop production function of wheat using experimental data of the Golestan Province, Iran. Irrigation and Drainage. (ICID), 58: 445-455.

Letey J, Dinar A. 1986. Simulated crop production functions for several crops when irrigated with saline waters. Hilgardia, 54: 1-32.

Letey J, Dinar A, Knapp KC. 1985. Crop-water production function model for saline irrigation waters.

Maas EV. Hoffman GJ. 1977. Crop salt tolerance current assessment. Journal of Irrigation and Drainage Engineering, 103 (2): 115-134.

Oster JD, Stottlmyer DE, Arpaia ML. 2007. Salinity and water effects on 'Hass' Avacado yields. J. Am. Soc. Hort. Sci. 132(2): 253-261.

Oweis, T.Y. and Hachum, A.Y. 2003. Improving water productivity in the dry areas of west Asia and north Africa. CAB International, water productivity in agriculture: Limits and Opportunities for improvement (eds J.W. Kijne, R. Barker and D. Molden).

Rhoades JD, Kandidah A, Mashali AM. 1992. The use of saline waters for crop production . FAO Irrigation and Drainage Paper, No. 48.

Russo D, Bakker D. 1986. Crop water production functions for sweet corn and cotton irrigated with saline waters. Soil Sci. Soc. Am. J., 51: 1554-1562.

Sankhayan PL. 1988. Introduction to the Economics of Agricultural Production. Prentice-Hall of India, Private Limited New Delhi, pp. 131.

Sepaskhah, A. R. and Akbari, D. 2005. Deficit irrigation planning under variable seasonal rainfall. Biosystem Engineering, 92(1): 97-106.

Shani U, Dudley LM. 2001. Field studies of crop response to drought and salt stress. Soil Sci. Soc. Am. J., 65:1522-1528.

Stewart, J.I. and Hagan, R.M., 1973. Functions to predict effects of crop water deficits. J. of Irrigation and Drainage. 99(4):421-439.

Tanji KK. (editor). 1990. Agricultural salinity assessment and management. ASCE Monograph, No 71.

Zhang, H. and Oweis, T., 1999. Water- yield relations and optimal irrigation scheduling of wheat in the Mediterranea region. Agric. Water Manage. 38, 195-211.

Zhang, H., Wang, X., You, M., Liu, C. 1999. Water-yield relations and water – use efficiency of winter wheat in the North China plain. Irrg. Sci. 19, 37-45.

4

The Nature of Rainfall at a Typical Semi-Arid Tropical Ecotope in Southern Africa and Options for Sustainable Crop Production

J. Mzezewa[1] and E.T. Gwata[2]
[1]Department of Soil Science, University of Venda
[2]Department of Plant Production, University of Venda
South Africa

1. Introduction

Climate plays an important role in crop biomass production. Extreme climatic conditions and high seasonal variability of climatic parameters could adversely affect productivity (Li et al., 2006) since rainfall determines the crop yields and the choice of the crops that can be grown. The pattern and amount of rainfall are among the most important factors that affect cropping systems. The analysis of rainfall records for long periods provides information about rainfall patterns and variability (Lazaro et al., 2001). Drought mitigation can be planned by understanding daily rainfall behaviour (Aghajani, 2007). Dry spell analysis assists in estimating the probability of intra-season drought in order to adjust water management practices (Tesfaye and Walker, 2004; Kumar and Rao, 2005).

It is important to know the likely durations and probability of wet spells particularly at critical times during the growing season (Dennet, 1987; Sivakumar, 1992). Probability distributions are used widely in understanding the rainfall pattern (Abdullah and Al-Mazroui, 1998). The normal distribution is one of the most important and widely used parameter in rainfall analysis (Kwaku and Duke, 2007). Despite the wide applicability of the normal distribution, there are many instances when observed rainfall distributions are neither normal nor symmetrical. Rainfall can be abnormally distributed (Stephens, 1974) except in wet regions (Edwards et al., 1983). Jackson (1977) observed that annual rainfall distributions are markedly skewed in semi-arid areas and the assumption of normal frequency distribution for such areas is inappropriate. Rainfall can also be described by other distributions such as Gamma distribution (Abdullah and Al-Mazroui, 1998; Aksoy, 2000; Garcia et al., 2007), the log-Pearson type III distribution (Chin-Yu, 2005), the Weibull and Gumbel distributions (Tilahun, 2006).

In semi-arid areas, marginal and erratic rainfall, exacerbated by high runoff and evaporation losses, constrain crop production. The ecotope at Thohoyandou experiences these conditions. By definition, an ecotope is a homogenous piece of land with a unique combination of climate, topographic and soil characteristics (Hensley et al. 2000). In order to understand the feasibility of using a water harvesting system at the ecotope, rainfall analysis and the identification of prevailing rainfall patterns is required (Dennet, 1987; Rappold, 2005). The major objectives of this chapter are to outline (i) an analysis of long-term (1983-

2005) rainfall data recorded for the ecotope at Thohoyandou in the Limpopo River basin (LRB) (South Africa) in order to provide a basis for future management of crop production and (ii) viable options for sustainable crop production at the location and similar agro-ecological areas.

2. Analyses of rainfall distribution patterns

Rainfall distribution patterns can change over a given period. This is expected particularly when taking into account the global climate change. This climate change impacts negatively on agricultural activities (Rosenweig and Parry, 1994; Rosenberg 1992). The nature of the changes in rainfall distribution patterns may also depend on the geographical location of the area of interest.

2.1 Geographical location of the ecotpe
The ecotope used in this study is located at the University of Venda (21° 58′ S, 30° 26′ E; 596 m above sea level) at Thohoyandou in the LRB. The location falls in the lowveld of the greater LRB which is situated in the southern part of southern Africa between about 20° and 26° S latitude and 25 and 35° E longitude. The greater LRB encroaches over the national borders of four different countries namely Botswana, Mozambique, South Africa and Zimbabwe. According to the Koppen Classification, the basin is predominantly semi-arid.

The daily temperatures at Thohoyandou vary from about 25°C to 40° C in summer and between approximately 12°C and 26° C in winter. Rainfall is highly seasonal with 95% occurring between October and March, often with a mid-season dry spell during critical periods of crop growth (FAO, 2009). Mid-season drought often leads to crop failure and low yields (Beukes et al., 1999). The average rainfall is about 800 mm but varies temporarily. The soils at the ecotope are predominantly deep (>150 cm), red and well drained clays with an apedal structure. Clay content is generally high (60 %) and soil reaction is acidic (pH 5.0). The soils are formed *in situ* and classified locally as Hutton form (Soil Classification Working Group, 1991) equivalent to Rhodic Ferralsol (WRB, 2006).

2.2 Statistical analysis of rainfall data
The records of daily rainfall data and reference potential evapotranspiration records were obtained from local institutions (the National Weather Service and Research Council) and used for determining annual and monthly totals. The years or consecutive months with missing data were not included in the calculations of averages. This was so for 1982, 2006 and 2007. Consequently, a 23 year rainfall data set (1983-2005) was analyzed following the standard procedure for analysing rainfall data for agricultural purposes which involves summarising the daily data to obtain monthly totals and then annual totals (Abeyasekera et al., 1983). This was done partly to reduce the volume of data in subsequent analyses. The approaches used in data analysis were similar to those used by Belachew (2002) and Tilahun (2006).

2.3 Rainfall distribution pattern and probability distribution models
The identification of the probability distributions of annual and monthly rainfall data was also important. The observed distributions were fitted to theoretical probability distributions by comparing the frequencies (Tilahun, 2006). Data normality was tested using

skewness and kurtosis coefficients and probability distributions were evaluated with the aid of probability plots and curve fitting using Minitab 14 statistical software (Minitab Inc., 2004)). The goodness-of-fit tests were based on the Anderson-Darling (AD) test (Stephens, 1974) which measures how well the data follow a particular distribution pattern. The p-value with the greatest magnitude was considered to be the best fit. Where the p-values were equal, the smallest AD value was then used to decide the best fit. Rainfall data from October to March (which is the cropping season at the ecotope) were considered in fitting distributions. In addition, four probability distribution models namely the normal, lognormal, Gamma and Weibull distributions were tested (Table 1).

Distribution	Probability density function	Parameter description
Normal	$f_{(x)} = n(x; \mu, \sigma) = \dfrac{1}{\sigma\sqrt{2\Pi}}$ $\exp\left[-\dfrac{1}{2}\left(\dfrac{x-\mu}{\sigma}\right)^2\right] for -\infty \le x \le \infty$	μ= mean of the population x σ= standard deviation of the population x
Lognormal	$f(x) = \dfrac{\exp\left(-\dfrac{1}{2}\left(\dfrac{\ln x - \mu}{\sigma}\right)^2\right)}{x\sigma\sqrt{2\pi}}$	σ = the standard deviation of ln x μ = the mean of ln x
Gamma	for $0 \le x \le \infty$ $f_{(x)} = \dfrac{1}{\beta^\alpha \Gamma_{(\alpha)}} \alpha^{x-1} e^{-x/\beta}$	α is the scale parameter β is the shape parameter of the distribution $\Gamma_{(\alpha)}$ = the normalising factor
Weibull	for $0 \le x \le \infty$ $f_{(x)} = \dfrac{\alpha}{\beta}\left(\dfrac{x}{\beta}\right)^{\alpha-1} \exp\left(-\left[\dfrac{x}{\beta}\right]^\alpha\right)$	α = the scale parameter β = the shape parameter of the distribution

Table 1. Probability distribution models used for testing the rainfall distribution patterns at the ecotope.

The results of the analysis showed that there was variation among the AD values depending on the distribution model applied (Table 2). Based on the AD goodness-of-fit and p-values, the normal distribution model was the best in describing the annual rainfall at the ecotope. This was in agreement with the observation reported previously for a semi-arid environment in Kenya at which the annual rainfall approximated a normal distribution (Rowntree, 1989). In contrast, 50% of the monthly rainfall patterns during the cropping season (October to March) were described best by the lognormal theoretical distribution model (Table 3). This suggested that for a given ecotope, the distribution model that best describes the annual rainfall pattern is not necessarily identical to that describing the monthly rainfall pattern. In addition, the results also suggested that there is no constant best distribution function for rainfall distribution for all the months at an ecotope. In a similar study in Ethiopia, Tilahun (2006) found that most of the monthly rainfall data sets were best described by the lognormal distribution while annual rainfall distribution patterns fitted either the Weibull or the Gumbel or the Gamma distribution models.

	Normal				Lognormal				Gamma				Weibull			
	AD	μ	σ	The p-values	AD	μ	σ	p-value	AD	α	β	p-value	AD	α	β	The p-values
Oct	1.168	64.38	56.41	<0.005	0.308	3.775	0.9904	0.533	0.250	45.14	1.420	>0.250	0.269	68.86	1.216	>250
Nov	1.529	97.89	77.14	<0.005	0.370	4.292	0.8297	0.397	0.466	52.59	1.861	>0.250	0.555	107.80	1.385	0.154
Dec	0.912	140.00	89.59	0.017	0.289	4.735	0.6718	0.584	0.419	54.17	2.583	>0.250	0.474	157.80	1.695	0.232
Jan	0.595	135.00	100.20	0.109	0.489	4.578	0.9126	0.200	0.341	80.48	1.677	>0.250	0.316	147.90	1.384	>250
Feb	0.905	133.90	117.10	0.017	0.539	4.375	1.2060	0.149	0.389	122.30	1.094	>0.250	0.404	137.70	1.078	>0.250
Mar	0.958	91.63	79.50	0.013	0.995	4.086	1.0910	0.010	0.519	70.57	1.298	0.218	0.478	97.06	1.185	0.228
Annu-al	0.339	781.50	248.10	0.468	0.818	6.603	0.3711	0.029	0.586	89.43	8.738	0.144	0.361	867.30	3.634	>0.250

AD = Anderson-Darling statistic

Table 2. Goodness-of-fit values and parameters of theoretical probability distributions fitted to annual and monthly rainfall data

Rainfall Month	Best DistributionModel
October	Lognormal
November	Lognormal
December	Lognormal
January	Weibull
February	Weibull
March	Gamma

Table 3. The best-fit distribution models for the respective monthly rainfall patterns at the semi-arid ecotope in the Limpopo River basin.

2.4 The exceedance probability of annual and monthly rainfall

Apart from establishing the rainfall distribution pattern, the exceedance probability of annual and monthly rainfall, was calculated. By definition, it is the probability that a given amount of rainfall is exceeded in a specific unit period. The probability of exceedance of annual and monthly rainfall was calculated from the respective rainfall distribution parameters as obtained from testing the probability distribution models described in section 2.3. This information is useful in selecting crops since each crop has a specific water requirement to take it through the growth cycle (Rappold, 2005). The information may also be critical in designing appropriate water storage facilities for supplementary irrigation in future. In order for such facilities to be efficient, they need to be constructed in proportion to the amount of water that can be expected during a rainfall event (Schiettecatte, 2005). However, the chances of implementing such strategies in many parts of Africa inhabited by smallholder farmers remain remote and highly unlikely.

At the ecotope, the probability of exceeding various amounts of annual rainfall diminished as the threshold rainfall amount increased. For example, there was 94 % chance of receiving annual rainfall >400 mm whilst the chance of having >1 500 mm of rainfall was zero (Table 4). There was 47 % probability of exceeding 800 mm of annual rainfall. During February, there was a 72% chance of receiving rainfall ≥ 5.0 mm yet the probability of receiving rainfall > 500.0 mm diminished to 2.0% (Table 5). The mean annual rainfall for the site was about 781.0 mm (Table 6).

Annual rainfall (mm)	Probability of Exceedance (%)
>400.0	94.0
>600.0	77.0
>800.0	47.0
>1 000.0	19.0
>1 200.0	5.0
>1 500.0	0.0

Table 4. The probability of receiving annual rainfall exceeding specific amounts ranging from 400.0 mm to 1 500.0 mm.

Month	Monthly Rainfall (mm)					
	5.0	50.0	100.0	200.0	500.0	600.0
	Probability of exceedance (%)					
Oct	99.0	44.0	20.0	6.0	1.0	0.0
Nov	100.0	68.0	35.0	11.0	1.0	1.0
Dec	100.0	89.0	58.0	20.0	1.0	1.0
Jan	99.0	80.0	56.0	22.0	0.0	0.0
Feb	72.0	71.0	49.0	22.0	2.0	0.0
Mar	97.0	63.0	35.0	9.0	0.0	0.0

Table 5. The probability of receiving monthly rainfall exceeding specific amounts ranging from 5.0 mm to 600.0 mm

Parameter	Mean	SD	CV	Min	Max	SC	KC
Oct	64.38	56.41	1.14	3.5	243.8	1.64	3.31
Nov	97.89	77.14	1.27	7.3	296.5	1.40	1.20
Dec	139.95	89.59	1.56	31.3	331.7	0.78	-0.68
Jan	134.96	100.24	1.35	12.0	420.0	1.02	1.29
Feb	133.87	117.07	1.14	5.3	420.7	0.87	-0.02
Mar	91.63	79.50	1.15	5.7	361.2	1.87	5.15
Apr	39.87	53.94	0.74	0.1	249.2	2.57	8.55
May	17.74	35.35	0.50	0.0	162.7	3.53	13.74
Jun	12.41	14.10	0.88	0.0	45.9	1.00	-0.16
Jul	8.00	15.29	0.52	0.0	67.3	3.01	10.37
Aug	7.87	10.87	0.72	0.0	48.5	2.67	8.60
Sept	25.29	38.11	0.66	0.0	162.5	2.58	7.35
Annual	781.47	248.07	3.15	281.2	1239.3	-0.15	-0.14

CV = Coefficient of variation; SD = Standard deviation; SC = Skewness coefficient; KC = Kurtosis coefficient.

Table 6. Statistical parameters for mean monthly and annual rainfall data (1983-2005).

2.5 The probability of dry spells

Equally important was determining the probability of dry spells. A dry day was defined as a day receiving less than 1.0 mm of rainfall. A dry spell was a sequence of dry days bracketed by wet days on both sides (Kumar and Rao, 2005). The analysis of the frequency of dry spells was adapted from the method of Belachew (2002). In this approach, a period of Y years of records, the number of times i that a dry spell of duration t days occurs, was counted on a monthly basis; then the number of times I that a dry spell of duration longer than or equal to t occurs was computed through accumulation. The consecutive dry days (1d, 2d, 3d ...) were prepared from historical data. The probabilities of occurrence of consecutive dry days were estimated by taking into account the number of days in a given month n. The total possible number of days, N, for each month over the analysis period was computed as, $N = n*Y$, hence the probability p that a dry spell equal or longer than t days was computed as:

$$p = \frac{I}{N}$$

The distribution of daily rainfall totals by amount and frequency was obtained by using frequency analysis of historic daily rainfall data. The statistical parameters (namely the mean, standard deviation, coefficient of variation, coefficient of skewness and coefficient of kurtosis) for both the annual and monthly rainfall were also determined (Table 6). For a normal distribution, the skewness and kurtosis coefficients should be zero or near zero. This approach can be confirmed further by determining the p-values using the AD test.

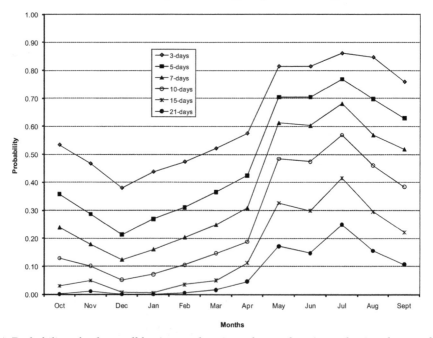

Fig. 1. Probability of a dry spell lasting ≥n days in each month estimated using the raw data from 1983-2005 at the semi-arid ecotope. (n=3, 5, 7, 15, 21).

The occurrence of dry spells has particular relevance to rain-fed agriculture, as rainfall water is one of the major requirements for plant life in rain-fed agriculture (Belachew, 2002; Rockstrom et al., 2002). The probability of occurrence of dry spells of various durations varied from month to month (Fig. 1). December had the lowest probabilities of occurrence of dry spells of all durations. Generally the occurrence of dry spells of all durations decreased between October and March. This period coincides with the rainy season in the region (Lynch et al., 2001). The probability of having a dry spell increased with the reduction in the duration of the dry spell. In other words, there were more chances of having a 3d dry spell than a 10 or 21d dry spell. For instance, in December, there was 20% probability of having a dry spell lasting five days but 0% probability of having a dry spell lasting 21d (Fig. 1). This trend was in agreement with observations reported in literature (Sivakumar, 1992; Aghajani , 2007).

2.6 The variability of rainfall

The rainfall data for the ecotope indicated that monthly rainfall was strongly skewed to the right (high positive values of skewness coefficients) and highly leptokurtic, a phenomenon common in semi-arid regions. The yearly rainfall analysis indicated that the mean annual rainfall (781.0 mm) at the ecotope was accompanied by a high (248.0 mm) standard deviation. In addition, the coefficient of variation of the annual rainfall was high (315 %) indicating high variability of rainfall from year to year. The monthly rainfall analysis indicated that the site receives about 80% of annual rainfall during the period October to March (Fig. 2). This was in agreement with the observation that most of the rainfall amount in the region occurs between October and March (Landman and Klopper, 1998). However, this rainfall amount received at the the ecotope should be approached with caution partly because of the relatively high potential evapotranspiration. The effective rainfall is low. The coefficient of variation for the monthly rainfall was high (156 %) confirming the high variability in the monthly rainfall at the location. This result was consistent with the findings by Tyson (1986) who reported a similar rainfall pattern in the interior regions of South Africa.

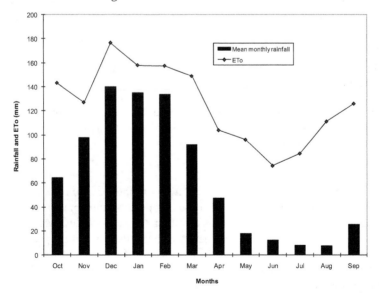

Fig. 2. Mean monthly rainfall and potential evapotranspiration (ETo) over a 23-year period at the semi-arid ecotope.

2.7 Agro-climatic zonation
An aridity index represents climatic aridity and is used to determine the adequacy of rainfall in satisfying the water needs of the crop (Tsiros et al., 2008). Mzezewa et al. (2010) reported that the aridity index for the study area was 0.52. Based on the UNESCO (1979) classification criteria, the study area is on the borderline case between semi-arid and sub-humid since it receives low rainfall and experiences high evapotranspiration. At the semi-arid ecotope at Thohoyandou, the potential evapotranspiration was consistently higher than the mean monthly rainfall throughout the year (Fig. 2) indicating that rainfall was not effective at the study site. Moreover, research conducted elsewhere in semi-arid regions showed that approximately 70 % of annual rainfall is lost due to evaporation from the soil (Hoffman, 1990; Jalota and Prihar, 1990; Botha et al., 2003). Therefore, the adoption of effective practices that maximise the utilization of rainfall and minimize water loss is imperative so as to achieve sustainable crop production at the ecotope.

3. Options for crop production
Information on rainfall amount and variability is important in deciding on the choice of crops and cultivars to produce at the ecotope. The rainfall threshold under rain-fed conditions varies from one crop species to the other and depends on a combination of several environmental factors such as soil type and diurnal temperatures. For instance, the water retention properties of sandy soils would be considerably different from those of clay soils while high day temperatures encourage evapotranspiration. Although the mean annual rainfall at the ecotope was in excess of 700 mm, there was high variability (315%) in the quantity of rainfall received. Nonetheless, the high evaporative demand as indicated by the Aridity Index (0.52) of the ecotope means that most of the rain is not available for crop use. On the other hand, the optimum rainfall for a staple cereal such as corn (Zea mays) is considered to range between 500 and 800 mm (Ovuka and Linqvist, 2000) but in crop production, the distribution of the rainfall is critical.

This spatial and temporal variability in annual rainfall experienced in the area imposes several major challenges for crop growers particularly small-holder farmers. Therefore, their crop choices must take into consideration the challenges imposed by moisture stress during the cropping season. In this regard, the use of drought tolerant crop technologies as a means for achieving sustainable crop production in the area is merited. Such crops include specific legumes and small-grain cereals.

3.1 Cropping systems
The ecotope is an agro-ecological representation of the prevailing conditions in the Limpopo basin. The crop production at the ecotope is dependent on the agricultural systems in the area. The systems consist of either the large-scale sector or small-scale sector (also called the small-holder sector). The latter is more prevalent and is characterized by resource-poor farmers in possession of small land holdings averaging approximately 0.4 ha per household. The pressure of limited land for cultivating crops, insufficient production inputs as well as the subsistence form of crop production practiced by typical smallholder farmers largely influence the types of crops they grow. The cropping systems typically consist of rain-fed cereals such as corn grown either in a

monocrop or intercropped with a variety of minor crops such as peanut (*Arachis hypogeae*), common bean (*Phaseolus vulgaris*), cowpea (*Vigna unguiculata*). Corn is the staple cereal. In some sub-regions of the LRB such as the dry-land area of Sekhukhune district, the farmers grow small grain cereals (especially millets) intercropped with tapery bean (*Phaseolus acutifolius*). In contrast, crop production in the large-scale sector is more intensive, highly mechanized and commercially oriented.

3.2 Drought tolerant legumes

The specific legume species recommended for production under these frequent moisture deficit conditions include chickpea (*Cicer arietinum*), cowpea (*Vigna unguiculata*) pigeonpea (*Cajanus cajan*) and tapery bean (*Phaseolus acutifolius*). These legumes offer to the farmers numerous advantages ranging from promiscuous (non-specific or tropical) type of nodulation to high potential for income generation (Table 7). Their ability to form effective nodules with soil rhizobia ubiquitous in African soils enables them to be sown without treating the seed with commercial inoculants. The legumes improve soil fertility through biological nitrogen fixation (Mapfumo et al., 1999; Serraj et al.,1999; Shisanya 2002) and provide food for human consumption.

	Agronomic Aspect	Chickpea	Cowpea	Pigeonpea	Tapery bean
(i)	Requirement for commercial inoculants for seed at planting time	None	None	None	None
(ii)	Requirement for production inputs	Minimal	Minimal	Minimal	Minimal
(iii)	Tolerance to drought	High	Very High	Very High	High
(iv)	Yield potential (approx.)	Very High (5.0 t ha^{-1})	High (4.0 t ha^{-1})	High (3.0 t ha^{-1})	Medium to High (2.0 t ha^{-1})
(v)	Amount of N$_2$-fixed (approx.)	High (124 kg ha^{-1})	Medium (120 kg ha^{-1})	Very High (166 kg ha^{-1})	Not Available
(vi)	Ability to ratoon and produce another harvest	None	None	Yes	None
(vii)	Requirement for expertise in seed production	None to Very Low	None to Very Low	None to Very Low	None to Very Low
(viii)	Household utilization	Yes	Yes	Yes	Yes
(ix)	Storability	High	Low	High	Medium
(x)	Use as stock feed	Yes	Yes	Yes	Yes
(xi)	Demand in domestic or regional or international markets	Very High	Very High	Very High	Very High

Table 7. Summary of important agronomic traits of four tropical legumes suitable for production by small-holder farmers at the ecotope.

3.2.1 Mechanisms of drought tolerance in the legumes

The legumes employ a variety of mechanisms to avoid the effects of moisture stress. One of such mechanisms is drought avoidance in which the legume plant maintains a relatively high tissue water potential despite a soil moisture deficit. The legume plant can also escape the effect of drought stress by completing its life cycle before the occurrence of soil and plant water deficit. In drought tolerance, the plants are able to withstand water deficit periods with low tissue water potential (Mitra 2001). According to Turner (1986), plants that utilize the drought tolerance mechanism are able to maintain turgor by accumulating compatible solutes in the cell environment (osmotic adjustment), increased cell elasticity, decreased cell volume and resistance to desiccation through protoplasmic resistance. Indeterminate legume types with sequential flowering are capable of recovering from drought spells producing a compensatory flush of pods when soil moisture is restored. In addition, the long tapered root associated with most leguminous species facilitates moisture acquisition from deep layers of the soil.

Chickpea is unique partly because it is a post-rainy season (planted in April/May at the ecotope) crop which thrives on the residual moisture after other field crops are harvested. Under receding soil moisture conditions, the chickpea deep root system directly contributes to the grain yield (Kashiwagi et al., 2006). Because of the lack of competition for space with other crops in the field then, it is advantageous for the small-holder farmers who have limited land for cultivation and no irrigation facilities for producing off-season crops during the dry season (May to October). Chickpea is also tolerant to drought, producing higher yields in winter than in summer (Saxena et al., 1993; Katerji, et al., 2001; Kumar et al., 2001; Sabaghpour et al., 2006). On the other hand, tolerance to drought in both cowpea (Elhers and Hall 1997; Singh et al., 1999; Singh and Matsui, 2002) and pigeonpea (Gwata and Siambi, 2009; Kumar et al. 2011; Sekhar et al. 2010) have been reported. Apart from its ability to develop deep roots, cowpea can have reduced leaf size with thick cuticles that reduce water loss (Graham and Vance 2003). The crop utilizes a combination of avoidance, escape and tolerance mechanisms. In tapery bean, the preliminary results obtained from a study aimed at evaluating the potential of the crop at the ecotope, showed that it matures early, within three months after planting. According to Beaver et al. (2003), selecting for early maturity, efficiency in nutrient partitioning to the reproductive structures as well as phenotypic plasticity in bean is useful in adapting the crop to drought conditions. In addition, tapery bean is grown widely in the dry-land area of the Sekhukhune district in Limpopo (Mariga *pers. comm.*, 2011) and the semi-arid areas in eastern Kenya (Shisanya, 2002) indicating its adaptation to drought prone areas.

3.2.2 Agronomic performance of the legumes

In a study designed to evaluate a typical drought tolerant legume, pigeonpea, for two seasons for both adaptation and yield potential at the ecotope under rainfed conditions, 60% of the twenty genotypes flowered early and matured within the cropping season indicating their adaptation to the area (Mogashoa and Gwata 2009). At least five of the cultivars obtained grain yields >1.5 t/ha over the two year period. The highest mean yield (1.9 t/ha) was four-fold higher than that attained by the check cultivar. In another study conducted under similar rain-fed conditions in a semi-arid environment in Malawi, pigeonpea attained considerably high yields (Gwata and Siambi, 2009). The authors reported a high grain yield (3.0 t/ha) for the genotype ICEAP 01480/32 (Table 8).This indicated the potential of the crop under the semi-arid conditions and a viable option for legume farmers in the semi-arid agro-ecological zones. Since biological nitrogen fixation is adversely affected by heat and depleted soil moisture conditions (Zahran 1999; Serraj et al., 1999), the relatively high grain yield observed for the grain legumes indicated that their nitrogen fixation mechanisms were not suppressed under the prevailing agro-ecological conditions. Therefore, these legumes were adapted to the agro-ecological conditions at the ecotope and similar environments in the region. However, the subsistence small-holder farmers in the area tend to prefer planting corn prior to any other crop at the beginning of the cropping season in spite of the high risks paused by either mid-season or end-of-season moisture stress every year (Fig. 3). The sustainable production of corn, by small-holder growers is constrained by highly unpredictable rainfall, high temperatures and low soil fertility among other factors. In our view, small grain cereals are more appropriate than corn for the predominantly arid conditions at the ecotope and similar agro-ecological areas in the region.

Fig. 3. Contrast in sensitivity to mid-season moisture stress between corn (in the foreground) and pigeonpea (in the background).

	Cultivar	Grain Yield (t/ha)
(i)	ICEAP 01144/3	2.7 a
(ii)	ICEAP 01160/15	2.4 a
(iii)	ICEAP 01480/32	3.0 a
(iv)	ICEAP 01162/21	2.6 a
(v)	ICEAP 01167/11	2.2 a
(vi)	ICEAP 01514/15	2.9 a
(vii)	*Royes	1.0 b
(viii)	**Mutawa Juni	1.1 b

Table 8. Grain yield of pigeonpea germplasm evaluated under rain-fed conditions in Malawi. (*Source*: Adapted from Gwata and Siambi, 2009).

Means in the column followed by the same letter are not significantly different at the 0.05 probability level by Tukey's test. *Commercial cultivar in Malawi; **Unimproved traditional landrace popular in Malawi.

3.3 Drought tolerant small-grain cereals
In many parts of Africa, small grain cereals such as sorghum (*Sorghum bicolor*), pearl millet (*Pennisetum glaucum*) and finger millet (*Eleusine coracana*) are useful for both human consumption and animal feed. For instance, sorghum is regarded as the major source of food in many countries in sub-Saharan Africa such as Chad, Sudan and Tanzania (Bucheyeki et al., 2010). In parts of southern Africa, sorghum grain is used for brewing commercial alcoholic and non-alcoholic beverages (Mushonga et al 1993). The stover is used for fodder, fencing, thatching as well as fuel purposes (Rai et al., 1999). Similarly, pearl millet is used for food and feed (Rai et al., 1999). On the other hand, finger millet is used widely for human consumption throughout southern Africa (Mushonga *et al.*, 1993; Mnyenyembe, 1993). In contrast to the other cereals, finger millet is rich in the protein fraction eleusinin and considerable amounts of cystine, methionine and tryptophan. It also contains a range of minerals such as calcium, copper and manganese. Partly because of its slow digestion, it is preferred by some end-users. The crop residue also provides readily digestible nutrients in livestock feeds.

3.3.1 Mechanisms of drought tolerance in the small grain cereals
These small grain cereals cope with moisture deficit in a variety of ways. For instance, pearl millet can withstand periods of moisture stress and still produce biomass and grain because of its high water use efficiency as opposed to increased water uptake efficiency by a deep root system (Zegada-Lizarazu and Lijima 2005). Similarly, the adaptive features of sorghum enable the crop to grow well in agro-ecological conditions where other staple cereals such as corn would not be suitable (Haussmann et al., 2000). These features include a highly branched rooting system, considerable amounts of silica in the root endodermis, ability to form tillers and capability of rolling the leaves to reduce water loss. In finger millet, drought tolerance was attributed to the ability to synthesize stress proteins (Umar et al., 1995).

3.3.2 Agronomic performance of the small-grain cereals
The performance of the small-grain cereals at the ecotope have not yet been investigated adequately. However, studies conducted elsewhere in marginal tropical zones in southern Africa reported moderately high grain yield potentials. For instance, in Zambia, finger millet attained 3.0 t/ha (Agrawal et al., 1993; Mnyenyembe 1993). In comparison, sorghum attained similar high yields in the African sub-continent (Table 9). According to Gari (2001), these small grain cereals, particularly pearl millet, have superior adaptation to drought and poor soils, providing a reliable harvest under such conditions, with minimal inputs. Moreover, rotating these cereals with the legumes enhances their productivity since they benefit from biological nitrogen fixation by the legumes (Batiano and Ntare 2000). Alternatively, the cereals can be intercropped with the legumes especially cowpea and pigeonpea. The benefits of cereal x legume intercrops in sub-Saharan Africa have been documented widely (Stoop and Staveren 1981; Ntare 1989; Klaij and Ntare 1995; Batiano and Ntare 2000).

Crop	Yield Potential (t/ha)	Country (Major Agro-ecological region)	Source
Finger millet	3.3	Malawi (southern Africa)	Mnyenyembe, 1993
	3.0	Zambia (southern Africa)	Agrawal et al., 1993
Sorghum	2.8	Chad (north/west Africa)	Yapi et al.,1998
	3.0	Mali (north Africa)	Shetty et al.,1991
	1.8	South Africa (southern Africa)	Olembo et al., 2010
	3.5	Tanzania (east Africa)	Bucheyeki et al.,1991
Pearl Millet	2.6	Mali (north Africa)	Wilson et al., 2008
	2.0	Mali (north Africa)	Shetty et al., 1991
	2.7	Zambia (southern Africa)	Wilson et al., 2008

Table 9. Potential yield of three small grain-cereals observed in various African countries.

3.4 Use of short-duration crops
Another sustainable option for the farmers in the region represented by the ecotope is to utilize short-duration (or short-season) crop technologies. The types are able to flower and mature before the on-set of mid-season moisture stress, thus escaping drought. Among tropical legumes, short-duration cultivars have been developed for cowpea (Gwathmey and Hall 1992; Hall and Patel 1992; Mligo and Singh 2006) and pigeonpea (Gwata and Siambi 2009; Silim and Omanga 2001). Short-duration cowpeas that can attain 2.0 t/ha within 60 to 70 d after planting have been reported (Ehlers and Hall 1997). In chickpea, extra-short duration genotypes were reported for the semi-arid tropics (Kumar et al., 2001). These genotypes require < 100 d from sowing to maturity and would be ideal for the limited moisture conditions during the post-rainy season at the ecotope.

3.5 Use of field water conservation and harvesting techniques
In-field water harvesting (IRWH) technique described by various researchers (Hensley et al., 2000; Li et al., 2000; Mzezewa et al., 2011) promotes rainfall runoff on a 2.0 m wide no-till strip between crop rows and collects the runoff water in basins where it infiltrates into the soil profile. The in-field runoff captured in the water basins, can result in increased crop yields. In 2003, Botha et al., demonstrated that the soil moisture captured in this way was responsible for increased crop yields on smallholder farms in Thaba Nchu, South Africa. In a similar study utilizing this water harvesting technique, both corn and sunflower attained up to 50% higher yields compared to conventional production systems on duplex soils in the Free State province in South Africa (Hensley et al., 2000). There has been increasing interest recently in South Africa of making crop production less risky and sustainable in semi-arid ecotopes through in-field rainwater water harvesting (Botha et al., 2003). In a study conducted at the ecotope, Mzezewa et al., (2011), observed that IRWH resulted in significantly higher water use, water use efficiency and precipitation use efficiency for both sunflower and cowpea in comparison with the conventional tillage system. However, despite the increase in crop yields, considerable losses of moisture through evaporation were observed. Li et al., (2000) successfully used gravel-mulched furrows to reduce evaporation and increase the grain yield of corn. Alternatively, living soil cover (that is using live or green mulches) between crop plants was proposed as a sustainable option for

minimizing water loss from the soil surface. According to Aladesanwa and Adigun (2008), the living mulches can be integrated in an intercrop in order to derive some of the benefits associated with intercropping such as increased yield under conditions of moisture stress, weed suppression, reducing the risks of drought and pests as well as achieving adequate crop diversity. Nonetheless, the economic potential of adapting mulches depends on an array of factors such as the cost of the complementary changes (for example availability of mulch adapted field equipment) as well as land availability.

4. Concluding remarks

The statistical analysis of rainfall data at the ecotope revealed that the rainfall is low and highly unreliable, thus constraining crop production. Soil moisture which is available for crop growth only for limited periods at the location can be lost by evaporation. Sustainable production of cereals and legumes at the ecotope and similar agro-ecological environments hinges on adopting drought tolerant or short-duration cultivars as well as water conservation practices such as in-field rainwater harvesting. Information from the combination of climatic analysis and crop cultivar performance needs to be considered in selecting appropriate planting dates.

5. References

Abdullah, M.A. and Al-Mazoui, M.A. (1998). Climatological study of the south western region of Saudi Arabia. I. Rainfall analysis. Clim. Res. 9, 213-223.

Abeyasekera, S.., Seneviratne, E.K., Leaker, A. and Stern, R.D. (1983). The analysis of daily rainfall for agricultural purposes. J. Natn. Sci. Counc. Sri Lanka 11(2), 165-183.

AghajanI, G.H. (2007). Agronomical Analysis of the characteristics of the precipitation (Case study: Sazevar, Iran). Pakistan J. Biol. Sci. 10 (8), 1353-1358.

Agrawal, B.L., Siame, J.A. and Uprichard, G.T. (1993). Status of finger millet (*Eleusine coracana* Gaertn.) in Zambia, 1993. In: Riley, K.W., Gupta, S.C., Seetharam, A. and Mushonga, J.N. (ed.). Advances in small millets. pp. 21-28. New Delhi: Oxford and IBH.

Aksoy, H. (2000). Use of gamma distributions in hydrological analysis. *Turk. J. Eng. Environ. Sci.* 24, 419-428.

Aladesanwa, R.D. and Adigun, A.W. (2008). Evaluation of sweet potato (Ipomea batatas) live mulch at different spacings for weed suppression and yield response of corn (Zea mays L.) in southwestern Nigeria. Crop Protection 27, 968-975.

Batiano, A. and Ntare, B.R. (2000). Rotation and nitrogen fertilizer effects on pearl millet, cowpea and groundnut yield and chemical properties in a sandy soil in the semi arid tropics, West Africa. J. Agric. Res. 134, 277 - 284.

Beaver, J.S., Rosas ,J. C., Myers J., Acosta J., Kelly J.D., Nchimbi-Msolla S., Misangu R., Bokosi J., Temple S., Arnaud-Santana E and Coyne, D. P. (2003). Contributions of the Bean/Cowpea CRSP to cultivar and germplasm development in common bean. Field Crops 82, 87 -1 02.

Belachew, A. (2000). Dry-spell analysis for studying the sustainability of rain-fed agriculture in Ethiopia: The case of the Arbaminch area. Proc. 8th Nile 2002 Conf. 22 June 2002,

Addis Ababa, Ethiopia.. International Commission on Irrigation and Drainage (ICDID).

Beukes, D.J., Bennie, A.T.P. and Hensley, M. (1999). Optimization of soil water use in the dry cropproduction areas of South Africa. In: N van Duivenbooden, M Pala, C Studer and CL Bielders (eds.)Proc. of the 1998 (Niger) (April 26-30 April) and 1999 (Jordan) (May 9-13) Workshops of the Optimizing Soil Water Use (OSWU) Consortium, entitled: Efficient Soil Water Use: The Key to Sustainable Crop Production in the Dry Areas of West Asia, and North and Sub-Saharan Africa. Aleppo, Syria: ICARDA; and Patancheru, India: ICRISAT. pp 165-191.

Botha, J.J., Van Rensburg, L.D., Anderson, J.J., Hensley, M., Macheli, M.S., Van Staden, P.P., Kundlande, G., Groenewald, D.G. and Baiphethi, M.N. (2003). Water Conservation Techniques on Small Plots in Semi-Arid Areas to Enhance Rainfall Use Efficiency, Food Security, and Sustainable Crop Production. WRC Report No. 1176/1/03. Water Research Commission, Pretoria, South Africa.

Bucheyeki, T.L., Shenkalwa, E.M., Mapunda, T.X. and Matata L.W. (2010). Yield performance and adaptation of four sorghum cultivars in Igunga and Nzega districts of Tanzania. Comm. Biom. Crop Sci. 5, 4 - 10.

Chin-Yu, L. (2005). Application of rainfall frequency analysis on studying rainfall distribution characteristics of Chia-Nan Plain area in southern Taiwan. Crop, Environ. Bioinf. 2, 31-38.

Dennett, M.D. (1987). Variation of rainfall-the background to soil and water management in dryland regions. Soil Use Manage. 3(2), 47-51.

Edwards, K.A., Classen, G.A. and Schroten, E.H.J. (1983). The Water Resource in Tropical Africa and its Exploitation. ILCA Research Report No. 6. Available online: http://ww.fao.org/wairdocs/ilri/x5524e/x5524e00.htm#contents (Accessed on March 6, 2009).

Ehlers, J.D. and Hall, A.E. (1997). Cowpea (*Vigna unguiculata* (L.) Walp.). Field Crops Res. 53, 187-204.

FAO (2009). *Climate and rainfall.* http://wwww.fao.org/wairdocs/ilri/x5524e/x5524e03.htm (Accessed 26 February 2009).

Garcia, M., Raes, D., Jacobsen, S.E. and Michel, T. (2007). Agroclimatic constraints for rainfed agriculture in the Bolivian Altipano. *J. Arid Environ.* 71 (1), 109-121.

Garí, J.A.(2001). Review of the African millet diversity. FAO - Food and Agriculture Organisation of the United Nations, Paper for the International workshop on fonio, food security and livelihood among the rural poor in West Africa. IPGRI / IFAD, Bamako, Mali, 19-22 November 2001.

Graham, P.H. and Vance, C.P. (2003). Legumes: importance and constraints to greater use. Plant Physiol. 131, 872-877.

Gwata, E.T. and Siambi, M. (2009). Genetic enhancement of pigeonpea for high latitude areas in Southern Africa. Afr. J. Biotech. 8, 4413-4417.

Gwata, E.T. (2010). Potential impact of edible tropical legumes on crop productivity in the small-holder sector in Sub-Saharan Africa. J. Food Agric. Environ. 8, 939-944.

Gwathmey, C.O. and Hall, A.E. (1992). Adaptation to midseason drought of cowpea of cowpea genorypes with contrasting senescence traits. Crop Sci. 32, 773-778.

Hall, A.E. and Patel, P.N. (1985) Breeding for resistance to drought and heat. *In*: S.R. Singh and K.O. Rachie (eds). Cowpea Research, Production and Utilization. John Wiley and Sons, New York, pp. 135-151.

Haussmann, B.I.G., Obilana, A.B., Oyiecho, P.O., Blum, A., Schipprack, W. and Geiger, H.H. (2000). Yield and yield stability of four population types of grain sorghum in a semi-arid area of Kenya. Crop Sci. 40,319-329.

Hensley, M., Botha, J.J., Anderson, J.J., Van Staden, P.P. and Du Toit, A. (2000). Optimizing Rainfall Use efficiency for Developing Farmers with Limited Access to Irrigation Water. WRC Report No. 878/1/00. Water Research Commission, Pretoria, South Africa.

Hoffman, J.E. (1990). The Influence of Soil Cultivation Practices on the Water Balance of a Avalon Soil with Wheat at Bethlehem. M.Sc. Agric. Thesis, Univ. Free State, Bloemfontein, South Africa.

Ismail, A.M., Hall, A.E., Ehlers, J.D. (2000). Delayed-leaf senescence and heat-tolerance traits mainly are independently expressed in cowpea. Crop Sci. 40,1049–1055.

Jackson, I.J. (1977). Climate, Water and Agriculture in the Tropics. Longman, New York.

Jalota, S.K. and Prihar, S.S. (1990). Bare soil evaporation in relation to tillage. *Adv. Soil Sci.* 12, 187-216.

Kashiwagi, J., Krishnamurthy, L., Crouch, J.H. and Serraj, R. (2006). Variability of root characteristics and their contributions to seed yield in chickpea (Cicer arietinum L) under terminal drought stress. Field Crops Res. 95, 171-181.

Katerji, N., van Hoorn, J.W., Hamdy, A., Mastrorilli, M., Oweise, T. and Malhotra, R.S. (2001). Response to soil salinity of two chickpea varieties differing in drought tolerance. Agricultural Water Management 50, 83-96.

Klaij, M.C. and Ntare, B.R. (1995). Rotation and tillage effects on yield of pearl millet (Pennisetum glaucum) and cowpea (Vigna unguiculata) and aspects of crop water balance and soil fertility in a semi-arid tropical environment. J. Agric. Sci. 124, 39 - 44.

Kumar, R.R., Karjol, K. and Naik, G.R. (2011). Variation of sensitivity to drought stress in pigeonpea [Cajanus cajan (L.) Millsp.] during seed germination and early seedling growth. World J. Sci. Technol. 1, 11-18

Kumar, J., Pannu, R.K. and Rao, B.V. (2001). Development of a short-duration chickpea for the subtropics. Int. Chickpea Pigeonpea Newsl. 8, 7-8.

Kumar, K.K. and Rao, T.V.R. (2005). Dry and wet spells at Campina Grande-PB. Revista Brasileira de Meteorologia 20 (1), 71-74.

Kwaku, X.S. and Duke, O. (2007). Characterization and frequency analysis of one day annual maximum and two five consecutive days' maximum rainfall of Accra, Ghana. ARPN J. Eng. Appl.Sci. 2 (5), 27-31.

Landman, W.A. and Klopper, E. (1998). 15-year simulation of the December to March rainfall season of the 1980s and 1990s using canonical correlation analysis (CCA). Water SA 24 (4), 281-286.

Lazaro, R., Rodrigo, F.S., Gutierrez, L., Domingo, F, and Puigdefafregas, J. (2001). Analysis of a 30-year rainfall record (1967-1997) in semi-arid SE Spain for implications on vegetation. J. Arid Environ. 48, 373-395.

Li, X.Y., Gong, J.D., Wei, X.H. (2000). In-situ rainwater harvesting and gravel mulch combination for corn production in the dry semi-arid region of China. Journal of Arid Environments 46, 371-382.

Li, X.Y., Shi, P.J., Sun, Y.L., Tang, J. and Yang, Z.P. (2006). Influence of various in-situ rainwater harvesting methods on soil moisture and growth of *Tamarix ramosissima* in the semiarid loess region of China. For. Ecol. Manage. 23 (1), 143-148.

Lynch, S.D., Zulu, J.T., King, K.N. and Knoesen, D.M. (2001). The analysis of 74 years of rainfall recorded by the Irwins on two farms south of Potchefstroom. Water S *A*27 (4), 559-564.

Mapfumo, P., Giller, K.E., Mpepereki, S. and Mafongoya, P.L. (1999). Dinitrogen fixation by pigeonpea of different maturity types on granitic sandy soils in Zimbabwe. Symbiosis 27,305-318.

Minitab Inc. (2004). Minitab Student Release 14 Software. State College, PA.

Mitra, J. (2001). Genetics and genetic improvement of drought resistance in crop plants. Current Sci. 80, 758-763.

Mligo, J.K. and Singh, B.B. (2007). Registration of 'Vuli-1' Cowpea. Crop Sci 47,437-438.

Mogashoa, K.E. and Gwata, E.T. (2009). Preliminary indicators of adaptation of pigeonpea [*Cajanus cajan* (L.) Millsp.] to Limpopo (South Africa). In: Proc. Afr. Crop Sci. Soc. 9th Conf., Cape Town, RSA. (Abstract, pp. 141).

Mushonga, J.N., Muza, F.R., and Dhliwayo, H.H. (1993). Development, Current and Future Research Strategies on Finger Millet in Zimbabwe. In: Riley, K.W., Gupta, S.C., Seetharam, A. and Mushonga, J.N. (ed.). Advances in small millets. pp. 11-19. New Delhi: Oxford and IBH.

Mnyenyembe, P.H. (1993). Past and present research on finger millet in Malawi. In: Riley, K.W., Gupta, S.C., Seetharam, A. and Mushonga, J.N. (ed.). Advances in small millets. pp. 29-37. New Delhi: Oxford and IBH.

Mzezewa, J., Gwata, E.T. and van Rensburg, L.D. (2011). Yield and seasonal water productivity of sunflower as affected by tillage and cropping systems under dryland conditions in the Limpopo Province of South Africa. Agric. Water Management 98, 1641-1648.

Mzezewa, J., Misi, T. and van Rensburg, L.D. (2010). Characterisation of rainfall at a semi-arid ecotope in the Limpopo province (South Africa) and its implications for sustainable crop production. Water SA 36 (1), 19-26.

Ntare, B.R. (1989). Intercropping morphologically different cowpeas with pearl millet in a short season environment in the Sahel. Exp. Agric. 26, 41 - 47.

Olembo, K.N., Mmboyi F, Kiplagat S, Sitieney JK and Oyugi FK. (2010). Sorghum breeding in sub-Saharan Africa: the success stories. ABSF, Nairobi, Kenya. pp. 37

Ovuka, M. and Linqvist, S. (2000). Rainfall variability in Murang'a District, Kenya: Meteorological data and farmers' perception. Geografiska Ann. *82 A* 1, 107-119.

Rai, K.N., Murty, D.S., Andrews, D.J. and Bramel-Cox, P.J. (1999). Genetic enhancement of peral millet and sorgum for the semi-arid tropics of Asia and Africa. Genome 42, 617-628.

Rappold, G.H. (2005). Precipitation analysis and agricultural water availability in the Southern Highlands of Yemen. Hydrol. Proc. 19 (12), 2437-2449.

Rockstrom, J., Jennie, B. and Fox, P. (2002). Rainwater management for increased productivity among small-holder farmers in drought prone environments. Phys. Chem. Earth, Parts A/B/C 27 (11-22), 949-959.

Rosenberg, N.J. (1992). Adaptation of agriculture to climate change. Clim. Change 21: 385-405

Rosenweig, C. and Parry, M.L. (1994). Potential impact of climate change on world food supply. Nature 367, 133-138.

Rowntree, K.M. 1989. Rainfall characteristics, rainfall reliability and the definition of drought: Baringo District, Kenya. S. Afr. Geogr. J. 71 (2), 74-80.

Sabaghpour, S.H., Mahmodi, A.A., Saeed, A., Kamel, M. and Malhotra, R.S. (2006). Study on chickpea drought tolerance lines under dryland condition of Iran. Indian J. Crop Sci. 1: 70-73

Saxena, N.P., Krishnamurthy, L. and Johansen, C. (1993). Registration of a drought resistant chickpea germplasm. Crop Sci. 33, 1424.

Schiettecatte, W., Ouessar, M., Gabriels, D., Tange, S., Heirman, S. and Abdelli, F. (2005). Impact of water harvesting techniques on soil and water conservation: a case study on a micro catchment in southeastern Tunisia. J. Arid Environ.61 (2), 297-313.

Sekhar, K., Priyanka, B., Reddy, V.D., and Rao K.V. 2010. Isolation and characterization of a pigeonpea cyclophilin (CcCYP) gene, and its over-expression in Arabidopsis confers multiple abiotic stress tolerance. Plant Cell Environ. 33, 1324–1338.

Serraj, R., Vadez, V., Denison, R.F. and Sinclair, T.R. (1999).Involvement of ureides in nitrogen fixation inhibition in soybean. Plant Physiol. 119, 289–296.

Shetty, S.V.R., Beninati, N.E, and Beckerman, S.R. 1991. Strengthening sorghum and pearl millet research in Mali. Patancheru 502 324, Andhra Pradesh, India: International Crops Research Institute for the Semi-Arid Tropics.

Shisanya, C.A. (2002). Improvement of drought adapted tepary bean (Phaseolus acutifolius) yield through biological nitrogen fixation in semi-arid SE-Kenya. Europ. J. Agron. 16, 13-24.

Silim, S.N., Omanga, P.A. (2001). The response of short-duration pigeonpea lines to variation in temperature under field conditions in Kenya. Field Crops Res. 72, 97-108.

Singh, B.B., Mai-Kodomi, Y. and Terao, T. (1999). A simple screening method for drought tolerance in cowpea. Indian J. Genet. 59,211 – 220.

Singh, K.B., Malhotra R.S., Saxena M.C. and Bejiga, G. (1997). Superiority of winter sowing over traditional spring sowing of chickpea in the Mediterranean region. Agron. J. 89,112-118.

Singh, B.B., Matsui, T. (2002). Cowpea varieties for drought tolerance. In: Fatokun CA, Tarawali SA, Singh BB, Kormawa PM, Tamo M. (Eds.), Challenges and Opportunities for Enhancing Sustainable Cowpea Production, World Cowpea Conference III Proc. 4–8 September. International Institute of Tropical Agriculture, Ibadan, Nigeria, pp. 287-300.

Sivakumar, M.V.K. (1992). Empirical analysis of dry spells for agricultural applications in West Africa. J. Clim. 5, 532-539.

The Nature of Rainfall at a Typical Semi-Arid Tropical Ecotope in Southern Africa
and Options for Sustainable Crop Production

91

Soil Classification Working Group (1991). Soil Classification- A Taxonomic System for South Africa. Department of Agricultural Development, Pretoria, South Africa.

Stephens, M.A. (1974). EDF Statistics for goodness of fit and some comparisons. J. Am. Stat. Assoc. 69, 730-737.

Stoop, W.A. and Staveren, J.P.V. (1981). Effects of cowpea in cereal rotations on subsequent crop yields under semi-arid conditions in Upper-Volta. In: Biological Nitrogen Fixation Technology for Tropical Agriculture (eds.) P.C. Graham and S.C. Harris. Cali, Colombia, Centro Internancional de Agricultura Tropical.

Subbarao, G. V., Chauhan Y.S. and Johansen, C. (2000). Patterns of osmotic adjustment in pigeonpea - its importance as a mechanism of drought resistance. Europ. J. Agron. 12, 239-249.

Tesfaye, K. and Walker, S. (2004). Matching of crop and environment for optimal use: the case of Ethiopia. Phys. Chem. Earth Parts A/B/C 29 (15-18), 1061-1067.

Tilahun, K. (2006). The characterisation of rainfall in the arid and semi-arid regions of Ethiopia. Water SA 32, 429-436.

Tsiros, E., Kanellou, E., Domenikiotis, C. and Dalezios, N.R. (2008). The role of satellite derived vegetation health index and aridity index in agroclimatic zoning. Geophysical Research Abstracts, 10, EGU2008-A-07223, 2008 SRef-ID: 1607-7962/gra/EGU2008-A-07223. EGU General Assembly 2008.

Turner, N.C. (1986). Crop water deficits: a decade of progress. Adv. Agron. 39, 1-59.

Tyson, P.D. (1986). Climate change and variability in southern Africa. Oxford University Press, Cape Town, South Africa. 220pp.

Umar, S., Prasad, T.G. and Kumar, M.U. (1995). Genetic variability and recovery growth and synthesis of stress proteins in response to polyethylene glycol and salt stress in finger millet. Annals Bot. 76, 43-49. UNESCO (1979). Map of the World Distribution of Arid Regions. Accompanied by explanatory notes. UNESCO, Paris, France. MAB Technical Notes 17- 54.

Van Boxtel, J., Singh, B.B., Thottappilly, G. and Maule, A.J. (2000). Resistance of Vigna unguiculata (L.) Walp. breeding lines to blackeye cowpea mosaic and cowpea aphid borne mosaic poty virus isolates under experimental conditions. J. Plant Dis. and Protec. 107,197–204.

Wilson, J.P, Sanogo, M.D., Nutsugah, S.K., Angarawai, I., Fofana, A, Traore H, Ahmadou, I, and Muuka, F.P. (2008) Evaluation of pearl millet for yield and downy mildew resistance across seven countries in sub-Saharan Africa. Afric.J. Agric. Res. 3, 371-378.

WRB (2006). A Framework for International Classification, Correlation and Communication. World Soil Resources Rep.103. IUSS/ISRIC/FAO.

Yapi, A., Dehala, G., Ngawara, K., and Abdallah, I. (1998) Sorghum S 35 in Chad - Adoption and Benefits . In: Bantilan, M.C.S., and Joshi, P.K. (eds.) Assessing joint research impacts: proceedings of an International Workshop on Joint Impact Assessment of NARS/ICRISAT Technologies for the Semi-Arid Tropics, 2-4 Dec 1996, ICRISAT, Patancheru, India. pp. 16-25

Zahran, H.H. (1999). Rhizobium-legume symbiosis and nitrogen fixation under severe conditions and in an arid climate. Microbiol Molec. Biol. Rev. 63, 968-989.

Zegada-Lizarazu, W. and Lijima, M. (2005). Deep water uptake ability and water use efficiency in pearl millet in comparison to other millet species. Plant Prod. Sci. 8, 454-460.

5

Impact of Soil Fertility Replenishment Agroforestry Technology Adoption on the Livelihoods and Food Security of Smallholder Farmers in Central and Southern Malawi

Paxie W. Chirwa[1] and Ann F. Quinion[2]
[1]University of Pretoria, Faculty of Natural & Agricultural Sciences
[2]45 Hanby Ave Westerville
[1]South Africa,
[2]USA

1. Introduction

The expanding information on agroforestry research and development around the globe shows that agroforestry is being promoted and implemented as a means to improve agricultural production for smallholder farmers with limited labor, financial, and land capital (Ajayi et al., 2007). In Africa, and particularly southern Africa, the main constraint to agricultural productivity is soil nutrient deficiency (Scoones & Toulmin, 1999; Sanchez *et al.*, 1997). For this reason, agroforestry research in the region has focused on soil fertility replenishment (SFR) technologies over the years and the adoption and scaling-up of these practices is the main thrust of the ongoing on farm research (Akinnifesi *et al.*, 2008; Akinnifesi et al, 2010). SFR encompasses a range of agroforestry practices aimed at increasing crop productivity through growing trees (usually nitrogen-fixing), popularized as *fertilizer tree* systems, directly on agricultural land. Fertiliser tree systems involve soil fertility replenishment through on-farm management of nitrogen-fixing trees (Akinnifesi et al 2010; Mafongoya et al., 2006). Fertiliser tree systems capitalise on biological N fixation by legumes to capture atmospheric N and make it available to crops. Most importantly, is the growing of trees in intimate association with crops in space or time to benefit from complementarity of resource use (Akinnifesi et al., 2010; Gathumbi et al., 2002). The different fertiliser tree systems that have been developed and promoted in southern Africa (see Akinnifesi et al, 2008; 2010) over the last two decades are briefly discussed below. The type of soil fertility replenishment (SFR) or fertilizer tree system appropriate for a particular setting is determined by a battery of ecological and social factors.

1.1 Intercropping

Intercropping is the simultaneous cultivation of two or more crops on the same field; usually, involving maize as the main crop in southern Africa and other agronomic crops as risk crops. In agroforestry based intercropping systems, species such as pigeon pea (*Cajanus cajan*), *Tephrosia vogelii*, *Faidherbia albida*, *Leucaena leucocephala*, and *Gliricidia sepium* are prominent. *Gliricidia* is a coppicing legume native to Central America with a foliage nitrogen content of up to 4% (Kwesiga *et al.*, 2003). It is currently being used in the intercropping technologies throughout southern Africa (Böhringer, 2001; Chirwa et al. 2003). In the intercropping system, *Gliricidia* is planted along with the maize crop. The trees are pruned at crop planting and again at first weeding and the pruned biomass is incorporated into the soil. The advantage of this system is that, because of its coppicing ability, the trees can be maintained for 15 to 20 years (Akinnifesi *et al.*, 2007), eliminating the need to plant each year, as is the case in the relay cropping system. However, it takes 2 to 3 seasons of intercropping before there is a significant positive response in maize yield (Böhringer, 2001; Chirwa *et al.*, 2003) and the technology is labor intensive because of the required pruning (Kwesiga *et al.*, 2003).

The benefits of intercropping on maize yields have proved to be highly substantial. Akinnifesi *et al.* (2006) reported soil fertility levels in *Gliricidia*/maize systems to be significantly greater than sole maize. In the second cropping season, maize yields in the intercropping plots were twice that of sole maize plots. Additionally, maize yields in the intercropping systems maintained an average of 3.8 MT ha^{-1} over a ten year period, compared to an average 1.2 MT ha^{-1} in the sole maize plots (Akinnifesi *et al.*, 2006). Results from Makoka Research Station in southern Malawi showed that by the fourth year, maize yields in the intercropping system were double those of the controls (sole maize) (Kwesiga *et al.*, 2003). Table 1, adapted from Kwesiga *et al.* (2003), illustrates the potential yield benefits of the intercropping technology.

Recom-mended fertilizer (%)	1992-1993		1993-1994		1994-1995		1995-1996		1996-1997	
	SM	G/M	SM	G/M	SM	G/M	SM	G/M	SM	G/M
MT ha^{-1}										
0	2.0	1.60	1.20	2.50	1.10	2.10	1.07	4.72	0.56	3.28
25	3.4	3.10	1.60	3.00	2.20	2.90	3.49	6.34	2.11	4.23
50	4.2	4.00	2.40	3.20	2.40	2.90	4.23	6.70	1.89	4.39

SM=sole maize, G/M= Gliricida/maize intercropping recommended fertilizer rates: 96 kg N and 40 kg P ha^{-1}.
Source: Kwesiga, *et al.*, 2003

Table 1. Maize grain yields from a Gliricidia/maize intercropping system with different levels of fertilizer from 1992 to 1997 at Makoka, Malawi.

1.2 Relay cropping

Relay cropping is a system whereby nitrogen-fixing trees, shrubs, or legumes such as *Sesbania sesban, Tephrosia vogelii, S. macrantha, Crotalaria spp.*, or perennial pigeon pea (*Cajanus cajan*), are grown as annuals and planted 3 to 5 weeks after the food crop. Staggering, or relaying, the agroforestry species and crop plantings reduces competition (Akinnifesi *et al.*, 2007; Kwesiga *et al.*, 2003). The agroforestry species are allowed to grow and develop beyond the main crop harvest. At the beginning of the, second season they are felled and the woody stems are collected for use as fuel while the remaining biomass is incorporated into the soil as green manure. Early reports reviewed by Snapp *et al.*, (1998) indicated that after 10 months of growth, *Sesbania* produced 30 to 60 kg N ha^{-1} and 2 to 3 MT ha$\eth^1$ of leafy biomass, plus valuable fuelwood from the stems. In southern Malawi, Phiri *et al.*, (1999) found a significant influence of *Sesbania* relay cropping on maize yields at various landscape positions. In another study, tree biomass production averaged 1 to 2.5 MT ha$\eth^1$ for *T. vogelii,* and 1.8 to 4.0 MT ha$\eth^1$ for *S. sesban* and a corresponding average maize grain yield of 2 MT ha$\eth^1$ (Kwesiga *et al.*, 2003). Relay cropping is suitable for areas of high population density and small farm sizes because it does not require farmers to sacrifice land to fallow. The drawback of this system is that the trees are felled and must therefore be re-planted each year. Furthermore, the technology relies on late-season rainfall in order for the trees to become fully established (Böhringer, 2001).

1.3 Improved fallow
Traditionally, farmers practiced rotational cultivation and allowed agricultural plots to lie in fallow for several years in order to replenish soil nutrients (Kanyama-Phiri *et al.*, 2000; Snapp *et al.*, 1998). With increasing populations and decreasing land holdings, many smallholder farmers can no longer afford to remove land from cultivation. For this reason, improved fallow technology has emerged as a promising alternative to traditional fallows. In an improved fallow, fast-growing, nitrogen fixing species such as *Sesbania sesban* , *Tephrosia vogelii, Gliricidia sepium,* and *Leucaena leucocephala* are grown for 2 to 3 years in the fallow plot after which, they are felled. The leaf matter can then be incorporated into the soil as green manure, and the woody stems can be used for fuel wood or construction materials. Farmers have also intensified this practice by intercropping during the first year of tree growth (Böhringer, 2001). Improved fallows are being used extensively in Eastern Zambia (Ajayi & Kwesiga, 2003; Ajayi *et al.*, 2003) as well as in parts of Malawi, Kenya, Zimbabwe, and Tanzania (Kwesiga *et al.*, 2003; Place *et al.*, 2003). Improved fallows are perhaps the most widely adopted SFR practice in southern Africa. Kwesiga *et al.* (2003) estimated that by 1998 over 14 000 farmers were experimenting with improved fallows in eastern Zambia, and that by 2006 a total of 400 000 farmers in southern Africa would be using the technology. In trials at Chipata, Zambia, maize yields increased from 2.0 MT ha$\eth^1$ in an un-fallowed plot to 5.6 MT ha$\eth^1$ after a 2 year *S. sesban* fallow (Kwesiga *et al.*, 2003). The same study also reported yield increases of 191% after a 2 year *T. vogelii* fallow and a 155% yield increase following a 2 year fallow with *C. cajan* (Kwesiga *et al.*, 2003). Despite the shorter fallow period, compared to traditional fallows, the success of improved fallow technology depends, in part, on the farmer's ability to remove land from crop production for a period of 2 to 3 years. In places where landholdings are small, fallows may not be a viable option for farmers. Other constraints include water availability, especially during tree establishment, and pests in the case of *Sesbania* (Böhringer, 2001). For this reason, intercropping and relay cropping have become the dominant SFR practices in central and southern Malawi (Kwesiga *et al.*, 2003; Thangata & Alavalapati, 2003).

1.4 Biomass transfer
In the biomass transfer technology, green manure is mulched and/or incorporated into agricultural soils. Biomass transfer is common in Zimbabwe, Tanzania, western Kenya, and northern Zambia where green biomass is grown in *dambos* (shallow, seasonally waterlogged wetlands) or on sloping land and areas that are unsuitable for agricultural production and where labor is not a limiting factor (Kwesiga *et al.*, 2003; Place *et al.*, 2003). The technology is

labor intensive as the mulch must be collected, transported to the agricultural field, and then incorporated into the soils. The amount and cost of labor associated with biomass transfer is the major limiting factor to the technology (Kuntashula et al., 2004). The advantage of this technology is that it allows for continuous cultivation as the incorporated green manure provides sustained soil nutrient replenishment (Place et al., 2003). Typically, *Tithonia diversifolia, Leucaena leucocephala, Senna spectabilis, Gliricidia sepium,* and *Tephrosia vogelii* are the most prominent species used in biomass transfer systems (Place et al., 2003). The technology has been reported to increase maize yields by up to 114% (Place et al., 2003). A compilation of independent studies in Malawi showed that green manures increased maize yields by 115.8%, when compared to unfertilized maize (Ajayi et al., 2007). Similarly, Ajayi et al. (2007) reported that incorporating 3.4 MT haõ¹ of dry weight of *Gliricidia* manure produced up to 3 MT haõ¹ of maize. Aside from the common use in maize production, biomass transfer is an important technology used in dambo cultivation of high-value cash crops, such as vegetables (Kwesiga et al., 2003). In addition to soil fertility and increased crop production, agroforestry provides other ecological and economic products and services including, but not limited to: wood production, pest management, and carbon sequestration.

1.5 Wood production for construction and energy
One of the most important products of SFR, to the smallholder farmer, is woody biomass production. Wood, for both fuel and construction, is critical to the livelihoods of rural farmers. An estimated 85% of the rural population in developing countries depends on woodlands and forests to sustain their livelihoods (Dixon et al., 2001). As population pressures and deforestation rates increase, there is an increasing demand for wood, but a decreasing supply. In Tanzania, for example, deforestation rates caused by activities associated with agriculture, illegal harvesting, and expanding settlements have reached 91 000 ha per year (Meghji, 2003). In Malawi, high population pressures have stressed the natural resources base, and especially the forest and woodland resources. The country's wood demand was evaluated to exceed the available supply by one third (Malawi, 2002; MEAD, 2002). Additionally, Malawi's forest cover decreased by 2.5 million ha between 1972 and 1992 and the current rate of deforestation is approximately 2.8% per year (MEAD, 2002). As a result of these trends, those who rely on wood for fuel, construction, and other livelihood activities are spending more time collecting and transporting wood to the detriment of other important household activities. Considering that fertilizer tree systems have been shown to produce up to 10 MT of woody biomass per hectare (Kwesiga & Coe, 1994), it is easy to see that the secondary benefit of wood production by agroforestry trees is an important, positive externality to these technologies. Two important species for wood production include *Sesbania sesban* and *Gliricidia sepium. S. sesban* produces a high volume of woody biomass in a short time, making it ideal for fuelwood production (AFT, 2008). In eastern Zambia, a *Sesbania sesban* improved fallow produced over 10 MT haõ ¹ (Kwesiga et al., 1999). Kwesiga & Coe (1994) reported fuelwood harvests of 15 and 21 MT haõ¹ following 2 and 3 year *Sesbania* fallows, respectively. Furthermore, Franzel et al. (2002) reported that a 2-year *Sesbania* fallow resulted in 15 MT of fuelwood. The woody biomass of *Gliricidia sepium* is suitable for both fuel and construction. As fuel, the wood of *G. sepium* burns slowly and with little smoke. Alternatively, the hard, durable wood is termite resistant and is used in fence, home, and tool construction (AFT, 2008). Chirwa et

al., (2003) reported that G. sepium, when grown in an unpruned woodlot, or as an improved fallow, produced 22 MT ha^{-1} yr^{-1} of fuelwood. The same study reported fuelwood production amounts of 1 MT ha^{-1} after a 2 year Gliricida/maize intercrop and 3.3 and 5.0 MT ha^{-1} after 3 years of Gliricida/maize/pigeon pea and Gliricidia/maize intercrop, respectively (Chirwa et al., 2003). A 5 year Gliricidia rotational woodlot in Tanzania was found to produce over 30 MT of woody biomass (Kimaro et al., 2007). Faidherbia albida and Leucaena leucocephala are two other SFR species planted in the southern Africa region that are managed for the dual purpose of soil fertility and woody biomass production (AFT, 2008).

1.6 Environmental services
1.6.1 Pest management
Another added benefit to some SFR agroforestry species is a pest management quality. Striga (S. asiatica and S. hermonthica) is a parasitic plant that thrives in nutrient starved soils (Ajayi et al., 2007; Berner et al., 1995; Gacheru & Rao, 2001; Sileshi et al., 2008). It attacks several of the major food crops, including maize, millet, rice, and sorghum. Seedlings attach to the roots of the host plant where they continue to grow underground for four to seven weeks; it is during this period that they cause the most damage (Berner et al., 1995). A single Striga plant can produce over 50 000 seeds and these seeds can remain viable in the soil for 10 to 14 years (Berner et al., 1995; Gacheru & Rao, 2001). Yield losses of 32% to 50% and 18% to 42% from Striga infestations have been reported in on-station trials in Kenya and Tanzania, respectively (Massawe et al., 2001). For smallholder, subsistence farmers, losses can be up to 100% with heavy infestation (Berner et al., 1995; Gacheru & Rao, 2001; Massawe et al., 2001).

High populations have necessitated the use of continuous cultivation. This leads to soil nutrient depletion and has caused an increase in the severity and spread of Striga infestations (Gacheru & Rao, 2001). Several agroforestry species have shown potential in combating Striga. For example, on moderately-infested sites in western Kenya, Desmodium distortum, Sesbania sesban, Sesbania cinerascuns, Crotalaria grahamiana, and Tephrosia vogelii fallows were found to decrease Striga by 40% to 72% and increase maize yields by 224% to 316% when compared to continuous maize plots (Gacheru & Rao, 2005). Additionally, Kwesiga et al. (1999) found less than 6 Striga plants 100 m^{-2} following 3 year Sesbania fallows in two experiments from Zambia. This is in stark contrast to the 1532 and 195 Striga plants 100 m^{-2} found in two experiments of continuously cultivated and unfertilized maize (Kwesiga et al., 1999).

Tephrosia vogelii has also been found to be effective as both a repellant and insecticide against Callosobruchus maculates, the main pest infecting stored cowpea. In a laboratory study conducted by Boeke et al. (2004), beetles exposed to tubes treated with T. vogelii powder laid fewer eggs in the first 24 hour period than beetles in the control. The T. vogelii powder was also found to reduce the parent beetle lifespan (Boeke et al., 2004). Another study reported that the juice of T. vogelii was effective in managing maize stem borer (Chilo partellus) populations in southern Tanzania and northern Zambia (Abate et al., 2000). Similarly, in Uganda, the presence of T. vogelii plants in sweet potato fields was reported to protect the potatoes from mole and rat damage (Abate et al., 2000). The dry, crushed Tephrosia vogelii leaves are also documented to be effective against lice, fleas, tics, and as a molluscicide (AFT, 2008).

1.6.2 Carbon sequestration

The Kyoto Protocol recognizes agroforestry as a greenhouse gas mitigation strategy and allows industrialized nations to purchase carbon credits from developing countries (Orlando *et al.*, 2002). In this context, agroforestry not only plays a part in mitigating the effects of global climate change through carbon sequestration (Ajayi *et al.*, 2007; Ajayi & Matakala, 2006), but also has the potential to contribute to farmer incomes through the sale of carbon credits (Takimoto *et al.*, 2008). Several initiatives have recently been developed to support and encourage farmers who adopt land use practices that render environmental services (Ajayi *et al.*, 2007). While there is increasing interest in the global warming mitigation potential of agroforestry, research has lagged behind in quantifying this potential for various systems (Albrecht & Kandji, 2003; Makumba *et al.*, 2007). While the volume of research on agroforestry and climate regulation is limited, there have been a few studies that reveal the carbon sequestration potential for some systems. For example, a *Gliricidia*/maize intercropping system in Malawi was found to sequester between 123 and 149 MT of C ha^{-1} in the first 0 to 200 cm of soil through a combination of root turnover and pruning application (Ajayi *et al.*, 2007; Makumba *et al.*, 2007). In a separate report, Montagnini & Nair (2004) estimated that the potential carbon sequestration for smallholder agroforestry systems in the tropics range from 1.5 to 3.5 MT ha^{-1} of C yr^{-1}. Albrecht & Kandji (2003) have calculated the carbon sequestration potential to be between 12 and 228 MT ha^{-1} for similar systems. Between fuel and pole wood production, pesticide qualities, and climate regulation, it is clear that agroforestry offers benefits beyond improved soil characteristics and crop yields. Table 2, adapted from Ajayi *et al.* (2007), highlights some of the private and social benefits of SFR technologies.

	Private	Social
Benefit	Yield increase	Carbon sequestration
	Stakes for tobacco curing	Suppresses noxious weeds
	Improved fuel wood availability	Improved soil structure, reduced erosion and run-off
	Fodder	Promotes biodiversity
	Bio-pesticide	Potential for community income diversity
	Suppresses weeds	
	Improved soil structure, reduced erosion and run-off	
	Diversification of farm production (cash crops)	

Source: Adapted from Ajayi, et al., (2007)

Table 2. Benefits of SFR Technologies

2. Case study of SFR technology in central and southern Malawi

There are a variety of agroforestry technology options that are being researched, tested, and adopted throughout the world. The type of SFR technology that is acceptable, appropriate, and sustainable to a particular setting is determined by a battery of ecological (climate, soil and terrain characteristics) and societal factors such as available land and labor and institutional support and regulations. As a result of the various ecological and social boundaries in the study area, the respondents in this study used a combination of one or more of the following SFR technologies: intercropping, relay cropping, improved fallow, and biomass transfer. This case study's main objective was to investigate the link between SFR adoption and poverty reduction in farming households of central and southern Malawi by assessing food security, asset status, and household activities and income. Specific objectives were as follows: (i) evaluate changes in food security resulting from increased yields associated with SFR adoption; (ii) determine if there is a cause-and-effect relationship between SFR adoption and household assets as an indication of improved wealth and (iii) determine if SFR adoption has allowed households to diversify their activities and income.

2.1 Study areas
Forty-eight percent of the land area in Malawi is under cultivation. However, only 32% of this is classified as suitable land for rain fed agriculture (Malawi, 2002). Agricultural land increased from 3 million ha to 4.5 million ha between 1976 and 1990 while the average land holding size decreased from 1.53 ha in 1968/1969 to 0.8 ha in 2000 (Malawi, 2002). In order to achieve its various goals and objectives, the Ministry of Agriculture established a National Rural Development Programme that divides the country into various management units. There are eight Agricultural Development Divisions (ADD) within the country and each ADD is divided into several Rural Development Project (RDP) areas. The RDPs are further divided into Extension Planning Areas (EPA) and then finally into smaller Sections. The study was conducted in Rural Development Programmes in two districts of Malawi, Kasungu (S 13°2'0", E 33°29'0") in the central region and Machinga (S14° 58' 00", E35° 31' 00") in the southern region (Fig 1a). Within the Kasungu RDP the Chipala EPA was chosen and interviews were carried out in three different Sections. In Machinga ADD, interviews were conducted in Mikhole Section within Nanyumba EPA (Figure 1b).

2.1.1 Farming activities and food production
The primary source of food in Kasungu is from local crop production, with availability being lowest between January and February. Apart from tobacco (*Nicotiana tabacum*) the cash crop, maize (*Zea mays*) is the most important food crop and is cultivated by an estimated 95.9% of the population of Kasungu District (Malawi National Statistical Office, 2005). Groundnuts (*Arachis hypogaea*), rice (*Oryza sativa*), pulses and, to a lesser extent cassava (*Manihot esculenta*) are also important crops with 55.3%, 48%, 41% and 12.3% of the population cultivating these crops respectively (Malawi National Statistical Office, 2005). In Machinga, most households are subsistence farmers whose main crops are maize, cassava, and rice (MVAC, 2005). Almost 98% of households in Machinga cultivate maize, 67.7% grow pulses, 38.6% grow groundnuts, 42.1% cultivate rice, and 26.6% cultivate cassava (Malawi National Statistical Office, 2005). Traditionally, households engage in a multiple cropping of maize/pulse farming system (Msuku et al., s.d.). Relay planting, the inclusion of N-fixing legumes, and incorporation of crop residues into the soil are common soil fertility management practices (Msuku et al., s.d.). Most farmers cannot afford inorganic fertilizers: consequently, only about 20% of households use fertilizers, pesticides, or improved seed (Msuku et al., s.d.). While most plots are intercropped, tobacco is grown in pure stands (Msuku et al., s.d.). Tobacco is an important cash crop for those who cultivate it, but it is grown by only about 22% of the population (Malawi National Statistical Office, 2005).

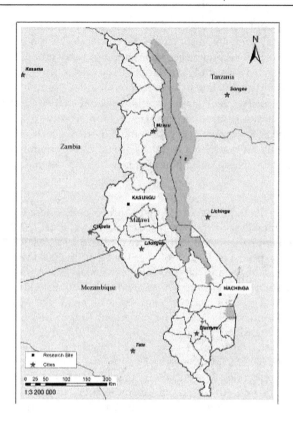

Fig. 1.a. Map of location of Kasungu and Machinga Agricultural Development Division

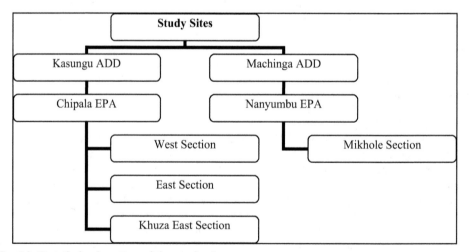

Fig. 1.b. Detailed subsection of the study sites

2.1.2 Wealth and income

According to the Malawi Baseline Livelihood Profiles (MVAC, 2005), wealth in the Kasungu region is heavily reliant on access to food and credit. Households with access to credit are more likely to have a larger land holding from which they can cultivate and harvest a higher crop yield. These households may also be able to purchase livestock such as cattle which can be used for meat, milk, farming, or sold for cash in times of stress. Overall, those considered "better-off" cultivate twice as much land, may own as many as 10 head of cattle, and/or own twice as many goats and chickens as those considered "poor" (MVAC, 2005). Crop sales are the primary source of income in the region, with tobacco constituting 65% to 85% of the average household income (MVAC, 2005). Approximately 64% of the population of Kasungu District grows tobacco (Malawi National Statistical Office, 2005). It is the most important cash crop in the region with an estimated 45% of the yearly tobacco sales in the country coming from Kasungu ADD, and 60% of this comes from Kasungu RDP (Mwasikakata, 2003). Among the poor, cash and in kind wages from ganyu[1] work are the second most important source of income, while for those households considered to be either middle or better-off, food crops and livestock sales are the secondary sources of income (MVAC, 2005). The average landholding size throughout the region is 0.4 ha (Msuku et al., s.d.; MVAC, 2005). Households who, according to the Malawi Baseline Livelihood Profiles, are considered "poor" own between 0.4 ha and 1.0 ha, while for those deemed to be better-off, land holdings average between 1.2 ha and 2.4 ha (MVAC, 2005). Despite low soil fertility and market access problems, crop sales are the most important source of income in Machinga. Crops such as groundnuts (Arachis hypogaea), sweet potatoes (Ipomoea batats), and soya beans (Glycine max) are sold mainly in the local markets (MVAC, 2005). Other major income sources include labor and firewood sale. Even after household production and in kind payments, poor households still face a 33% food deficit. According to the Malawi Integrated Household Survey of 2005, Machinga District has a 73.7% poverty rate and only 36.7% of the population has an adequate food supply. The sandy soils, poor infrastructure, and high population density, make poverty relief and hunger alleviation especially challenging.

2.2 Methodology

The sites, communities, and individual households were selected using purposive sampling strategies (Babbie and Mouton 2001) based on information provided by the project staff and local extension officers. In total, 131 household interviews were conducted, 65 from Kasungu and 66 from Machinga. Farmers were selected on the basis of length of SFR and/or agroforestry technology use; having been adopters of the fertilizer tree technologies for at least 5 years.

2.2.1 Data analysis

Household characteristics such as number of household members and landholding size were summarized using descriptive statistics. Frequency tables and descriptive statistics were used to identify and evaluate trends in the agroforestry technology use, crop production, shocks, assets, and income. Sign and Signed Rank Non-parametric (also called

---

[1] Ganyu refers to casual labor or piecework and is paid for with either cash or in kind upon completion of the job

Wilcoxon Matched Pairs test) analysis was used to test for a change in the crop yield and asset variables between pre- and post-adoption (Clewer and Scarisbrick 2006). The test for equality of proportions was used to examine the probability of an increase in income amount, number and type of income sources, and maize yields as a result of the technologies adoption. Chi-square analysis test was used to determine if there was an influence of the addition of agroforestry related activities on both the amount and number of income sources

3. Results and discussion

3.1 The relationship between the adoption of SFR technologies and food security

The majority of respondents (65%) reported an increase in maize yield due to SFR use with an average total yield increase of 381.5 kg in Kasungu and 241.7 kg in Machinga (Table 3). The difference between sites was likely due to the fact that respondents in Machinga cultivate much smaller areas. The results confirm what the existing literature has already established, that integrated soil fertility technologies do cause a significant increase in crop production (Ajayi et al., 2007; Akinnifesi et al., 2006; Kwesiga et al., 2003; Phiri et al., 1999). Some of the studies have even shown increases of over 100% (Ajayi et al., 2007; Phiri et al. 1999; Place et al., 2003) for various agroforestry technologies. However, the respondents only provided information about the amount of yield increase with no reference to any baseline information pertaining to yields per hectare and so the reported increases cannot be extrapolated to kg per hectare; making it difficult to directly compare the production at the two sites. It should also be mentioned that no data was collected on the use of inorganic fertilizers among the respondents. At the time of interview, the government fertilizer subsidy program supplied 100 kg of inorganic fertilizer to approximate 50% of the smallholder farming sector (Malawi Ministry of Agriculture and Food Security, 2008) and it is likely that some of the respondents in this study were recipients of these subsidies. The use of both organic and inorganic fertilizer options are complementary and will contribute to increasing crop yields.

Crop	Kasungu	Machinga
Maize*	381.5 (192.4)	241.7 (126)
Cassava*	188.2 (92.75)	50
Vegetables	34.1 (28.60)	17.1 (6.98)

* Indicates that differences between sites are significant at $p<0.05$

Table 3. Mean increases (kg) (and SD) of crops in Kasungu and Machinga districts

The other two crops that also showed significant increases in yield since adoption of SFR technologies were cassava and vegetables (Table 3). Respondents at both sites grew vegetables for consumption and sale. However, the sale of vegetables was a much more common source of income in Kasungu than in Machinga (Table 4). The use of biomass transfer in dambo (wetlands) cultivation of high value cash crops such as vegetables has been shown to provide a potential net profit of US$700 to US$1000 per hectare (Ajayi et al., 2006). Ajayi and Matakala (2006) reported that in Zambia the use of Leucaena biomass in cabbage cultivation resulted in a net profit of US$5 469 per hectare. It is likely that, when compared to Machinga, the larger land holdings in Kasungu has allowed the more prevalent

use of biomass transfer, and resulted in the production of larger quantities of cash crops
(vegetables) which has also contributed to a more diversified income portfolio.

Crop	Kasungu		Machinga	
	%	Mean Rank (SD)	%	Mean Rank (SD)
Groundnuts	86.2	2.11 (1.22)	92.4	1.54 (0.79)
Cassava	64.6	2.95 (1.46)	30.3	3.05 (1.05)
Potato	56.9	3.59 (1.36)	25.8	3.47 (0.94)
Maize	52.3	2.76 (1.07)	3.0	2.0 (0)
Vegetables	46.2	3.07 (1.55)	7.6	2.2 (1.3)
Tobacco	43.1	1.50 (0.88)	50.1	1.82 (1.07)
Cotton	18.5	2.25 (1.48)	3.0	5.0 (0)
Pulses	15.4	3.90 (1.10)	45.5	3.13 (1.67)
Millett	4.6	3.33 (2.52)	0	-
Rice	0	-	65.2	2.60 (1.00)
Sorghum	0	-	10.6	3.57 (1.13)

A rank of 1 is considered the most important

Table 4. Percent (%) of respondents cultivating, and mean ranking of, cash crops in Kasungu
and Machinga districts

3.2 SFR technology adoption and the impact on household assets

Table 5 shows the ownership of assets at the two study sites. Ownership of bed mats,
bicycles, radios, goats, and chickens increased significantly ($p < 0.05$) between pre- and
post-SFR adoption. The results show that the majority of respondents (85% to 100%)
attributed an increase in asset ownership to SFR use. Assets increased both in number
(purchasing additional chickens, for example) and in diversity (for example, purchasing a
first radio). However, it was not possible to determine if asset status was directly
correlated to the number of years since adoption. This would have required taking asset
inventories at regular intervals, e.g. annually, over time. It is therefore impossible to
determine if there is a relationship between assets and years of SFR use. It can however,
be said that there is a significant change in asset status between pre- and post-adoption.
Studies from Ellis *et al.* (2003) and the Malawi Baseline Livelihood Profiles (MVAC, 2005),
found that changes in livestock ownership may indicate a change in wealth. Through
wealth-ranking exercises in Zomba and Dedza districts of Malawi, Ellis *et al.* (2003) found
that households considered to be "well-off" owned, among other things: 5 or more cattle,
3 to 5 goats, and at least one bicycle. Similarly, using livestock ownership as one indicator
of wealth, the Malawi Baseline Livelihood Profiles (MVAC, 2005) reported that in
Kasungu district those considered poor owned zero to 5 goats or chickens, those in the
middle wealth bracket owned zero to 3 cattle and up to 6 goats and chickens, and those
considered better-off owned 3 to 10 cattle and 5 to 10 goats and/or chickens. The same
study reported that for Machinga district, households classified as poor owned 4 to 6
chickens, those in the middle owned 1 to 4 goats and/or 4 to 6 chickens, and the better-off
households owned up to 15 goats and 15 or more chickens.

3.3 SFR adoption and diversity of income among households
3.3.1 Seasonal income generating activities
Crop sales were the most common and most important sources of income at both sites (Table 6). This is consistent with the Malawi Baseline Livelihood Profiles (MVAC, 2005) which reported that crop sale is the largest source of income in both the Kasungu /Lilongwe Plain and Phalombe Plain and Lake Chilwa Basin (which includes the Machinga site) areas. The majority of crop sales occur between the months of May and September (Figure 2). This is expected since these are the months during which most agronomic crops are harvested

Asset	Kasungu (n=65)	Machinga (n=66)	Total (n=131)
Iron Roof	9.2	6.1	7.6
Radio*	61.5	39.4	-
Bicycle	46.2	53	49.6
Bank Account*	18.5	0	-
Bed Mats	100	100	100
Goats	27.7	28.8	28.2
Chickens	64.6	48.5	56.5
Cattle	1.5	0	0.7
Other	32.3	7.6	19.8

* Indicates a significant difference (Fisher's exact p-value<0.05) between the sites and means could not be pooled

Table 5. Percent of respondents reporting asset ownership

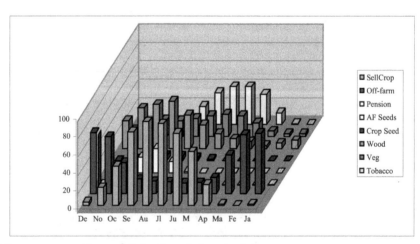

Fig. 2. Income sources by month for the whole sample

The study found off-farm wages to be especially important sources of income between November and March (Figure 2), coinciding with the annual food shortages and hunger periods experienced by many subsistence farmers. These months are also times of high labor demand as it is during these months that land preparation, land clearing, ridging, and vegetables are harvested (MVAC, 2005). Farmers are therefore faced with the dilemma of hiring out labor for payment or working on their own plots. If household labor resources are constrained, the loss of an active household member to off-farm labor may be to the detriment of household land preparation and subsequent crop production (Place et al., 2007). However, the additional income (either cash or in-kind) from off-farm labor may be more critical to meeting the household's immediate needs.

Tobacco sale in Kasungu was the most highly ranked cash crop (Table 6) although it was not as common an income source as would have been expected based on the literature. The MVAC report (2005) also identified tobacco as the most important cash crop in Kasungu district, accounting for 65-85% of the income across all wealth groups. The high ranking of tobacco in the present study supports the results from Mwasikakata's (2003) study which found nearly 45% of the yearly tobacco sales in Malawi come from Kasungu ADD, and 60% of this comes from Kasungu RDP (Mwasikakata, 2003).

Income Source	Machinga (n=66)		Kasungu (n=65)		Whole Sample (n=131)	
	%	Average Rank (SD)	%	Average Rank (SD)	%	Average Rank (SD)
Sell Crops	95.5	1.51 (0.76)	89.2	2.60 (1.38)	92.4	2.03 (1.22)
Off-farm wages	71.2	1.87 (0.85)	90.7	2.56 (1.59)	81	2.25 (1.35)
Sell Wood	33.3	2.77 (0.61)	72.3	3.43 (1.49)	52.6	3.22 (1.32)
Sell AF seeds	6.1	2.75 (0.50)	80.0	3.48 (1.49)	42.7	3.43 (1.45)
Tobacco	44.0	2.10 (1.01)	40.0	3.08 (2.12)	42	2.56 (1.69)
Sell Vegetables	12.1	3.13 (1.25)	47.7	3.58 (1.43)	29.7	3.49 (1.39)
Sell Maize	3	2.50 (0.71)	52.3	2.94 (1.32)	27.5	2.92 (1.29)
Sell crop seeds	0		10.7	5.00 (1.83)	5.3	5.00 (1.83)
Other	0		3.1	2.00 (0.00)	1.5	2.00 (0.00)
Pension#	0		1.5	2.00	0.76	2.00
Sell Other#	0		1.5	6.00	0.76	6.00

Rank of 1 indicates most important source of income
#Indicates only one household reported this sources of income

Table 6. Percent of respondents reporting and average ranking of various income sources

3.3.2 Income diversity as a result of SFR adoption

With agroforestry adoption come other income generating opportunities. For example, wood from the agroforestry species can be sold for fuel or construction materials; seeds can be collected and sold; and if increased crop yields produce a surplus, those crops can also be sold (see Table 6). It was therefore hypothesized that SFR use would promote the diversification of income generating activities (IGAs). Responses from Machinga showed that there was no significant diversification of income sources. While a few households did report an increase in the number of income sources, the majority said the number had remained the same. The results from Kasungu, however, showed a significant number of respondents reporting an increase in the number of household income sources. Since Kasungu respondents were cultivating larger plots, and more respondents used improved fallows and biomass transfer than in Machinga, they may have more income generating resources available to them. For example, the use of improved fallows requires more land than relay cropping and the resulting woody biomass yield will be greater in a plot that is dedicated to an improved fallow than in a plot where woody growth shares the same space as food crops. Therefore, the resulting volume of saleable wood will be greater from an improved fallow than from a relay cropping system.

3.4 The impact of SFR adoption on household vulnerability and coping strategies

Vulnerability is the potential to be adversely affected by an event or change and is a robust function of the interaction between and among natural or environmental variability, socio-economic processes, and policy (Eriksen *et al.*, 2005).The vulnerability context refers to external shocks, trends, and seasonality over which people have little or no control (DFID, 1999). While some changes to these external forces can have a positive influence in reducing vulnerability, many interactions among external shocks, trends, and seasonal processes provide a positive feedback into increased vulnerability. In this study, while some households demonstrated a positive change in income, crop yield, and assets, this does not appear to have been significant enough to allow for any substantial reduction in vulnerability, except for perhaps a shorter annual hunger period, the significance of which should not be ignored. It is difficult to separate the effects of hunger, illness, labor shortage, and crop loss as the presence of one can directly affect another. Case studies from an investigation of SFR livelihood impacts conducted by Place *et al.* (2007) in western Kenya revealed that shocks and coping strategies were key causes of poverty. Therefore, this study looked for any changes in the household's ability to cope with shocks as an indication of increased security and decreased vulnerability. Hunger is by far the most prevalent shock or crisis facing smallholder farmers, as illustrated by the fact that all of the respondents in this study were still vulnerable to several months of food insecurity each year. It was hypothesized that if SFR adoption had enabled households to increase crop production and diversify their livelihoods, then they would also have been able to invest in various adaptation and coping strategies that would mitigate the adverse effects of any shock or crisis that arose. Despite the gains in food security, brought about by a significant increase in crop yields, a marked decrease in hunger periods (Figure 3), and in some cases a more diversified income portfolio and asset inventory, there is still an obvious lag in household security, the ability to absorb and cope with shocks, and overall improved welfare.

When households live on the margin of survival, livelihood strategies focus more on addressing immediate needs and surviving shocks than progressing out of poverty (Eriksen *et al.*, 2005). The results revealed that where households were able to increase their income, the added income was reinvested into activities that support the household's immediate needs (Table 7), rather than investing in any form of insurance. In a study of household budgets in western Kenya, David (1997) found that up to 87% of all household expenditure went towards purchasing food and non-food necessities, while only 7% went towards farm inputs such as hired labor, fertilizer, and seed. This study agrees with David's (1997) conclusion that resource-poor farmers have little or no savings and households give priority to investments which yield short-term returns.

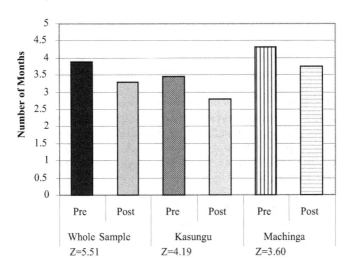

Fig. 3. Average number of hunger months before (Pre) and after (Post) SFR adoption. All differences are significant at p<0.05

Allocation	Kasungu (n=48)	Machinga (n=30)	Whole Sample (n=78)
Savings	16.7	6.7	12.8
Pay Debts	93.8	96.7	94.8
Purchase Household Items	100	100	100
Purchase Food	97.9	100	98.7
Purchase Agricultural Supplies	97.9	96.7	97.4
Medical Fees	47.9	53.3	50
School Fees*	35.4	13.3	-

Values are the percent of "yes" responses from those who reported an increase in income.
* Indicates a significant difference (Fisher's exact p-value<0.05) between the sites and means could not be pooled

Table 7. Percent of respondents reporting various allocations of additional income

Households allocated income to the purchase of household items, agricultural supplies, and food. These items have an immediate and direct effect on the wellbeing and security of household members. Investing additional income or resources into savings, non-essential assets, or school fees are investments that have long-term implications to the household's well-being, but may be at the expense of immediate needs. It was not expected that households in this study would have become fully food self-sufficient, but rather that they would have been able to spend less money to meet immediate needs and be able to put more income towards non-essential investments, such as savings, or school fees. The results show that households who had seen an increase in income are able to allocate income to a variety of areas, though they still rely heavily on purchasing food and non-food necessities. A full economic analysis at the household level would be necessary to determine if households have realized any significant financial relief since SFR adoption. The use of various coping strategies provides another indication of a household's vulnerability. Ideally, households would have some form of insurance or "safety net" to rely on in difficult times. In the absence of formal security measures however, households are likely to sell productive assets, reallocate time to increase income, or a previously non-working member may enter in the labor market (Jacoby & Skoufias, 1997; Skoufias, 2003) in response to unexpected challenges. It is not surprising then that in this study, households at both sites relied heavily on selling assets, crops, and labor as a response strategy. When households choose to sell their physical assets or crops as a coping mechanism in response to a shock, they may be able to mitigate the immediate effects of the crisis, but to the detriment of future stability. This observation is supported by Skoufias (2003) who observed that poor households may be forced to use coping strategies that ultimately prevent movement out of poverty.

4. Conclusion

This case study has confirmed that agroforestry, and specifically integrated soil fertility replenishment technologies have the ability to increase crop production and provide additional income. This acknowledgement points to the conclusion that farmers have an understanding of the importance of soil fertility and the currently low soil nutritional status. Other studies have also found that even in the absence of knowledge about the chemical or structural properties of soils, farmers are keenly aware of, and have noticed detrimental changes in various aspects of their local environments such as rainfall patterns, and soil performance over time and soil analysis consistently supports farmer perceptions of soil fertility (Desbiez et al., 2004; Mairura et al., 2007; Murage et al., 2000; Thomas et al., 2007).

Hunger months have decreased, and in many cases, income has increased. However, the respondents in the two study areas still live on the margins of survival. This study revealed that while food security is paramount to sustaining the livelihoods of smallholder farmers, livelihood security and poverty reduction depend on more than increased food production. SFR technologies are fulfilling their primary role as a means to food security, but their adoption does not lead to significant livelihood improvements. Achieving lasting impacts requires that initiatives take an integrated approach and address not only household food production, but the multifaceted dynamics of social institutions, markets/economy, and policy. However, it is apparent that despite the repeated confirmation of the challenges associated with land, labor, seed and training, little has been done to find solutions to these issues.

While agroforestry alone cannot completely bring households out of poverty, it can play a significant role by improving food security and providing additional income opportunities. Livelihood improvements will depend on several factors. First, market inefficiencies must be remedied and economic barriers must be broken down. Second, the challenges identified by the respondents, especially access to resources and training, need to be addressed in a participatory way that promotes education and empowerment. As these two issues are tackled, households will become better equipped to manage the complexities that arise from SFR adoption and livelihood diversification, such as managing crop surplus and additional income.

5. References

Abate, T., van Huis, A. & Ampofo, K.J.O. 2000. Pest management strategies in traditional agriculture: an African perspective. *Annual Review of Entomology* 45: 631-659.

Agroforestree Database (AFT). [Online]. 2008. Available at: http://www.worldagroforestry.org/Sites/TreeDBS/aft.asp [10, May 2008]

Ajayi, O.C., Akinnifesi, F., Mullila-Mitti, J., DeWolf, J.J., Matakala, P.W. & Kwesiga, F.R. 2006. *Adoption, Profitability, Impacts and Scaling-Up of Agroforestry Technologies in southern African countries.* ICRAFP [2006298]. Lilongwe, Malawi.

Ajayi, O.C., Akinnifesi, F., Sileshi, G., Chakeredza, S. & Matakala, P. 2007. Economic framework for integrating environmental stewardship into food security strategies in low-income countries: case of agroforestry in southern African region. *African Journal of Environmental Science and Technology* 1: 59-67.

Ajayi, O.C., Franzel, S., Kuntashula, E. & Kwesiga, F. 2003. Adoption of improved fallow technology for soil fertility management in Zambia: Empirical studies and emerging issues. *Agroforestry Systems* 59: 317-326.

Ajayi, O.C. & Kwesiga, F. 2003. Implications of local policies and institutions on the adoption of improved fallows in eastern Zambia. *Agroforestry Systems* 59: 327-336.

Ajayi, O.C. & Matakala, P. 2006. *Agroforestry can improve food security, farm diversification and income generation in Zambia Nairobi, Kenya.* World Agroforestry Centre ICRAF, Nairobi.

Akinnifesi FK, Ajayi OC, Sileshi G, Chirwa PW, & Chianu J. 2010. Fertilizer trees for sustainable food security in the maize-based production systems of East and Southern Africa. A review. *Agronomy for Sustainable Development* .30: 615–629

Akinnifesi, F.K., Ajayi, O.C., Sileshi, G., Kadzere, I. & Akinnifesi, A.I. 2007. *Domesticating and commercializing indigenous fruit and nut tree crops for food security and income generation in Sub-Saharan Africa.* New Crops International Symposium. (2007: Southampton). United Kingdom.

Akinnifesi, F.K., Chirwa, P.W., Ajayi, O.C., Sileshi, G., Matakala, P., Kwesiga, F.R., Harawa, H. & Makumba, W. 2008. Contributions to agroforestry research to livelihood of smallholder farmers in southern Africa: 1. Taking stock of the adaptation, adoption and impact of fertilizer tree options. *Agricultural Journal* 3:58-75.

Akinnifesi, F.K., Makumba, W. & Kwesiga, F.R. 2006. Sustainable maize production using *Gliricidia*/maize intercropping in southern Malawi. *Exploratory Agriculture* 42: 441-457.

Akinnifesi, F.K., Makumba, W., Sileshi, G., Ajayi, O.C. & Mweta, D. 2007. Synergistic effect of inorganic N and P fertilizers and organic inputs from *Gliricidia sepium* on productivity of intercropped maize in Southern Malawi. *Plant Soil* 294: 203-217.

Albrecht, A. & Kandji, S.T. 2003. Carbon sequestration in tropical agroforestry systems. *Agriculture, Ecosystems and Environment* 99: 15-27.

Babbie E & Mouton J. 2001. The Practice of Social Research. Oxford University Press, Oxford.

Berner, D.K., Kling, J.G. & Singh, B.B. 1995. *Striga* research and control a perspective from Africa. *Plant Disease* 70: 652-660.

Boeke, S.J., Baumgart, I.R., van Loon, J.A.A., van Huis, A., Dicke, M. & Kossou, D.K. 2004. Toxicity and repellence of African plants traditionally used for the protection of stored cowpea against *Callosobruchus maculates*. *Journal of Stored Products Research* 40: 423-438.

Böhringer, A. 2001. Facilitating the wider use of agroforestry for development in southern Africa. *Development in Practice* 11: 434-448.

Chirwa, P.W., Black, C.R., Ong, C.K. & Maghembe, J.A. 2003. Tree and crop productivity in *Gliricidia*/maize/pigeon pea cropping systems in southern Malawi. *Agroforestry Systems* 59: 265-277.

Clewer, A.G. & Scarisbrick, D.H. 2006. *Practical Statistics and Experimental Design for Plant and Crop Science*. Chichester: John Wiley & Sons, Ltd.

David, S. 1997. Household economy and traditional agroforestry systems in western Kenya. *Agriculture and Human Values* 14: 169-179.

Department for International Development (DFID). 1999. Sustainable Livelihoods Guidance Sheets [Online]. 1999. Available at: http://www.livelihoods.org/info/info_guidancesheets.html#1 [2007, October 31].

Desbiez, A., Mathews, R., Tripathi, B. & Ellis-Jones, J. 2004. Perceptions and assessment of soil fertility by farmers in the mid-hills of Nepal. *Agriculture, Ecosystems and Environment* 103:191-206.

Dixon, J., Gulliver, A. & Gibbon, D. 2001. *Farming Systems and Poverty Improving Farmers' Livelihoods in a Changing World*. Rome: FAO and World Bank.

Ellis, F., Kutengule, M. & Nyasulu, A. 2003. Livelihoods and rural poverty reduction in Malawi. *World Development* 31:1495-1510.

Eriksen, S.H., Brown, K. & Kelly, P.M. 2005. The dynamics of vulnerability: locating coping strategies in Kenya and Tanzania. *The Geographical Journal* 171: 287-305.

Franzel S., Phiri D. & Kwesiga F. 2002. Assessing the adoption potential of improved fallows in eastern Zambia, in S. Franzel & S.J. Scherr, (eds.). 2002. *Trees on the Farm: Assessing the Adoption Potential of Agroforestry Practices in Africa*. Wallingford, UK: CAB International.

Gacheru, E. & Rao, M.R. 2001. Managing *Striga* infestation on maize using organic and inorganic nutrient sources in western Kenya. *International Journal of Pest Management* 47: 233-239.

Gathumbi S., Ndufa J., Giller K.E., & Cadisch G. (2002) Do species mixtures increase above- and belowground resource capture in woody and herbaceous tropical legumes? Agron. J. 94, 518–526.

Jacoby, H.G. & Skoufias, E. 1997. Risk, financial markets, and human capital in a developing country. *Review of Economic Studies* 64: 311-335.

Kanyama-Phiri, G., Snapp, S., Kamanga, B. & Wellard, K. 2000. *Towards integrated soil fertility management in Malawi: incorporating participatory approaches in agricultural research.* Managing Africa's Soils No. 11. IIED, London.

Kimaro, A.A., Timmer, V.R., Mugasha, A.G., Chamshama, S.A.O. & Kimaro, D.A. 2007. Nutrient use efficiency and biomass production of tree species for rotational woodlot systems in semi-arid Morogoro, Tanzania. *Agroforestry Systems* 71: 175-184.

Kuntashula, E., Mafongoya, P.L., Sileshi, G. & Lungu, S. 2004. Potential of biomass transfer technologies in sustaining vegetable production in the wetlands (*dambos*) of eastern Zambia. *Exploratory Agriculture* 40: 37-51.

Kwesiga, F., Akinnifesi, F.K., Mafongoya, P.L., McDermott, M.H. & Agumya, A. 2003. Agroforestry research and development in southern Africa during the 1990s: Review and challenges ahead. *Agroforestry Systems* 59: 173-186.

Kwesiga, F. & Coe, R. 1994. The effect of short rotation *Sesbania sesban* planted fallows on maize yield. *Forest Ecology and Management* 64: 199-208.

Kwesiga, F.R., Franzel, S., Place, F., Phiri, D. & Simwanza, C.P. 1999. *Sesbania sesban* improved fallows in eastern Zambia: their inception, development, and farmer enthusiasm. *Agroforestry Systems* 47: 49-66.

Mafongoya P.L., Bationo A., Kihara J., & Waswa B.S. (2006) Appropriate technologies to replenish soil fertility in southern Africa, Nutr.Cycl. Agroecosys. 76,137–151.

Mairura, F.S., Mugendi, D.N., Mwanje, J.I., Ramisch, J.J., Mbugua, P.K. & Chianu, J.N. 2007. Integrating scientific and farmers' evaluation of soil quality indicators in Central Kenya. *Geoderma* 139: 134-143.

Makumba, W., Akinnifesi, F.K., Janssen, B. & Oenema, O. 2007. Long-term impact of gliricidia-maize intercropping system on carbon sequestration in southern Malawi. *Agriculture, Ecosystems and Environment* 118: 237-243.

Malawi. 2002. Malawi poverty reduction strategy paper. Lilongwe.

Malawi. 2005. Integrated household survey 2004-2005 Volume 1 household socio-economic characteristics. Zomba: National Statistical Office.

Malawi. 2008. Ministry of Agriculture and Food Security Agricultural Development Programme Project ID Number: MASIP/TA/ADP-SP 1901/14 Environmental and social impact assessment final draft report. Lilongwe.

Malawi Ministry of Mines, Natural Resources, and Environment, Environmental Affairs Department (MEAD). 2002. State of Environment Report for Malawi 2002. [Online]. Lilongwe. Available at:
http://www.malawi.gov.mw/Mines/EnviromentalAffairs/SOER2001se/soeindex .html. [2008, March 1].

Malawi Vulnerability Assessment Committee (MVAC). 2005. *Malawi Baseline Livelihood Profiles: Version 1.*

Massawe, C.R., Kaswende, J.S., Mbwaga, A.M. & Hella, J.P. 2001. *On-farm verification of maize/cowpea intercropping on the control of Striga under subsistence farming.* Seventh Eastern and Southern Africa Regional Maize Conference, Nairobi.

Meghji, Z.H. 2003. *Participatory forest management in Tanzania.* [Online]. XII World Forestry Congress, 2003. Quebec City, Canada. Available at:
http://www.fao.org/DOCREP/ARTICLE/WFC/XII/0813-C1.HTM. [2008 May 10].

Montagnini, F. & Nair, P.K.R. 2004. Carbon sequestration: An underexploited environmental benefit of agroforestry systems. *Agroforestry Systems* 61: 281-295.

Msuku, I.R., Lowole, M.W., Nzima, J. & Msango, A. s.d. *Agroforestry potentials for the land-use systems in the unimodal plateau of southern Africa Malawi*. Rapport AFRENA Report No. 5 Malawi Agroforestry Task Force, Malawi.

Murage, E.W., Karanja, N.K., Smithson, P.C. & Woomer, P.L. 2000. Diagnostic indicators of soil quality in productive and non-productive smallholders' fields of Kenya's Central Highlands. *Agriculture, Ecosystems and Environment* 79:1-8.

Mwasikakata, M. 2003. *Tobacco: an economic lifeline? The case of tobacco farming in the Kasungu Agricultural Development Division, Malawi*. Sectoral Activities Programme Working Paper 184. International Labour Office, Geneva.

Orlando, B., Baldock, D., Canger, S., Mackensen, J., Maginnis, S., Socorro, M., Rietbergen, S., Robledo, C. & Schneider, N. 2002. *Carbon, Forests and People: Towards the integrated management of carbon sequestration, the environment and sustainable livelihoods*. IUCN, Gland, Switzerland and Cambridge, U.K.

Phiri, A.D.K., Kanyana-Phiri, G.Y. & Snapp, S. 1999. Maize and sesbania production in relay cropping at three landscape positions in Malawi. *Agroforestry Systems* 47: 153-162.

Place, F., Adato, M., Hebinck, P. & Omosa, M. 2003. *The Impact of Agroforestry-Based Soil Fertility Replenishment Practices on the Poor in Western Kenya*. FCND Discussion Paper No. 160. International Food and Policy Research Institute, Washington, D.C.

Place, F., Adato, M. & Hebinck, P. 2007. Understanding rural poverty and investment in agriculture: An assessment of quantitative and qualitative research in western Kenya. *World Development* 35: 312-325.

Sanchez, P.A., Shepherd, K.D., Soule, M.J., Place, F.M., Buresh, R.J., Izac, A.N., Mokwunye, A.U., Kwesiga, F.R., Ndirity, C.G. & Woomer, P.L. 1997. Soil fertility replenishment in Africa: An investment in natural resource capital, in R.J. Buresh, P.A. Sanchez & F. Calhoun (eds.). 1997. *Replenishing Soil Fertility in Africa SSSA Special Publication Number 51*. Madison: SSSA.

Scoones, I. & Toulmin, C. 1999. *Policies for Soil Fertility Management in Africa*. Department for International Development, London.

Sileshi, G.W., Kuntashula, E., Matakala, P. & Nkunika, P.O. 2008. Farmers' perceptions of tree mortality, pests and pest management practices in agroforestry in Malawi, Mozambique, and Zambia. *Agroforestry Systems* 72: 87-101.

Skoufias, 2003. Economic crises and natural disasters: Coping strategies and policy implications. *World Development* 31: 1087-1102.

Snapp, S.S., Mafongoya, P.L. & Waddington, S. 1998. Organic matter technologies for integrated nutrient management in smallholder cropping systems of southern Africa. *Agriculture, Ecosystems, and Environment*. 71: 185-200.

Takimoto, A., Nair, P.K.R. & Nair, V.D. 2008. Carbon stock and sequestration potential of traditional and improved agroforestry systems in the West African Sahel. *Agriculture, Ecosystems, and Environment* 125: 159-166.

Thangata, P.H. & Alavalapati, J.R.R. 2003. Agroforestry adoption in southern Malawi: the case of mixed intercropping of *Gliricidia sepium* and maize. *Agricultural Systems* 78: 57-71.

Thomas, D.S.G., Twyman, C., Osbahr, H. & Hewitson, B. 2007. Adaptation to climate change and variability: farmer responses to intra-seasonal precipitation trends in South Africa. *Climate Change* 83: 301-322.

Water Deficit Condition Affecting Rice Production – Challenges and Prospects

Deivanai Subramanian
Department of Biotechnology, Faculty of Applied Sciences, AIMST University
Malaysia

1. Introduction

Drought is one of the inherent abiotic constraints that affect agricultural productivity worldwide. It is estimated that drought stress can potentially reduce nearly 20% of crop yield around the World (Bouman et al., 2005; Scheiermeier, 2008). Global climatic changes such as dry spell, heat waves and uneven precipitation patterns limit water availability for farming (Bates et al., 2008). However, water is needed at every phase of plant growth from seed germination to plant maturation (Athar & Ashraf, 2005; Chaves et al., 2003) and any degree of imbalance in the uptake would pose a serious threat to agriculture by adversely affecting the growth and grain yield (Wang et al., 2001). Further, water deficit stress could occur at any time of growing seasons; however severity of stress effect on productivity depends on distribution of rainfall.

Rice is the staple food for almost half of the world population. Rice farming is considered as one of the world's most sustainable and productive cropping system, as it is adapted to wide range of environment ranging from tropical low lands to mountains and from deep water swamp to uplands. In general, rice crop is semi aquatic and can thrive well in waterlogged soil and hence its production system relies on ample water supply. Based on the availability of water, rice can be grown in different ecological conditions such as low land rainfed, low land irrigated, deep water and upland. In global scenario, irrigated rice is considered as productive farming system and has accounted for 55% of total harvested area with a contribution of 75% of total productivity. Further, annual productivity of irrigated rice is estimated to be 5% more than that of rainfed rice (Fairhurt & Dobermann, 2002). Meanwhile, resource for irrigation has declined gradually over the past decades due to rapid urbanization and industrialization which exacerbates the problem of water scarcity (Gleick et al., 2002). Current rice production systems rely on ample supply of water and it is estimated that on average rice require 1900 liters of water to produce 1kg of grain. Even a short period of water deficit is highly sensitive to rice farming and rice productivity (O'Toole, 2004). Different developmental stages of rice such as tillering phase, panicle initiation and heading known to respond differently to drought stress (Botwright Acuna et al., 2008; Kamoshita et al., 2004), however, factors such as timing, intensity and duration of stress have detrimental effect on plant growth. Liu et al., (2006) reported that reproductive stage, especially during flowering, is more vulnerable to stress and cause spikelet sterility.

Rice-cropping system is considered as economic backbone of many Asians as 90% of the World total rice is grown and consumed in Asia. A small decrease in availability of water would drastically affect grain yield and imperil food security. Increasing crop tolerance to water scarcity would be the most economic approach to improve the productivity and to minimize agricultural use of fresh water resource. To fulfill this objective, a deeper understanding of the possible mechanisms under water stress environment is a must. Recent studies have shown that plants have evolved various morphological, physiological, biochemical and molecular mechanisms to cope up with adverse climatic effect. Conventional breeding strategies utilized those mechanisms to certain extent and achieved a linear improvement in yield by exploiting genetic variation of phenotypic traits in the germplasm (Atlin et al., 2006; Lafitte et al., 2006; Serraj, 2005). However, development of tolerant varieties to water deficit condition has been slow due to lack of understanding on the mechanism of drought tolerance. Several efforts have been made to improve crop productivity under water limiting environment. Munns (1993) suggested that physiological based approach could furnish a new insight to breeding programme by offering reliable key indicators for screening drought tolerant genotypes to improve crop yield over a range of environment. Nevertheless, very little information is available on magnitude of genetic variation associated with physiology of tolerance. Furthermore, plants evolve a wide spectrum of adaptive mechanisms from whole plant to molecular level which varies within species and cultivars to overcome drought stress (Bartels & Sunkar, 2005; Jones, 2004; Rampino et al., 2006; Yamaguchi Shinozaki and Shinozaki, 2005). To exploit genetic basis of physiological variations effectively, investigations on molecular basis of stress are indispensable. Rapid advancement in molecular techniques helps to elucidate the control mechanisms linked to stress perception and responses by dissecting yield and integrative traits influenced under stress. Number of drought inducible gene and gene products have been identified at transcriptional and translational level and most of the gene products may function in stress tolerance at cellular level (Umezawa et al., 2006; Yamaguchi Shinozaki & Shinozaki, 2005; Zang et al., 2004). Further, all these investigations have provided important clues for functional characterization of stress responsive gene and stress tolerance mechanism; thereby widen our knowledge in mitigation of drought stress. The present review addresses the recent advances of the adaptive strategies in rice and facilitates the development of enhanced tolerance by integrating functional genomics into breeding programs.

2. Physio morphological traits for yield under drought stress

In rice, a number of physio-morphological putative traits have been suggested to confer drought tolerance (Deivanai, et al., 2010). Indeed, root system architecture plays a primary constitutive role in acquisition of water and nutrient from the soil and maintains plant water status (Nguyen et al., 1997; Lafitte et al., 2001; Kato et al., 2006). Descriptors of root architecture such as rooting depth (Pantuwan et al., 1996; Wade et al., 1996), root density, root thickness, root distribution pattern (Lilley and Fukai, 1994; Fukai and Cooper, 1994) increases plant water uptake and avoid dehydration mechanism. However, water uptake efficiency is determined by the function of root length, soil type, root hydraulic conductance and transpiration demand (Nobel, 2005). Further, root system architecture is highly dynamic and its functions are affected by change in environmental conditions and soil types (Lafitte et al., 2001). Since root characteristics are invisible for direct selection, knowledge on

interrelationship between plant type and root architecture would be desirable to facilitate yield performance of rice varieties under different water regimes.
Traditional plant breeding has identified traits such as plant type (plant height, tiller number, leaf area, leaf area index), phenology (timing of germination, floral initiation, flowering and seed maturity) and canopy temperature for screening of genotypes to improve grain yield under stress (Turner, 1982; Cooper et al., 1999; Slafer, 2003; Turner, 2004; Reynold et al., 2005a). In fact, grain yield is a complex quantitative trait attributed by number of panicle/shoot/m^2, grains/spikelet/panicle and grain weight / panicle (Sinha and Kanna, 1975). Improvement of crop yield could be achieved by selecting the genotypes based on the intensity of genetic variation. Several investigations have observed significant genetic variation for root number, diameter, depth, branching, root to shoot ratio, water extraction and root penetration (Yadav et al., 1997; Lafitte et al., 2001; Price et al., 2002b; Pantuwan et al., 2002a & b; Nhan et al., 2006). Meanwhile, high heritability was also noticed for root thickness, root dry weight and root length density (Lafitte et al., 2001). Moderate heritability for grain yield was reported by Babu et al., 2003; Atlin et al., 2004; Lafitte et al., 2004b; Yue et al., 2005; Kumar et al., 2007. Though, several traits were considered for improving drought tolerance in rice, only few have contributed significantly towards grain yield under drought condition (Lafitte, 2003). It is because most of the traits that respond to stress are constitutive and are not adaptive to stress (Parry et al., 2005; Komoshita et al., 2008). Further, substantial differences in root system architecture and grain yield have been reported between species, cultivars and land races. Obviously, grain yield is characterized by yield potential of the genotypes under target environment and intrinsically linked with plant water status (Araus et al., 2003; Rizza et al., 2004; Blum, 2005), a relevant physiological measure of interest under drought prone environment. Plant water status is determined by a balance between water uptake through root system and demand by shoot. Since plant production is the function of water use (WU), water use efficiency (WUE) and harvest index (HI), it is therefore vital to understand its effect during defined developmental stage to design effective selection method to improve plant production under dry environment. Several direct measurements are recommended to assess plant water status and its physiological consequences, some widely used measures are mentioned in this review

2.1 Water use efficiency (WUE)

WUE provide the means of efficient use of water and serves as a breeding target in water saving agriculture. Traditionally defined as the ratio of dry matter produced per unit of water transpired, and constitute one of the key determinants in controlling plant production. It is also referred as "transpiration efficiency" and estimated from the measures of leaf gas exchange or by using carbon isotope discrimination. Mean while, carbon isotope discriminant method had contributed much in the study of WUE and provided large volume of data in relation to genetic diversity and plant breeding. Studies have shown that yield related traits such as biomass accumulation and transpiration rate are highly interlinked with stomatal control and leaf area (Cabuslay et al., 2002; Richards, 2000; Tardieu & Tuberosa, 2010). For example, when plant experienced mild water stress, it increased WUE by regulating stomatal conductivity and lower carbon isotope discriminant measures. In most of the drought related studies low measures of carbon isotope discrimination found to be associated with high WUE. Higher WUE in turn lower photosynthetic rate due to reduced rate of transpiration and consequently slower the rate of plant growth (Codon et al., 2004). It is obvious that increased WUE has resulted in smaller or short duration plants with reduced transpiration, biomass production and yield potential due to reduced water use. Blum (2005) pointed out that high WUE is largely a function of reduced WU and suggested that it can be used as a yardstick to measure irrigation efficiency in agriculture. Further, he proposed direct selection of plant type with moderate growth, short duration and reduce leaf area would promote WUE. Currently, agricultural sectors are slowly progressing due to use of genotypes with increased WUE and improved agronomic practices (Pereira et al., 2006; Richards et al., 2002).

2.2 Leaf water potential (LWP)
Leaf water potential (LWP), a measure of whole plant water status and has long been recognized as an indicator for dehydration avoidance (Pantuwan et al., 2002). When water deficit in leaf goes beyond a threshold level, stomata closes as a response to lower the rate of transpiration. According to Hsiao and Bradford (1982), stomata functions as safety valve to regulate water loss when tissue water status becomes too low, whereby minimize the severity of water deficiency in plants. Thus higher LWP is maintained by stomatal closure and varietal differences in stomatal response to water status have been reported by Jongdee et al., (1998), O'Toole & Cruz (1980) and Turner (1974), it is mainly due to differential capacity of hydraulic conductance among the genotypes. Further, genotype with deeper and thicker roots shown to extract more soil moisture effectively and maintains higher plant water status (Yoshida & Hasegawa, 1982). In a study, Sibounheuang et al., (2001) observed consistence in the performance of LWP and suggested a mighty relation between shoot water potential and internal plant water conductivity. Mean while, Kumar et al., (2004) reported a positive association between grain yield and relative water content with LWP.
Stomatal closure found to increase leaf temperature and measure of canopy temperature serves as an indirect measure of plant water status. Boonjung and Fukai (1996) as well as Pantuwan (2000) observed taller genotypes tend to have larger canopy than the shorter genotypes and they found smaller canopies had lower demand for water and were able to maintain higher LWP. The two parameters namely leaf rolling and leaf death are considered as determinant for estimating LWP. Earlier studies have reported higher estimates of visual scores and heritability for those two determinants while screening cultivars for drought tolerance.

2.3 Osmoregulation
Osmoregulation is receiving increasing recognition as an effective physiological mechanism of drought tolerance. Drought stress known to alter internal plant water status and lower water potential of cell environment (Babu et al., 1999). As a consequence, solutes are actively accumulated at high concentration within cells in order to maintain water potential (Blum, 1998; Kramer & Boyer, 1995). Such solutes are referred as osmolytes which include amino acids, sugars, polyols, quaternary ammonium and sulfoium compounds (Rontein et al., 2002). According to Blum (1998) the process of accumulation of solutes during the development of water shortage and enabling the plant to maintain hydrated state is termed as osmotic adjustment (OA), while buffering mechanism against dehydrated condition by addition or removal of solutes from cytoplasm is osmoregulation.
Generally, OA restores turgor pressure of both root and shoot, thereby permit stomatal conductivity to continue and help to sustain plant growth at time of transpirational demand

(Blum, 1996, 2005; Serraj & Sinclair, 2002; Stoop et al., 1996). Further, Blum (1996) found that these compounds stabilize macromolecular structures and membrane proteins. Nevertheless, the role of osmolytes are diverse and investigations on osmoprotectants have revealed that proline regulates cellular redox during stress (Kuznetsov & Shevyakova, 1999) while, manitol protects oxidation sensitive cellular structure by scavenging reactive oxygen species (Huang et al., 2000) whereas, trehalose and fructan function as membrane stability (Nakayama et al., 2000), aquaporin, a major intrinsic protein super family facilitate control of water permeability through membrane (Luu & Maurel, 2005).

A genetic variation of 0.3 – 0.5 MPa has been reported for OA in rice (Turner et al., 1986). Though OA enable to maintain water absorption and cellular turgor pressure, a number of contrasting reports have been published. For instance, Munns (1988) pointed that OA may not show positive effect on plant growth and grain yield. It was also argued that OA occurs in root facilitates elongation of root growth at the onset of soil drying and in turn assist the root to penetrate deeper in search of resources thereby enable the plant to extract more soil water. Serraj & Sinclair (2002) found that plants with higher root penetration capacity is capable of sustaining plant growth and they suggested investigation on OA with focus on roots and root tips would elucidate the role of OA in practical breeding.

2.4 Stay green

Retention of greenness in leaves helps the plants to live longer and increase crop productivity; hence this trait is considered as one of the key determinants for developing drought resistance in rice (Fukai & Cooper, 1995). Genotype possessing stay green trait maintain high photosynthetic activity often protects the plants from premature senescence (Campos et al., 2004) during the onset of stress. It is reported that stay green plants assimilate more nitrogen and retain high level of nitrogen content in the leaf, thereby retains photosynthetic capacity under water limited conditions (Borrell et al., 2001). Further it enhances the transpiration efficiency and enable the plants to use more water to ensure continuous availability of new assimilate thereby increase the grain filling and size. Studies have also shown that stay-green is positively associated with grain yield

3. Functional genomic approach

In the past decades, integration of physiological and biochemical studies on abiotic stresses have contributed extensively in identifying tolerance traits responsive to stresses (Richard, 2000). Conventional breeding techniques attempted to utilize genetic variations of those traits from varietal germplasm; however the success is limited by: i) complexity of stress responses, ii) low genetic variation of yield component under stress condition and iii) lack of suitable selection technique (Ashraf, 2010; Cushman & Bohnert, 2000). Recently engineering of drought tolerance in plants is being pursued as viable option as it seems to be more attractive and rapid approach for breeding drought tolerance. Since transfer of functional genes that are directly involved in drought mechanism by genetic engineering is complex, the success of present engineering strategies rely on understanding of key gene network and regulatory control of biological processes associated with drought stress (Seki, et al., 2003; Shinozaki et al., 2003). Further, the advent of high throughput genomic platforms has facilitated increasing number of tools and resources for elucidation of abiotic stress responses in plants. Furthermore, large scale genome sequencing projects has generated a number of sequence information from DNA microarrays, serial analysis of gene expression (SAGE), cDNA fragment sizing combined with amplified fragment length polymorphism (cDNA – AFLP), differential screening of cDNA libraries, expressed sequence tag (EST) sequencing, massive parallel signature sequencing (MPSS) etc., provides important clues for gene expression, functional characterization and identification of stress responsive genes (Bohnert et al., 2006, Parray et al., 2005; Umezawa et al., 2006).

More recently several stress inducible novel putative genes have been identified from wide range of tissue specific for genetic engineering. Meanwhile, sequencing project also enabled the development of molecular markers such as RFLPs, RAPDs, CAPS, PCR indels, AFLP, microsatellite such as SSR and SNP that are closely linked to target loci and facilitated mapping of quantitative traits loci (QTLs) for agronomically important attributes under drought stress (Lafitte et al., 2004; Talame et al., 2004). Subsequently those markers were used for marker assisted selection (MAS), a powerful tool for indirect selection of complex traits at early stage. However the efficiency of MAS depends on the distance between observed QTL and marker loci and also the magnitude of additive variance explained by QTL (Blumwald, et al., 2004). Thus the key role of functional genomics is to deduce the biological function of gene and gene products through genomic approaches like genetic mapping, transcriptional profiling and proteomics.

3.1 Quantitative trait loci (QTL) to dissect drought stress related traits

Progress in genomic sequencing of rice has facilitated a range of approaches to identify molecular markers (Nguyen et al., 1997), which examines the inheritance pattern of QTL in response to drought stress. Further these markers explore chromosome regions controlling genetic variations of physiological traits and play a central role in construction of linkage map using QTL analysis. QTL mapping approach has successfully identified a number of genetic regions that are expected to be associated with drought response, such as plant height and flowering time (Ishimaru et al., 2004; Li et al., 2003), root architecture (Courtois et al., 2003; Kamoshita et al., 2002; Price et al., 2002c; Tuberosa et al., 2003; Venuprasad et al., 2002; Zheng et al., 2003), root penetration (Johnson et al., 2000; Nguyen et al., 2004; Zheng et al., 2000), stay green phenotype (Jiang et al., 2004). OA, RWC and leaf rolling (Robin et al., 2003; Zhang et al., 2001), LWP (Yan Ying et al., 2008), leaf drying (Lafitte et al., 2004), yield and yield component under drought stress (Babu et al., 2003; Bernier et al., 2008; Campos et al., 2004; Kamoshita et al., 2008; Kumar *et al.*, 2007; Lafitte et al., 2004; Lanceras et al., 2004). Most of the mapping populations were derived from cross between upland japonica and low land indica cultivars (Courtois et al., 2003) under varying water regimes, further these studies identified the location of drought tolerant traits as well yield and yield components under stress on chromosomal regions. For example, Babu et al., (2003) documented the location of drought tolerant QTLs in rice in the regions of chromosome 1 (plant water status), chromosome 3 (biomass yield under stress), chromosome 4 (root morphological traits) and chromosome 9 (RWC and delayed flowering due to stress). Whereas another group (Lanceras et al., 2004) mapped physio-morphological traits and confirmed the region of QTL traits in chromosome 3 (Grain yield, biomass and delayed flowering), chromosome 4 (grain yield) and chromosome 8 (biomass yield and spikelet sterility). In both the studies, the traits such as biomass, yield and yield components were found to be common and located at the same intervals on chromosome 3, 4 and 9. In another study, Bernier et al., (2007) while evaluating F_3 progenies derived from a cross between two upland rice cultivars under field trials for two years, identified a stable QTL for drought tolerance on chromosome 12 linked to grain yield, biomass production, harvest index, plant height and early flowering. Wealth of information thus generated on QTL over last two decades helps to develop strategies for marker assisted selection (MAS) to screen drought tolerant strains at early stage of growth. Several studies have been conducted under both well watered and drought stressed conditions and a significant progress has been made during last few years in marker assisted selection (Jearakongman, 2005; Price et al., 2002c; Shen et al., 2001; Yue et al., 2005; Steel et al., 2006). One of the classic achievement of MAS is the release of first ever highly drought tolerant Indian rice variety "Birsa Vikas Dhan III (PY,84)", characterized by early maturity, good quality high grain yield (Steele, 2009). Though several putative loci for drought stress tolerance have been identified, the contribution of MAS to development of drought stress tolerant cultivars has been marginal. The effectiveness of MAS is challenged by number of factors such as: i) accuracy and preciseness of phenotyping of traits, ii) very large genetic x environment interaction component wherein the QTL established in one environment often disappear in another and iii) poorly defined genetic architecture of polygene controlling yield (Ashraf, 2010; Cattivelli et al., 2008).

The efficiency of MAS could be improved through map based cloning (MBC), where specific QTL is introgressed into sensitive cultivars through pyramiding, a process of combining several genes together into single genotype (Steele et al., 2006). To date most plant QTLs have been cloned by positional cloning approach using functional markers with a goal to identify candidate gene responsible for drought stress and to manipulate the target trait more directly. Most of the comprehensive physical map reported from the earlier studies covers a region of 35-64 cM resolution which may contain several hundred to few thousand genes, need to be fine mapped. High density genetic map using single nucleotide polymorphism (SNP) and other marker may precisely detect QTL genes associated with drought tolerance. Further, fine mapping of specific QTL regions could be achieved by developing near isogenic lines (Nguyen et al., 2004). However, identification and characterization of QTL genes involved in regulatory network is still remain challenging.

3.2 Transcript profiling

ESTs provide a direct approach for discovering genes in response to stress, with the advent of high throughput trancriptome studies, several ESTs have been generated from different cDNA libraries (viz., full length, normalized and subtracted) which offered a foundation in deciphering the role of regulatory network in stressed tissues derived at various developmental stages (Goff, 1999). Numerous putative genes respond to dehydration stress have been categorized by EST based gene expression profiling (Shinozaki et al., 2000; Yamaguchi – Shinozaki & Shinozaki, 2006). Some of these genes that are induced during stress protects the plant cell directly while the others involved in signaling cascades with diverse pathways, suggesting the complexity of the mechanism involved in sensing and responding to multifarious stresses (Bartels & Sunkar, 2005; Blumwald et al., 2004; Bray et al., 2000). Further, the plasticity of plant response to water limited conditions is governed by a number of transcription factors (TFs) which can modulates and regulates various stress inducible genes either independently or constitutively. More than 50 different TFs have been identified and characterized, are found to be member of large multigene families such as bZIP- (Martinez-Garcia et al., 1998), MYB (Jin & Martin, 1999), MYC, AP2/ERF (Riechmann,

& Meyerowitz., 1998) , NAC (Kikuchi et al., 2000) and WRKY (Dong et al., 2003). Many genes that respond to multiple stresses are induced by abscisic acid (ABA); a phytohormone which acts as a key signaling intermediate in controlling the expression of stress related genes. A detail examination of ABA regulated genes has shown the presence of both ABA dependent / independent regulatory systems in ABA biosynthesis (Vinocur & Altmam, 2005; Yamaguchi-Shinozaki & Shinozaki, 2005). Further, the transcriptional factors such as MYC and MYB function as activator in one of the ABA-dependent regulatory system, while a cis-acting element known as drought responsive element factor/ C-repeat (DREB/CRT) is involved in ABA-independent regulatory system (Shinozaki et al., 2000). Several experimental approaches on genomic analysis have reported different TFs that are associated with stress responsive gene induction are presented in Table 1.

TF family	Gene category	Gene name	Physiological response	Reference
AP2/ERF	DREB1/CBF	OstDREB1A	Stomatal closure	Ito et al., 2006; Oh et al., 2005; Xiao et al., 2009
NAC	SNAC	AtSNAC1	Stomatal closure	Rabbani et al., 2003
bZIP	AREB/ABF	AtABF3	Reduced leaf rolling and wilting	Oh et al., 2005
TFIII-A Zinc finger	ZFP252	OsZFP252	Proline and sugar accumulation	Xu et al., 2008
NF-Y (A, B, C)	NF-YB	AtNF-YB1 ZmNF-YB2	High photosynthesis	Nelson et al.,2007
WRKY Zinc finger	WRKY	OsWRKY11	Reduced leaf wilting and slow water loss	Wu et al., 2009
EAR Zinc finger	Zat10/STZ	AtZat10	High spikelet fertility and grain yield	Xiao et al., 2009

Table 1. Genome wide transcriptome analysis of drought response in rice

To date, substantial amount of published works have examined the mechanism of plant response to various environmental changes and demonstrated that some transcription factors significantly overlaps with the expression of gene that are induced in response to different stress (Chen & Murata, 2008; Seki et al., 2001). Further, overexpression of one or more transcription factors confirmed the activation of TF regulons that modulate a wide range of signaling pathways in achieving tolerance under multiple stress conditions (Umezawa et al., 2006). For example, it enables the regulation of key enzymes in the biosynthesis of compatible solutes such as proline (Ito et al., 2006; Zhu et al., 1998), glycinebetaine (Quan et al., 2004), variety of sugars and sugar alcohol, viz., trehalose (Garg et al., 2002), manitol (Abebe et al., 2003), galactinol and raffinose. Transgenic rice plants produced by overexpression of transcription factors help to understand and manipulate the responses of plant stress. Considerable progress has been made in developing transgenic rice strains that are tolerant to drought stress and the results are summarized in Table 2.

Gene	Gene action	Phenotype	Reference
Adc	Arginine decarboxylase (Polyamine synthesis)	Drought resistance	Capell et al., 2004
codA	Choline oxidase (glycinebetaine synthesis)	Recovery from a week long stress	Mohanty et al., 2003
COX	Choline oxidase (glycinebetaine synthesis)	Stress tolerance	Su et al., 2006
HVA1	Group 3LEA protein (late embryogenesis abundant)	Dehydration avoidance and cell membrane stability	Babu et al., 2004
OCPI1	Chymotrypsin inhibitor like 1 (proteinase inhibitor gene)	Stress tolerance	Huang et al., 2007
OsLE A3-1	LEA protein (late embryogenesis abundant)	Drought resistance for yield in the field	Xiao et al., 2007
P5CS	Pyrroline carboxylate synthase (proline synthesis	Reduced oxidative stress under osmotic stress	Hong et al., 2000
P5CS	Pyrroline carboxylate synthase (proline synthesis)	Increased biomass production under drought stress	Zhu et al., 1998
RWC3	Aquaporin (water channel protein)	Stress response	Huang et al., 2007
TPS	Trehalose-6-phosphate synthase (Trehalose synthesis)	Drought tolerance	Jang et al., 2003
TP	Trehalose-6-phosphatase (Trehalose synthesis)	Drought tolerance	Lee et al., 2003

Table 2. Achievements made in overexpression of candidate genes through transgenic approach in rice

3.3 Proteomics

In general, abiotic stresses cause considerable dysfunction in proteins, proteomic approaches focused on protein changes in response to stresses and explore the functional network of protein. A global protein expression profile can be investigated using two dimensional polyacrylamide gel electrophoresis (2DE) technique coupled with protein identification by mass spectrometry (MS). This technique facilitated identification of new proteins of interest and elucidated the expression profile, post translational modification, interactions and *de novo* synthesis of proteins (Peck, 2005). Salekdeh et al., (2002) investigated the drought responsiveness using lowland indica (IR62266) and upland japonica (CT9993). They quantify nearly 1000 rice leaf proteins out of which 42 responded to stress. Similar such studies have identified several proteins based on its function and classified them into two groups. One group of proteins such as heat shock proteins (HSPs), late embryogenesis abundant protein (LEA), Dehydrin proteins, RuBisCo and reactive oxygen species (ROS) play a direct role in protecting the plant cells against stress by involving in osmotic adjustment, chaperon like activity and scavenging of reactive oxygen species. The second group of proteins are viz., mitogen activated protein kinase-MAPK and calcium-dependent protein kinases-CDPK (Ludwig et al., 2004), salt overlay sensitive-SOS kinases (Zhu, 2001), phospholipases (Frank et al., 2000) and transcriptional factors (Choi et al., 2000) are actively involved in signaling cascades and transcriptional control. Overexpression of signaling factors known to control a broad range of downstream events and has resulted in superior tolerance (Umezawa et al., 2006)

Huge amount of data generated from proteomic studies provided useful information on individual enzymes and transporters that are involved in stress responsive network including protein modifications, interactions and *de novo* synthesis. Further the information would be helpful in developing biomarker for molecular cloning. However, the application of a proteomic approach at the whole cell level is limited by several factors such as protein abundance, size, hydrophobicity and other electrophoretic properties (Parker *et al.*, 2006; Timperio *et al.*, 2008). Moreover, low abundance proteins including regulatory proteins and rare membrane proteins are out of scope of most proteomic techniques, it is due to chemical heterogeneity of proteins associated with diverse functions within a cell. The limitation could be resolved using comprehensive protein extraction protocol for proteome analysis.

4. Conclusion and future prospects

The present review summarizes the achievements of breeding enhanced tolerance towards water deficit condition in rice. In early 1980's large number of studies on drought stress have identified some morphological features such as strong root system, short stature plants, reduced leaf area and limited tillering ability were capable of maintaining high plant water status to enhance drought avoidance. Among those features, the most important is the root system as it ensures extraction of soil water from greater depth in upland regions and maintains high LWP during stress. While, in rainfed lowland regions, soil forms hard encrustation and inhibit root penetration when the available soil moisture is exhausted. The genotypes with dense and thick roots were suggested to improve selection efficiency. Further, physiological studies in the past have provided knowledge on complex network of drought stress related traits and suggested relevant drought related determinants such as WUE, osmotic potential, utilization of stem reserve, dry matter production, etc., that could be used to achieve high potential (Blum, 2005).

Conventional breeding program effectively utilized those traits and achieved drought tolerance by generating reasonable number of cultivars. However, the program is limited by lack of appropriate screening technique. It is because drought stress can occur at any time of the developmental stage; starting from sowing to grain maturity. Generally it is widely accepted that stress at reproductive phase is critical and deserves attention. However, genotypic response that contributes to drought stress avoidance/ tolerance is largely depends on the genetic mechanism of tolerance in the target environment (Fukai and cooper, 1995).

With the advancement in biotechnological tool, the genetic basis of drought tolerance has received considerable attention. Gene governing quantitative traits were identified using a variety of molecular markers and their loci controlling drought tolerance were mapped on the chromosomal regions. Mapping of QTL has resulted in greater understanding of genetic phenomenon of drought tolerance traits. Despite the significant progress in cereal genomics, the QTL approach has not widely practiced in marker assisted breeding and still remains a major challenge. The efficiency of MAS is hampered by the complexity of gene governing grain yield, epistatic interactions and epigenetic variation among QTLs. However, Tuberosa and Salvi (2006) opinioned that conscious selection of mapping population and careful introgression of specific alleles from genotype to other, through pyramiding could bring success to MAS.

In the last decade, transgenic and functional genomic approaches offered a reliable promise in identifying stress responsive genes, pathways and deciphering the mechanism of stress tolerance. Further it enabled to solve several essential key questions associated with stress tolerance through gene expression profiling and engineering of tolerant traits. A large number of functionally characterized genes, transcriptional factors and promoters were introduced by such methods to enhance tolerance against abiotic stresses. Although several reports have highlighted the significance of this approach (Cattivelli, et al., 2008; Kamoshita et al., 2008; Umezawa et al., 2006; Vij and Tyagi, 2007; Yang et al., 2010), introgression of genomic portion often associated with undesirable agronomic traits and only very few field screening and genetic transformations have resulted in improved grain yield under drought condition. It is likely due to the fact that interaction between number of edaphic and climatic factors poses difficulty in screening of stress tolerance (Ashraf et al., 2008). Further it is anticipated that transcriptional regulation as well as post transcriptional gene plays a major role in determining the tolerance against various stress. Therefore while using functional genomics, it is important to consider the phenomenon of regulatory network including siRNA and miRNA to fine tune the expression of genes associated with stress responses (Yang et al., 2010). Indeed, the post genomic era has offered a great potential to increase the efficiency of breeding by determining the phenotype more precisely. Further, the challenge in stabilizing high yield under drought condition could be achieved in near future by integrating plant breeding with multidisciplinary approach based on plant physiology, functional genomics and by adapting comprehensive screening technique.

5. References

Abebe, T., Guenzi, A.C., Martin, B & Cushman, J.C. (2003). Tolerance of manitol accumulating trangenic sheat to water stress and salinity. *Plant physiology*, 131, 1748-1755. www.plantphysiol.org/cgi/doi/10.1104/pp.102.003616.

Araus, J.L., Villegas, D., Aparicio, N., Garcia del Moral, L.F., El Hani, S., Rharrabti, Y., Ferrio, J.P., & Royo, C. (2003). Environmental factors determining carbon isotope discrimination and yield in durum wheat under Mediterranean conditions. Crop Science, 43, 1, 170- 180. doi: 10.2135/cropsci2003.1700

Ashraf, M. (2010). Inducing drought tolerance in plants: recent advances. Biotechnology Advances. 28, 169-183. doi:10.1016/j.biotechadv.2009.11.005

Ashraf, M., Athar, H.R., Harris, P.J.C., Kwon, T.R. 2008. Some prospective strategies for improving crop salt tolerance. *Advances in Agronomy*. 97, 45-110. doi:10.1016/S0065-2113(07)00002-8

Athar, H.R., & Ashraf, M. (2005). Photosynthesis under drought stress. In: *Handbook of photosynthesis* (Ed.): Pessarakli, M. CRC Press, Taylor and Fransis group. New York. Pp 793-804. doi: 10.1093/aob/mcj017

Atlin, G. N., Lafitte, H. R., Venuprasad, R., Kumar, R., & Jongdee, B. (2004). Heritability of mean grain yield under reproductive-stage drought stress and correlations across stress levels in sets of selected and unselected rice lines in the Philippines, Thailand, and India: implications for drought tolerance breeding. p. 85-87. In: D. Poland, M. Sawkins, J. -M. Ribaut, and D. Hoisington (ed.) *Resilient crops for water limited environments*: Proceedings of a workshop held at Cuernavaca, Mexico, 24-28 May 2004. CIMMYT, Mexico D.F.
http://apps.cimmyt.org/english/docs/proceedings/resilient/resil_crops.pdf

Atlin, G.N., Lafitte, H.R., Tao, D., Laza, M., Amante, M., & Courtois, B. (2006). Developing rice cultivars for high-fertility upland systems in the Asian tropics. *Field Crops Research*. 97, 1, 43-52. doi:10.1016/j.fcr.2005.08.014

Babu, R.C., Nguyen, B. D., Chamarerk, V. P., Shanmugasundaram, P., Chezhian, P., Jeyaprakash, S.K., Ganesh, A., Palchamy, S., Sadasivam, S., Sarkarung, S., Wade, L. J., & Nguyen, H. T. (2003). Genetic analysis of drought resistance in rice by molecular markers. *Crop Science*, 43, 1457- 1469.
https://www.soils.org/publications/cs/pdfs/43/4/1457

Babu, R.C., Pathan, M. S., Blum, A., & Nguyen, H.T. (1999). Comparison of measurement methods of osmotic adjustment in rice cultivars. *Crop Science*, 39, 150-158.
http://www.plantstress.com/Methods/OA_methods.pdf

Babu, R.C., Zhang, J., Blum, A., Ho, T.H.D., Wu, R., & Nguyen, H.T. (2004). HVA1, a LEA gene from barley confers dehydration tolerance in transgenic rice (*Oryza sativa* L.) via cell membrane protection. Plant Science 166, 855–862.

Bartels, D., & Sunkar, R. (2005). Drought and salt tolerance in plants. *Critical Reviews in Plant Science*, 24, 23–58. doi:10.1080/07352680590910410

Bates, B.C., Kundzewicz, Z.W., Wu, S., & Palutikof, J.P., Editors. (2008). In: *Climate change and water*. Technical Paper of the Inter-governmental Panel on Climate Change 65. Geneva (Switzerland): IPCC Secretariat. p 105 http://www.ipcc.ch/pdf/technical-papers/climate-change-water-en.pdf

Bernier, J., Kumar, A., Venuprasad, R., Spaner, D., & Atlin, G. (2007). A large effect QTL for grain yield under reproductive stage drought stress in upland rice. *Crop Science*, 47, 507-18. doi: 10.2135/cropsci2006.07.0495

Bernier, J., Kumar, A., Serraj, R., Spaner, D., & Atlin, G. (2008). Review: breeding upland rice for drought resistance. *Journal of Science Food and Agriculture*, 88, 927-39.
http://dx.doi.org/10.1002/jsfa.3153

Blum, A. (1996). Crop responses to drought and the interpretation of adaptation. *Plant Growth Regulation*, 20, 135-148 doi: 10.1007/BF00024010

Blum, A. (1998). Improving wheat grain filling under stress by stem reserve mobilization. Euphytica, 100, 77-83 doi: 10.1023/A:1018303922482

Blum, A. (2005). Drought resistance, water-use efficiency, and yield potential: are they compatible, dissonant, or mutually exclusive? Australian Journal of Agricultural Research, 56, 11, 1159-1168. http://dx.doi.org/10.1071/AR05069

Blumwald, E., Grover, A., & Good, A.G. (2004). Breeding for abiotic stress resistance: challenges and opportunities. In: New directions for a diverse planet. Proceeding of the 4th international Crop Science Congress, 26-Sep – 1 Oct 2004, Brisbane, Australia. http://cropscience.org.au/icsc2004/pdf/1953_blumwalde.pdf.

Bohnert, H.J., Gong, Q., Li, P., & Ma, S. (2006). Unraveling abiotic stress tolerance mechanisms – getting genomics going. Current Opinion in Plant Biology, 9, 180-188. doi:10.1016/j.pbi.2006.01.003

Boonjung, H., & Fukai, S. (1996). Effects of soil water deficit at different growth stages on rice growth and yield under upland conditions. 1. Growth during drought. Field Crops Research, 48, 1, 37-45. doi:10.1016/0378-4290(96)00039-1

Bouman, B.A.M., Peng, S., Castaoeda, A.R., & Visperas, R.M. (2005). Yield and water use of irrigated tropical rice system. Agricultural water Management, 74, 2, 87-105. doi:10.1016/j.agwat.2004.11.007

Borrell, A., Hammer, G., & Van Oosterom, E. (2001). Stay green; a consequence of the balance between supply and demand for nitrogen during grain filling. Annals of Applied Biology, 138, 91-95. doi: 10.1111/j.1744-7348.2001.tb00088.x

Botwright Acuna, T.L., Lafitte, H.R., & Wade, L.J. (2008). Genotype and environment interactions for grain yield of upland rice backcross lines in diverse hydrological environments. Field Crops Research, 108, 2, 117-125. doi:10.1016/j.fcr.2008.04.003

Bray, E., Bailey-Serres, J., Weretilnyk, E. (2000). Responses to abiotic stresses. In Biochemistry and Molecular Biology of Plants. (Edt.) Buchanan B, Gruissem W, Jones R Rockville: American Society of Plant Biologists, 1158-1203.

Cabuslay, G.S., Ito, O. & Alejar, A.A. (2002). Physiological evaluation of responses of rice (Oryza sativa L.) to water deficit. Plant Science, 163, 815-827. doi:10.1016/S0168-9452(02)00217-0

Campos, H., Cooper, M., Habben, J.E., Edmeades, G.O., & Schussler, J.R. (2004). Improving drought tolerance in maize: a view from industry. Field Crops Research, 90, 19-34. doi:10.1016/j.fcr.2004.07.003

Capell, T., Bassie, L., & Christou, P. 2004. Modulation of the polyamine biosynthetic pathway in transgenic rice confers tolerance to drought stress. PNAS 101, 9909-9914.

Cattivelli, L., Rizza, F., Badeck, F.W., Mazzucotelli, E., Mastrangelo, A.N., Francia, E., Marè, C., Tondelli, A., & Stanca, A.M. (2008). Drought tolerance improvement in crop plants: An integrated view from breeding to genomics. Field Crops Research, 105, 1-14. doi:10.1016/j.fcr.2007.07.004

Chaves, M.M., Maroco, J.P., & Pereira, J.S. (2003). Understanding plant responses to drought – from genes to whole plant. Functional Plant Biology, 30, 239-264. doi:10.1071/FP02076

Chen, T.H., & Murata, N. (2008). Glycinebetaine: an effective protectant against abiotic stress in plants. *Trends Plant Science* 13: 499-505. doi:10.1016/j.tplants.2008.06.007

Choi H.I., Hong, J.H., Ha, J., Kang, J.Y., & Kim, S.Y. (2000). ABFs, a family of ABA-responsive element binding factors. *Journal of Biological Chemistry*, 275, 1723-30. doi: 10.1074/jbc.275.3.1723, http://www.jbc.org/content/275/3/1723.full.pdf+html

Codon, A.G., Richards, R.A., Rebetzke, G.J., & Farquhar, G.D. (2004). Breeding for high water use efficiency. Journal of Experimental Botany, 55, 407, 2447-2460. doi: 10.1093/jxb/erh277, http://jxb.oxfordjournals.org/content/55/407/2447.full

Cooper, M., Fukai, S., & Wade, L. J. (1999). How can breeding contribute to more productive and sustainable rainfed lowland rice systems? *Field Crops Research*, 64, 199–209. doi:10.1016/S0378-4290(99)00060-X

Courtois, B., Shen, L., Petalcorin, W., Carandang, S., Mauleon, R., & Li, Z. (2003). Locating QTLs controlling constitutive root traits in the rice population IAC 165 x Co39. *Euphytica*, 134, 335-345. doi: 10.1023/B:EUPH.0000004987.88718.d6

Cushman, J.C., & Bohnert, H. J. (2000). Genomic approaches to plant stress tolerance. Current Opinion in Plant Biology, 3, 117-124. doi:10.1016/S1369-5266(99)00052-7

Dong, J., Chen, C., & Chen, Z. (2003). Expression profiles of the Arabidopsis WRKY gene super-family during plant defense response, *Plant Molecular Biology*, 51, 21–37. DOI: 10.1023/A:1020780022549

Deivanai, S., Sheela Devi, S., & Sharrmila Rengeswari, P. (2010). Physiochemical traits as potential indicators for determining drought tolerance during active tillering stage in rice (*Oryza sativa* L.). *Pertanika Journal of Tropical Agricultural Science*, 33, 1, 61 – 70. ISSN: 1511-3701

Fairhurt, T.H., & Dobermann, A. (2002). Rice in the global food supply. In: *better crops for international*. Special supplementary, 3-7. http://www.ipni.net/ppiweb/bcropint.nsf/$webindex/0E477FFC43BD62DA8525 6BDC00722F62/$file/BCI-RICEp03.pdf

Frank, W., Munnik, T., Kerkmann, K., Salamini, F., Bartels, D. (2000). Water deficit triggers phospholipase D activity in the resurrection plant Craterostigma plantagineum. *Plant Cell*, 12, 111-24. http://www.plantcell.org/content/12/1/111.full.pdf+html

Fukai, S., & Cooper, M. (1995). Development of drought-resistant cultivars using physio-morphological traits in rice. *Field Crops Research*, 40, 67-86. doi:10.1016/0378-4290(94)00096-U

Garg, A.K., Kim, J.K., Owens, T.G., Ranwala, A.P., Choi, Y.D., Kochian, L.V., & Wu, R.J. (2002). Trehalose accumulation in rice plants confers high tolerance levels to different abiotic stresses. *Proceedings of National Academy of Sciences*, U.S.A., 106, 21425-21430. doi: 10.1073/pnas.252637799, http://www.pnas.org/content/99/25/15898.full

Gleick, P.H., Wolff, E.L & Chalecki, R.R. (2002). *The new economy of water*: The risks and benefits of globalization and privatization of fresh water: Pacific Institute for studies in Development, environment and Security. Oakland CA., 48pp. ISBN No. 1-893790-07-X

Goff, S.A. (1999). Rice as a model for cereal genomics. *Current Opinion in Plant Biology*, 2, 86–89. doi:10.1016/S1369-5266(99)80018-1

Hong, Z.L., Lakkineni, K, Zhang, Z.M., & Verma, D.P.S. (2000). Removal of feedback inhibition of delta (1) pyrroline-5-carboxylate synthetase results in increased proline accumulation and protection of plants from osmotic stress. *Plant Physiology,* 122, 1129-1136. doi: 10.1104/pp.122.4.1129, http://www.plantphysiol.org/content/122/4/1129.full.pdf+html

Hsiao, T. C., & Bradford, K. J. (1982). Physiological consequences of cellular water deficits. In H. M.Taylor, W. R. Jordan, and T. R. Sinclair, (Eds). *Efficient water use in crop production.* Agron.Sac. Am., Madison, Wisconsin.

Huang, J., Hirji, R., Adam, L., Rozwadowski, K. L., Hammerlindl, J. K., Keller, W. A., & Selvaraj, G. (2000). Genetic engineering of glycinebetaine production toward enhancing stress tolerance in plants: Metabolic limitations. *Plant Physiology,* 122747-756. http://www.ncbi.nlm.nih.gov/pmc/articles/PMC58910/

Huang, Y., Xiao, B., Xiong, L. (2007). Characterization of a stress responsive proteinase inhibitor gene with positive effect in improving drought resistance in rice. *Planta ,* 226, 73–85. DOI: 10.1007/s00425-006-0469-8

Ishimaru, K., Ono, K., & Kashiwagi, T. (2004). Identification of new gene controlling plant height in rice using the candidate gene strategy. *Planta,* 218, 388-395. DOI: 10.1007/s00425-003-1119-z

Ito, Y., Katsura, K., Maruyama, K., Taji, T., Kobayashi, M., Seki, M., Shinozaki, K & Yamaguchi-Shinozaki, K. (2006). Functional analysis of rice DREB1/CBF-type transcription factors involved in cold responsive gene expression in transgenic rice. Plant and Cell Physiology. 47, 141-153. doi: 10.1093/pcp/pci230 http://pcp.oxfordjournals.org/content/47/1/141.full

Jang, I.C., Oh, S.J., Seo, J.S., Choi, W.B., Song, S.I., Kim, C.H., Kim, Y.S., Seo, H.S., Choi, Y.D., Nahm, B.H., Kim, J.K. (2003). Expression of a bifunctional fusion of the Escherichia coli genes for trehalose-6-phosphate synthase and trehalose-6- phosphatase in transgenic rice plants increases trehalose accumulation and abiotic stress tolerance without stunting growth. *Plant Physiology,* 131, 516– 524. doi: 10.1104/pp.007237

Jearakongman, S. (2005). Validation and discovery of quantitative trait loci for drought tolerance in backcross introgression lines in rice (*Oryza sativa* L.) cultivar IR64. *Ph.D Thesis.* Kasetsart University.

Jiang, G.H., He, Y.Q., Xu, C.G., Li, X.H., & Zhang, Q. (2004). The genetic basis of stay green in rice analyzed in a population of doubled haploid lines derived from an indica by japonica cross. *Theoritical and Applied Genetics,* 108, 688-698. DOI: 10.1007/s00122-003-1465-z

Jin, H., & Martin, C. (1999). Multifunctionality and diversity within the plant MYB-gene family. *Plant Molecular Biology,* 41, 577-585. DOI: 10.1023/A:1006319732410

Johnson, W.C., Jackson, L.E., Ochoa, O., Van Wijik, R., Peleman, J., St. Clair, D.A., Michelmore, R.W. (2000). Lettuce: A shallow rooted crop and its wild progenitor differ at loci determining root architecture and deep soil water extraction. Theoritical and Applied Genetics, 101, 1066-1073. DOI: 10.1007/s001220051581

Jones, H. (2004). What is water use efficiency? In: *water use efficiency in plant biology,* Edited by M. A. Bacon. Oxford, Blackwell.

Jongdee, B., Fukai, S., & Cooper, M. (1998). Genotypic variation for grain yield of rice under water-deficit conditions. In: Michalk DL, Pratley JE, editors. *Proceedings of the 9th Australian Agronomy Conference*, Wagga Wagga, 1998. p 403-406.
http://www.regional.org.au/au/asa/1998/4/185jongdee.htm

Kamoshita, A., Wade, L., Ali, M.L., Pathan, M.S., Zhang, J., Sarkarung, S., & Nguyen, H.T. (2002). Mapping of QTLs for root morphology of a rice population adapted to rainfed lowland conditions. *Theoritical and Applied Genetics*, 104, 880-893.
DOI: 10.1007/s00122-001-0837-5

Kamoshita, A., Rodriguez, R., Yama auchi, A., & Wade, L. (2004). Genotypic variation in response of rainfed lowland to prolonged drought and re-watering. Plant Production Science, 7, 4, 406-420.
http://www.jstage.jst.go.jp/article/pps/7/4/406/_pdf

Kato, Y., Abe, J., Kamoshita, A., & Yamagishi, J. (2006). Genotypic variation in root growth angle in rice (Oryza sativa L.) and its association with deep root development in upland fields with different water regimes. *Plant and Soil*, 287, 117–129.
http://www.springerlink.com/content/1713808t97578w3h/fulltext.pdf

Kikuchi, K., Ueguchi-Tanaka, M., Yoshida, K.T., Nagato, Y., Matsusoka, M., Hirano, H.Y. (2000). Molecular analysis of the NAC gene family in rice. *Molecular General Genetics*, 262, 1047-1051. DOI: 10.1007/PL00008647

Komoshita, A., Babu, R.C., Manikanda Boopathi, N., & Fukai. S. (2008). Phenotypic and genotypic analysis of drought resistance traits for development of rice cultivars adapted to rainfed environments. Field Crop Research, 109, 1-23.
doi:10.1016/j.fcr.2008.06.010

Kramer, P.J., & Boyer, J.S. (1995). Water relation of plants and soils. Academic Press, San Diego. http://dspace.udel.edu:8080/dspace/handle/19716/2830

Kumar, R., Malaiya, S., & Srivastava, M.N. (2004). Evaluation of morpho-physiological traits associated with drought tolerance in rice. Indian Journal of Plant Physiology 9, 305-307. ISSN 0019-5502

Kumar, R., Venuprasad, R., & Atlin, G.N. (2007). Genetic analysis of rainfed lowland rice drought tolerance under naturally-occurring stress in eastern India: Heritability and QTL effects. *Field Crops Research*, 103, 42-52. doi:10.1016/j.fcr.2007.04.013

Kuznetsov, V.V., & Shevyakova, N.I. (1999). Proline under stress; Biological role, metabolism and regulation. *Russian Journal of Plant Physiology*, 46, 274-287. ISSN 1021-4437

Lafitte, R., Blum, A., & Atlin, G. (2003). Using secondary traits to help identify drought tolerance genotypes. In: Fischer, K.S., Lafitte, R, Fukai, S. Atlin, G., Hardy, B (Eds.) *Breeding rice for drought prone environments*. International Rice Research Institute, Los Bunos. The Phillippines, pp. 37- 48.
http://books.irri.org/9712201899_content.pdf

Lafitte, H.R., Champoux, M.C., McLaren, G., & O'Toole, J.C. (2001). Rice root morphological traits are related to isozyme group and adaptation. Field *Crops Research*, 71, 57–70.
doi:10.1016/S0378-4290(01)00150-2

Lafitte, H.R, Price, A.H., & Courtois, B. (2004). Yield response to water deficit in an upland rice mapping population: associations among traits and genetic markers. *Theoritical and Applied Genetics*, 109, 1237- 46. DOI: 10.1007/s00122-004-1731-8

Lafitte, H.R., Yongsheng, G., Yan, S., & Li, Z.K. (2006). Whole plant responses, key processes, and adaptation to drought stress: the case of rice. Journal of Experimental Botany, 58, 2, 169-175. doi: 10.1093/jxb/erl101. http://jxb.oxfordjournals.org/content/58/2/169.full.pdf+html

Lanceras, J.C., Pantuwan, G., Jongdee, B., & Toojinda, T., (2004). Quantitative trait loci associated with drought tolerance at reproductive stage in rice. *Plant Physiology*, 135, 384–399. doi: 10.1104/pp.103.035527 http://www.plantphysiol.org/content/135/1/384.full.pdf+html

Lee, S.B., Kwon, H.B., Kwon, S.J., Park, S.C., Jeong, M.J., Han, S.E., Byun, M.O., Daniell, H. (2003). Accumulation of trehalose within transgenic chloroplasts confers drought tolerance. *Molecular Breeding* 11, 1–13.

Li, Z.K., Yu, S.B., Lafitte, H.R., Huamg, N., Courtois, B., Hittalmani, S., Vijayakumar C.H.M., Liu, G.F., Wang, G.C., Shasidhar, H.E., Zhuang, J.Y., Zheng, K.L., Singh, V.P., Siddu, J.S., Srivantaneeyakul, S., & Kush, G.S. (2003). QTL x environment interaction in rice.1. Heading date plant height. *Theoretical and Applied Genetics*, 108, 141–153. DOI 10.1007/s00122-003-1401- http://www.springerlink.com/content/6592e7manw59mvrl/fulltext.pdf

Lilley, J.M., & Fukai, S. (1994). Effect of timing and severity of water deficit on four diverse rice cultivars. I. Rooting pattern and soil water extraction. *Field Crops Research*, 37, 205–213. doi:10.1016/0378-4290(94)90099-X

Liu, J.X., Liao, D.Q., Oane, R., Estenor, L., Yang, X.E., Li, Z.C., & Bennett, J. (2006). Genetic variation in the sensitivity of anther dehiscence to drought stress in rice. *Field Crops Research*, 97, 87-100. doi:10.1016/j.fcr.2005.08.019

Ludwig, A., Romeis, T., & Jones, J. D. (2004). CDPK mediated signaling pathways: specificity and cross-talk. *Journal of Experimental Botany*, 55:181-8. doi: 10.1093/jxb/erh008, http://jxb.oxfordjournals.org/content/55/395/181.full.pdf+html

Luu, D.T., & Maurel, C. (2005). Aquaporins in challenging environment: Molecular gears for adjusting plant water status. *Plant, Cell and Environmental*, 28, 85-96. DOI: 10.1111/j.1365-3040.2004.01295.x http://onlinelibrary.wiley.com/doi/10.1111/j.1365-3040.2004.01295.x/pdf

Martinez-Garcia, J.F., Moyano, E., Alcocer, M.J.C., & Martin, C. (1998). Two bZIP proteins from *Antirrhinum* flowers preferentially bind a hybrid C-box/G-box motif and help to define a new sub-family of bZIP transcription factors. *Plant Journal*, 13, 489-505. DOI: 10.1046/j.1365-313X.1998.00050.x http://onlinelibrary.wiley.com/doi/10.1046/j.1365-313X.1998.00050.x/pdf

Mohanty, C., Kathuria, H., Ferjani, A, Sakamoto, A., Mohanty, P., Marata, N., & Tyagi, A.K. (2003). Trangenics of an elite indica variety *Pusa Basmathi* 1 harboring the coda gene are highly tolerant to salt stress. *Theoretical and Applied Genetics*, 106. 51-57. DOI: 10.1007/s00122-002-1063-5

Munns, R. (1988). Why measure osmotic adjustment? *Australian Journal of Plant Physiology*, 15, 717-726. doi:10.1071/PP9880717

Munns, R. (1993). Physiological processes limiting plant growth in saline soils: some dogmas and hypotheses, *Plant Cell Environment*, 16, 15-24. DOI: 10.1111/j.1365-3040.1993.tb00840.x,

Nakayama, H., Yosida, K., Ono, H., Murooka, Y., & Shinmyo, A. (2000). Ectoine, the compatible solute of Halomonas elongata confers hyper osmotic tolerance in cultured tobacco cells. Plant physiology, 122, 1239-1247. doi:10.1104/pp.122.4. http://www.plantphysiol.org/content/122/4/1239.full.pdf+html

Nelson, D.E., Repetti, P.P., Adams, T.R., Creelman, R.A., Wu, J., Warner, D.C., Anstrom, D.C., Bensen, R.J., Castiglioni, P.O., Donnarummo, M.G., Hinchey, B.S., Kumimoto, R.W., Maszle, D.R., Canales, R.D., Krolokowsko, K.A., Dotson, S.B., Gutterson, N., ratcliffe, O.J., & Heard J.E. (2007). Plant nuclear factor Y (NF-Y) B subunits confer drought tolerance and lead to improved corn yields on water-limited acres. *Proceedings of National Academy of Sciences*. USA. 104, 16450-16455. doi:10.1073/pnas.0707193104,
http://www.pnas.org/content/104/42/16450.full.pdf+html

Nhan, D.Q., Thaw, S., Matsuo, N., Xuan, T.D., Hong, N.H., & Mochizuki, T. (2006). Evaluation of root penetration ability in rice using the wax-layers and the soil cake methods. *Journal of the Faculty of Agriculture*, Kyushu University, 51, 251-256. ISSN: 0023-6152

Nobel, P.S. (2005). Physiochemical and environmental plant physiology (3rd Ed.). Elsevier Amsterdam. ISBN: 978-0-12-520026-4

Nguyen, H.T., Babu, R.C., & Blum, A. (1997). Breeding for drought resistance in rice: Physiological and molecular genetics considerations. *Crop Sciences*, 37, 1426-1434. ISSN 0011-183X

Nguyen, T.T., Klueva, N., Chamareck, V., Aarti, A., Magpantay, G., Millena, A.C.M., Pathan, M.S., Nguyen, H.T. (2004). Saturation mapping of QTL region and identification of putative candidate genes for drought tolerance in rice. *Molecular Genetics Genomics*, 272, 35-46. DOI: 10.1007/s00438-004-1025-5

Oh, S.J., Song, S.I., Kim, Y.S., Jang, H.J., Kim, M., & Kim YK. (2005). Arabidopsis CBF3/DREB1A and ABF3 in transgenic rice increased tolerance to abiotic stress without stunting growth. Plant Physiology, 138, 341–351.
doi: 10.1104/pp.104.059147,
http://www.plantphysiol.org/content/138/1/341.full.pdf+html

O'Toole, J. C. & Cruz, R. T. (1980). Response of leaf water potential, stomatal resistance, and leaf rolling to water-stress. *Plant Physiology*, 65, 428-432. doi: 10.1104/pp.65.3.428, http://www.plantphysiol.org/content/65/3/428.short

O'Toole, J.C. (2004). Rice and water: the final frontier. Paper presented at the First International Conference on *Rice for the Future*, 31 August-2 September 2004, Bangkok, Thailand.

Pantuwan, G. (2000). Yield responses of rice (Oryza sativa L.) genotypes to water deficit in rainfed lowlands. *Ph.D Thesis*, University of Queensland.

Pantuwan, G., Fukau, S., Cooper, M., Rajatasereekul, S., & O'Toole, J.C. (2002a) Yield responses of rice (Oryza sativa L.) genotypes to different types of drought under rainfed lowlands, Part-1. Grain yield and yield components. Field Crops Research, 73, 153-168. doi:10.1016/S0378-4290(01)00187-3

Pantuwan, G., Fukai, S., Cooper, M., Rajatasereekul, S., & O'Toole, J.C. (2002b). Yield response of rice (*Oryza sativa* L.) genotypes to different types of drought under rainfed lowlands. 2. Selection of drought resistance genotypes. *Field Crops Research*. 73, 169-180. doi:10.1016/S0378-4290(01)00195-2

Pantuwan, G., Ingram, K.T., Sharma, P.K. (1996). Rice root system development under rainfed conditions. *Proceedings of the Thematic Conference on Stress Physiology, Rainfed Lowland Rice Research Consortium*, Lucknow, India, February 28–March 5, 1994. International Rice Research Centre, Manila, Philippines, pp 198–206. http://books.irri.org/8186789006_content.pdf

Parker, R., Flowers, T.J., Moore, A.L. & Harpham, N.V.J. (2006). An accurate and reproducible method for proteome profiling of the effects of salt stress in the rice leaf lamina. *Journal of Experimental Botany*. 57, 5, 1109-1118.

Parry, M.A.J., Flexas, J., & Medrano, H. (2005). Prospects for crop production under drought: research priorities and future direction. *Annals of Applied Biology*, 147, 211-226. http://www.uib.es/depart/dba/plantphysiology/Grup%20de%20recerca/Papers/2005/Parry05.pdf

Peck, S.C. (2005). Update on Proteomics in Arabidopsis. Where Do We Go From Here? *Plant Physiology* 138, 591-599. doi: 10.1104/pp.105.060285, http://www.plantphysiol.org/content/138/2/591.full.pdf+html

Pereira, J.S., Chaves, M.M., Caldeira, M.C., & Correia, A.V. (2006). Water availability and productivity. In: Morison. J, Croft. M, (Eds.) *Plant growth and climate change*. Blackwell publishers, London. Pp 118-145. DOI: 10.1002/9780470988695.ch6

Price, A.H., Steele, K.A., Moore, B.J., & Jones, R.G.W. (2002b). Upland rice grown in soil filled chambers and exposed to contrasting water deficit regimes II. Mapping QTLs for root morphology and distribution. *Field Crops Research*, 76, 25-43. doi:10.1016/S0378-4290(02)00010-2

Price, A.H., Cairns, J.E., Horton, P., Jones, H.G., & Griffiths, H. (2002c). Linking drought resistance mechanisms to drought avoidance in upland rice using a QTL approach: progress and new opportunities to integrate stomatal and mesophyll responses. *Journal of Experimental Botany*, 53, 989-1004. doi: 10.1093/jexbot/53.371.989 http://jxb.oxfordjournals.org/content/53/371/989.full.pdf+html

Quan, R., Shang, M., Zhang, H., Zhao, Y., & Zhang, J. (2004). Engineering of enhanced glycine betaine synthesis improves drought tolerance in maize. *Plant Biotechnology Journal*, 2, 477-486. doi:10.1111/j.1467-7652.2004.00093.x.

Rabbani, M.A., Maruyama, K., Abe, H., Khan, M.A., Katsura, K., Ito, Y., Yoshiwara, K., Seki, M., Shinozaki, K., & Yamaguchi-Shinozaki, K. (2003) Monitoring expression profiles of rice genes under cold, drought, and high-salinity stresses and abscisic acid application using cDNA microarray and RNA get-blot analyses. Plant Physiology, 133: 1755-1767. doi/10.1104/pp.103.025742

Rampino, P., Spano, G., pataleo, S., Mita, G., Napier, J.A., Difonzo, N., Shrewry, P. R., & Perrotta, C. (2006). Molecular analysis of durum wheat "Stay green mutant"; expression pattern of photosynthesis related genes. *Journal of Cereal Science*, 43, 160-168. doi:10.1016/j.jcs.2005.07.004

Reynolds, M.P., Mujeeb-Kazi, A., Sawkimns, M. (2005a). Prospects for utilizing plant adaptive mechanisms to improve wheat and other crops in drought and salinity prone environments. *Annals of Applied Biology*, 146, 155 -162. ISSN *0003-4746*

Riechmann, J.L., & Meyerowitz, E.M. (1998). The AP2/EREBP family of plant transcription factors. Biological Chemistry. 379, 633–646. DOI: 10.1515/bchm.1998.379.6.633

Richards, R.A. (2000). Selectable traits to increase crop photosynthesis and yield of grain crops. *Journal of Experimental Botany*, 51, 447-458.
doi: 10.1093/jexbot/51.suppl_1.447,
http://jxb.oxfordjournals.org/content/51/suppl_1/447.full.pdf+html

Richards, R.A., Rebetzke, G.J., Condon, A.G., & Van Herwaarden, A.F. (2002). Breeding opportunities for increasing efficiency of water use and crop yield in temperature cereals. *Crop Science*, 42, 111-121. doi: 10.2135/cropsci2002.1110

Rizza, F., Badeck, F.W., Cattivelli, L., Destri, O., Di Fonzo, N., Stanca, A.M., (2004). Use of water stress index to identify barley genotypes adapted to rainfed and irrigated conditions. *Crop Science*, 44, 2127-2137.
https://www.crops.org/publications/cs/pdfs/44/6/2127.

Robin, S., Pathan, M.S., Courtois, B., Lafitte, R., Carandang, S., Lanceras, S., Amante, M., Nguyen, H.T., & Li, Z. (2003). Mapping osmotic adjustment in an advanced back cross inbred population of rice. *Theoritical and Applied Genetics*, 107, 1288-1296. DOI: 10.1007/s00122-003-1360-7

Rontein, D., Basset, G., & Hanson, A.D. (2002). Metabolic engineering of osmoprotectant accumulation in plants. *Metabolic Engineering*, 49-56. doi:10.1006/mben.2001.0208

Salekdeh, G. H., Siopongco, J., Wade, L.J., Ghareyazie, B., & Bennett, J. (2002). A proteomic approach to analyzing drought- and salt-responsiveness in rice. Field Crops Research. 76, 2-3, 199-219. doi:10.1016/S0378-4290(02)00040-0

Scheiermeier Q. 2008. A long dry summer. *Nature* 452, 270-273. ISSN:0028-0836

Seki, M., Kamei, A., Yamaguchi-Shinozaki, K., & Shinozaki K. (2003). Molecular responses to drought, salinity and frost: common and different paths for plant protection. *Current Opinion in Biotechnology*, 14, 194-199. doi:10.1016/S0958-1669(03)00030-2

Seki, M., Narusaka, M., Abe, H., Kasuga, M., Yamaguchi-Shinozaki, K., Carninci, P., Hayashizaki, Y., & Shinozaki, K. (2001) Monitoring the expression patterns of 1300 Arabidopsis genes under drought and cold stresses by using a full-length cDNA microarray. *Plant Cell*, 13, 61-72. doi: 10.1105/tpc.13.1.61,
http://www.plantcell.org/content/13/1/61.full.pdf+html

Serraj, R., & Sinclair, T.R. (2002). Osmolyte accumulation: can it really help to increase a crop yield under drought conditions? *Plant, cell and Environment*, 25, 333-341.

Serraj, R. (2005). Genetic and management options to enhance drought resistance and water use efficiency in dryland agriculture. Selected talk. Interdrought-II: coping with drought. The 2nd International Conference on *Integrated Approaches to Sustain and*

Improve Plant Production under Drought Stress. Rome, Italy, 24-28 September 2005. University of Rome La Sapienza, Rome, Italy. Online at www.plantstress.com

Shen. L., Courtois, B., McNally, K.L., Robin, S., & Li, Z. (2001). Evaluation of near isogenic lines of rice introgressed with QTLs for root depth through marker aided selection. *Theoretical Applied Genetics,* 103, 75-83 DOI: 10.1007/s001220100538

Sinha, S. K., & Khanna, R. (1975). Physiological, biochemical and genetic basis of heterosis. *Advances in Agronomy,* 27, 123-170.

Shinozaki, K., & Yamaguchi-Shinozaki, K. (2000). Molecular responses to dehydration and low temperature differences and cross-talk between two stress signaling pathways. *Current Opinion in Plant Biology,* 3, 217-223. doi:10.1016/S1369-5266(00)80068-0

Shinozaki, K., Yamaguchi-Shinozaki, K., & Seki, M. (2003). Regulatory network of gene expression in the drought and cold stress responses. *Current Opinion in Biotechnology,* 6, 410-417. doi:10.1016/S1369-5266(03)00092-X

Sibounheuang, V., Basnayake, J., Fukai, S., & Cooper, M. (2001). Leaf water potential as drought resistance character in rice. In: Fukai S, Basnayake J. (Eds.), ACIAR proceedings 101: *Increased lowland rice production in the Mekong Region.* Australian Centre for International Agricultural research, Canberra. Pp. 86-95. ISBN 1-863-20-311-7

Slafer, G.A. (2003). Genetic basis of yield as viewed from crop physiologist's perspective. *Annals of Applied Biology,* 142, 117-128. DOI: 10.1111/j.1744-7348.2003.tb00237.x

Steele, K. (2009). Novel upland rice variety bred using marker assisted selection and client oriented breeding released in Jharkhand. India. Bangor University. http://www.cazs.bangor.ac.uk/ccstudio/WhatsNew/cazsWhatsNew3.php?ID=14

Steel, K.A., Price, A.H., Shashidhar, H.E., & Witcombe, J.R. (2006). Marker assisted selection to introgress rice QTLs controlling root traits into an Indian upland rice variety. *Theoretical and Applied Genetics,* 112, 208 -221. DOI: 10.1007/s00122-005-0110-4

Stoop, J.M.H., Williamson, J.D., & Pharr, D.M. (1996). Manitol metabolism in plants: A method for coping with stress. *Trends in Plant Science,* 1139 -1144. doi:10.1016/S1360-1385(96)80048-3

Su, J., Hirji, R., Zhang, L., He, C., Selvaraj, G., & Wu, R. (2006). Evaluation of the stress inducible production of choline oxidase in transgenic rice as a strategy for producing the stress protectant glycinebetaine. *Journal of Experimental Botany,* 57, 1129-1135. doi: 10.1093/jxb/erj133, http://jxb.oxfordjournals.org/content/57/5/1129.full.pdf+html

Talame, V., Sanguineti, M.C., Chiapparino, E., Bahri, H., Ben Salem, M., Forster, B.P., Ellis, R.P., Rhouma, S., Zoumarou, W., Waugh R., & Tuberosa, R., (2004). Identification of *Hordeum spontaneum* QTL alleles improving field performance of barley grown under rainfed conditions. *Annals of Applied Botany,* 144, 309-20. DOI: 10.1111/j.1744-7348.2004.tb00346.x

Tardieu, F. & Tuberosa, R. (2010). Dissection and modeling of abiotic tolerance plants. *Current opinion in Plant Biology,* 13, 206-212. doi:10.1016/j.pbi.2009.12.012

Timeperio, A.M., Egidi, M.G., & Zolla, L. (2008). Proteomics applied on plant abiotic stresses: Role of heat shock proteins (HSP). *Proteomics.* 71, 391-411. doi:10.1016/j.jprot.2008.07.005

Tuberosa, R., & Salvi, S. (2006). Genomics-based approaches to improve drought tolerance of crops. *Trends in Plant Science,* 11, 8, 405-412. doi:10.1016/j.tplants.2006.06.003

Tuberosa, R., Salvi, S., Sanguineti, M.C., Maccaferri, M., Giuliani, S., & Landi, P. (2003). Searching for quantitative trait loci controlling root traits in maize: a critical appraisal. *Plant and Soil,* 255, 35-54. DOI: 10.1023/A:1026146615248

Turner, N. C. (1974). Stomatal behavior and water status of maize, sorghum, and tobacco under field conditions. *Plant Physiology,* 53, 360-365. doi: 10.1104/pp.53.3.360, http://www.plantphysiol.org/content/53/3/360.short

Turner, N.C. (1982). The role of shoot characteristics in drought resistance of crop plants. In: *Drought resistance in crops with emphasis on rice,* IRRI, Los Banos, Philippines, Pp 115-134. http://books.irri.org/getpdf.htm?book=9711040786

Turner, N.C. 2004. Agronomic options for improving rainfall –use efficiency of crops in dryland farming systems. *Journal of Experimental Botany,* 55, 2413-2425. doi: 10.1093/jxb/erh154, http://jxb.oxfordjournals.org/content/55/407/2413.full.pdf+html

Turner, N.C., O'Toole, JC., Cruz, RT., Yambao, EB., Ahmad, S., Namuco, OS., & Dingkuhn, M. (1986). Responses of seven diverse rice cultivars to water deficit. II. Osmotic adjustment, leaf capacity, leaf retention, leaf death, stomatal conductance and photosynthesis. *Field crops Research,* 13, 273- 286. doi:10.1016/0378-4290(86)90028-6

Umezawa, T., Fujita, M., Fujita, Y., Yamaguchi-Shinozaki, K., & Shinozaki, K. (2006). Engineering drought tolerance in plants: discovering and tailoring genes to unlock the future. *Current Opinion in Biotechnology,* 17, 113–122. doi:10.1016/j.copbio.2006.02.002

Venuprasad, R., Shashidhar, H.E., Hittalmani, S., & Hemamalini, G.S. (2002). Tagging quantitative trait loci associated with grain yield and root morphological traits in rice (Oryza sativa L.) under contrasting moisture regimes. Euphytica, 128, 293-300. DOI: 10.1023/A:1021281428957

Vinocur, B., & Altmam, A. (2005). Recent advances in engineering plant tolerance to abiotic stress: achievement and limitations. *Current opinion in Biotechnology,* 16, 123-132. doi:10.1016/j.copbio.2005.02.001

Wade, L.J., McLaren, C.G., Samson, B.K., Regmi, K.R., & Sarkarung, S. (1996). The importance of site characterization for understanding genotype by environment interactions. *In:* M Cooper, GL Hammer, (Eds.) *Plant Adaptation and Crop Improvement,* CAB International, Wallingford, UK, pp 549–562.

Wang, W.X., Vinocur, B., Shoseyov, O., & Altman, A. (2001). Biotechnology of plant osmotic stress tolerance: physiological and molecular considerations. Acta Horticulturae, 560, 285-292. http://www.actahort.org/books/560/560_54.htm

Wu, X., Shroto, Y., Kishitani, S., Ito, Y, & Toriyama, K. (2009). Enhanced heat and drought tolerance in transgenic rice seedlings overexpressing OsWRKY11 under the control of HSP101 promoter. *Plant Cell Reporter,* 28, 21-30. DOI: 10.1007/s00299-008-0614-x

Xiao, B.Z., Chen, X., Xiang, C.B., Tang, N., Zhang, Q.F., & Xiong, L.Z. (2009). Evaluation of seven function known candidate genes for their effects on improving drought resistance of transgenic rice under field conditions. *Molecular Plant*, 2, 73-83. doi: 10.1093/mp/ssn068,
http://mplant.oxfordjournals.org/content/2/1/73.full.pdf+html

Xiao, B., Huang, Y., Tang, N., & Xiong, L. (2007). Overexpression of LEA gene in rice improves drought resistance under the field conditions. *Theoretical and Applied Genetics*, 115, 35-46. DOI: 10.1007/s00122-007-0538-9

Xu, D.O., Huang, J., Guo, S.Q., Yang, X., Bao, Y.M., Tang, H.J., & Zhang, H.S. (2008). Overexpression of a TFIIIA-type zinc finger protein gene ZFP252 enhances drought and salt tolerance in rice (*Oryza sativa* L.) *FEBS Letter.*, 582, 1037-1043. doi:10.1016/j.febslet.2008.02.052

Yadav, R., Courtois, B., Huang, N., & McLaren, G. (1997). Mapping genes controlling root morphology and root distribution in a double haploid population of rice. *Theoretical and Applied Genetics*, 619-32 DOI: 10.1007/s001220050459

Yamaguchi-Shinozaki, K., & Shinozaki, K. (2006). Transcriptional regulatory networks in cellular responses and tolerance to dehydration and cold stresses. *Annual Review of Plant Biology*, 57, 781-803 DOI: 10.1146/annurev.arplant.57.032905.105444

Yamaguchi-Shinozaki, K., & Shinozaki K. (2005). Organization of *cis* acting regulatory elements in osmotic- and cold-stress- responsive promoters. *Trends in Plant Science*, 10, 88–94. doi:10.1016/j.tplants.2004.12.012

Yan Ying, Q.U., Ping, M.U., Xue Qin, L., Yi Xin, T., Feng, W., & Hong Liang, Z. (2008). QTL mapping and correlation between leaf water potential and drought resistance in rice under upland and lowland environments. *Acta Agronomic Sinica*, 34, 2, 198-206. doi:10.1016/S1875-2780(08)60008-5

Yang, S., Vanderbeld, B., Wan, J., & Huang, Y. 2010. Narrowing down the targets: towards successful genetic engineering of drought tolerant crops. *Molecular Plant.* 3, 3, 469-490. doi: 10.1093/mp/ssq016,
http://mplant.oxfordjournals.org/content/3/3/469.full.pdf+html

Yoshida, S., & Hasegawa, S. (1982). The rice root system: its development and function. In: O'toole, J.C. (Ed), *Drought Resistance in Crops with the Emphasis on Rice*. International Rice Research Institute, Los Banos, Philippines, pp. 97-114.
http://hdl.handle.net/123456789/297

Yue, B., Xiong, L., Xue, W., Xing, Y., Luo, L., & Xu, C. (2005). Genetic analysis for drought resistance of rice at reproductive stage in field with different types of soil. *Theoretical and Applied Genetics*, 111, 1127-1136. DOI 10.1007/s00122-005-0040-1,
http://www.ncpgr.cn/papers/Yueb_TAG2005.pdf

Zhang, J., Zheng, H.G., Aarti, A., Pantuwan, G., Nguyen, T.T., Tripathy, J.N., Sarial, A.K., Robin, S., Babu, R.C., Nguyen, B.D., Sarkarung, S., Blum, A., & Nguyen, H.T. (2001). Locating genomic regions associated with components of drought resistance in rice: comparative mapping within and across species. *Theoritical and Applied Genetics*, 103, 19-20. DOI: 10.1007/s001220000534

Zhang, J.Z., Creelman, R.A., & Zhu, J.K. (2004). From laboratory to field. Using information from Arabidopsis to engineer salt, cold, and drought tolerance in crops. *Plant Physiology*, 135, 615–621. doi: 10.1104/pp.104.040295, http://www.plantphysiol.org/content/135/2/615.full

Zheng, H.G., Babu, R.C., Pathan, M.S., Ali, L., Huang, N., Courtois, B., & Nguyen, H.T. (2000). Quantitative trait loci for root penetration ability and root thickness in rice: Comparison of genetic backgrounds. *Genome*, 43, 53-61.

Zheng, B.S., Yang, L., Zhang, W.P., Mao, C.Z., Wu, Y.R., Yi, K.K., Liu, F.Y., & Wu, P.(2003). Mapping QTLs and candidate genes for rice root traits under different water supply conditions and comparative analysis across three populations. *Theoretical and Applied Genetics*, 107, 1505-1515. DOI: 10.1007/s00122-003-1390-1

Zhu, B., Su, J., Chang, M., Verma, D.P.S., Fan, Y.L., & Wu, R. (1998). Overexpression of a D1-pyrroline-5-carboxylate synthetase gene and analysis of tolerance to water and salt stress in transgenic rice. *Plant Science*, 139, 41-48. doi:10.1016/S0168-9452(98)00175-7

Zhu, J.K. (2001) Cell signaling under salt, water and cold stresses. *Current Opinion in Plant Biology*, 4, 401-6. doi:10.1016/S1369-5266(00)00192-8

7

Lowland Soils for Rice Cultivation in Ghana

M. M. Buri[1], R. N. Issaka[1], J. K. Senayah[1], H. Fujii[2] and T. Wakatsuki[3]

[1]CSIR - Soil Research Institute, Academy Post Office, Kwadaso- Kumasi, Ghana
[2]Japan International Research Center for Agricultural Sciences (JIRCAS),
[3]Faculty of Agriculture, Kinki University, Nara
Japan

1. Introduction

In Ghana lowlands mostly comprising floodplains and inland valleys occur throughout the whole country. These lowlands have been characterized to be heterogeneous in morphology, soil type, vegetation and hydrology. Lowland soils therefore occur across all the agro-ecological zones of the country. These agro-ecological zones include the drier Savannahs (Sudan, Guinea and Coastal) which cover the northern and coastal parts and the Forest which covers the western, central and eastern corridors of the country. According to estimates, the country has over one million hectors of lowlands (Wakatsuki et al, 2004a, b) which can be developed for effective and sustainable rice cultivation, beside the pockets of areas used for dry season vegetable production.

In the Savannah agro-ecological zones rice cultivation is common due to the presence of abundant poorly drained lowlands. These soils are deep sandy loams to clay loams and abundant water resources from the major rivers, comprising the Volta, Oti, Nasia, Daka, Kulda and their numerous tributaries that could be tapped for irrigation. The soils are developed over shale / mudstone, sandstone, granites and phyllites (Adu, 1995a, 1995b, 1969; Dedzoe et al, 2001a & b). They include *Kupela* (Vertisol) and *Brenyase* (Fluvisol) series occurring over granite, *Pale* series (Gleysol) occurring over Birimian schist, *Lapliki*, *Siare* and *Pani* series found on floodplains of major rivers and *Lima* and *Volta* series developed over shale and mudstone. The most commonly used soils for rain-fed rice production in the Savannah agro-ecological zone include the inland valley soils occurring within the shale / mudstone areas of the Voltaian Basin. Under irrigation, soils found on the old levees of the Volta and Nasia rivers, *Lapliki* series, are highly suitable.

The Savannah agro-ecological zones experience the lowest rainfall (< 1000mm) amounts for the year. There is only one growing season, commonly referred too as the rainy season. The rainy season starts when the rains commence from May/June and ends in September/October. A long dry period is experienced from November to April. Temperatures are normally high and quite uniform throughout the year. Average monthly temperatures range from 25° to 35° C.

The Forest agro-ecological zones on the other hand, experience relatively higher rainfall amounts (> 15000mm per annum) with a bimodal pattern. The major season rains occur between March and mid-July with a peak in May/June. There is a short dry spell from mid-July to mid-August. The minor rainy season starts from mid-August to about the end of

October with a peak in September. A long dry period is experienced from November to February with possibilities of occasional rains. The rains appear to be nearly well distributed throughout the year, with amounts considered adequate for crop production occurring in the two peaks. Temperatures are normally high throughout the year with very little variations. The mean monthly temperatures range from 25° C in July/August to 29° C in March/April.

Agriculture is the dominant land use for these lowlands. Across the country, the lowlands are commonly cultivated to rice and sometimes to vegetables. Within the forest ecology, maize, cocoa, oil palm, plantain and citrus are also cultivated in these areas, while maize, sorghum and millet are also cultivated within these lowlands in the Savannah agro-ecological zones. With increasing intensity in use, there is the need for the development of technologies that will ensure the sustainable use of these lowlands, particularly for rice production. This chapter is therefore describing the nature of lowland soils in Ghana, their suitability for rice production and possible effective measures that need to be put in place for their sustainable use. Possible areas for further research are also provided.

2. Soil types

2.1 Soils of Savannah agro-ecological zones
The most extensive lowland soils in the Savanna agro-ecological zones are *Lima* and *Volta* series, which originate from shale and mudstone. *Lima* series is the most extensive, followed by *Volta* and *Changnalili* series respectively (Dedzoe et al, 2001a & b; Senayah et al, 2001). These soils occupy a generally flat (0-3% slope), broad and very extensive lowland plains that are generally suitable for mechanization. In a catena, the fringes of the valley adjoining the upland is occupied by *Changnalili* series (Stagnic Plinthosol), followed by *Lima series* (*Endogleyi–Ferric Planosol*) and with the *Volta* series (*Dystric or Eutric Gleysol*) occurring closest to the stream bed. A general description (Senayah et al,) of some of these soil series are as given below.

2.1.1 Lima series (Endogleyi – Ferric Planosol)
The profile description of *Lima series* is presented in Table 1. *Lima* soils are deep (>140 cm) and imperfectly to poorly drained. At the peak of the wet season, they are flooded intermittently, depending on the duration or breaks in the rainfall. Topsoil textures are loam, silt loam or sandy loam and the underlying subsoil textures range from sandy clay loams to clays.

2.1.2 Volta series (Dystric or Eutric Gleysol)
Table 2 shows a typical profile description of the *Volta series*. *Volta soils* are also very deep (>150 cm) and poorly drained. They occur close to streams or in depressions as compared to Lima. Textures are heavier and they are flooded for much longer periods than Lima soils. Topsoil textures are mainly silt loams and silt clay loams and in the underlying subsoil, silt clays and clays. The limitation of this soil is the difficulty of working when it is wet or dry. When wet, it is very sticky and easily gets stuck to implements and very hard when dry.

2.1.3 Lapliki series (Abrupti-Stagnic Lixisol)
Lapliki series is developed from mixed alluvial deposits and occur above flood plains. Unlike the *Volta* and *Lima* series, it is seldom flooded. It could be used under irrigation. The soil is

moderately well to imperfectly drained and occurs on middle to lower slopes. *Lapliki* series has a topsoil of grayish brown to light grey sandy loam, which is usually less than 30 cm thick. This grades into brownish yellow compact sandy clay loam below 30 cm and in turn overlies several meters of yellowish brown, mottled red sandy clay loam or sandy clay.

Horizon	Depth (cm)	Description
Ap	0- 23	Dark grayish brown (10YR 4.5/2) to light brown (7.5YR 6/3); few brownish yellow mottles; sandy loam; weak fine and medium granular; loose
Eg	23 - 36	blocky; friable. yellowish brown (10YR 5/4); many (15%) distinct brownish yellow mottles; sandy clay loam; weak fine and medium sub-angular
Btg c	36 – 68	Light yellowish brown (10YR 6/4); many (20%) distinct brownish yellow and yellowish red mottles; clay loam to clay; weak to moderate fine, medium and coarse sub-angular blocky; slightly firm to firm; many (20%) iron and manganese dioxide concretions.
Btg	68 – 140	Light brownish grey (10YR 6/2); common (10%) distinct dark red mottles; clay; strong medium prismatic; very firm; common (10%) iron nodules

Table 1. Profile description of a typical *Lima* soil series (FAO/WRB: *Endogleyi-Ferric Planosol*)

Fig. 1.a. Typical profile of *Lima* soil series (FAO/WRB: *Endogleyi-Ferric Planosol*)

Horizon	Depth (cm)	Description
Apg	0 – 35	Dark grayish brown (10YR 4/2); few distinct brownish yellow mottles; silt clay loam; moderate fine and medium sub-angular blocky; friable to slightly firm; very few (<3%) iron concretions
Bwg 1	35 – 60	Grayish brown (10YR 5/2); common distinct brown mottles; silt clay; moderate fine and medium sub-angular blocky; slightly firm; few (3%) hard iron and manganese dioxide concretions
Bwg2	60 – 116	Grayish brown (10YR 5/2; common (10%) distinct dark red and yellowish brown mottles; clay; moderate fine and medium sub-angular blocky firm; common (10%) iron and manganese dioxide concretions

Table 2. Profile description of a typical *Volta* soil series (FAO/WRB: *Eutric Gleysol*)

Fig. 2.a. Typical profile of *Volta* soil series (FAO/WRB: *Eutric Gleysol*)

2.2 Soil of forest agro-ecological zones

Soils within the forest agro-ecology, on the other hand, are developed from the Lower Birimian rocks. The soils fall under the Akumadan – Bekwai / Oda Complex and Bekwai – Zongo / Oda Complex associations (Adu, 1992). The lower slope is occupied by imperfectly drained gravel–free yellow brown silty clay loams, *Kokofu series,* which are developed from colluvium from upslope, *Temang series* which are gray, poorly drained alluvial loamy sands and *Oda series* which are clays occupy the valley bottoms. Most of the major streams are flanked by low, almost flat (0 – 2%) alluvial terrace consisting of deep, yellowish brown, moderately well to imperfectly drained silty clay loams, *Kakum series.* The Bekwai – Zongo / Oda complex association also consists of Bekwai, Nzima, Kokofu, Temang, Oda and Kakum series but in addition has a large tract of seepage iron pan soils called *Zongo series.* *Zongo* soils consist of sandy loam topsoil overlying yellow brown, imperfectly drained, clay loams containing ironstone concretions and iron pan boulders in the subsoil. On the other hand, *Nzima* soils are characterized by a high content of stones and gravel in some places resulting from the break up during weathering of veins and stringers of quartz injected into the phyllite. This results in the formation of the *Mim series.* A brief but general description of these soils includes:

2.2.1 Mim series (Ferric Acrisol)

Mim series is a moderately well to well drained soil found on middle slopes of 5-8%. The topsoil is dark reddish brown sandy loam. This overlies many to abundant (40-80%) quartz and stones in a reddish brown clay loam soil. This soil differs from the *Nzima series* by the higher gravel and stone content in the subsoil. Its effective depth is determined by the amount of quartz gravel and stones, where it becomes so abundant that there is only little soil material which varies between 30 and 60cm depth.

2.2.2 Zongo series (Plinthosol)

Zongo series is a moderately well to imperfectly drained soil, found on middle to lower slopes. The topsoil is dark grey sandy loam. The underlying subsoil is pale brown sandy clay loam containing ironstone gravel from 40cm, which increases with depth from many (15-40%) to abundant (40-80%).

2.2.3 Kakum series (Gleyic Lixisol)

Kakum soils are very deep (> 150cm), imperfectly to moderately well drained, occurring on the slightly raised old alluvial flats along the banks of major rivers/streams. The profile consists of dark brown, weak granular friable sandy loam at the topsoil. The subsoil is yellowish brown and faintly mottled, strong brown friable clay loam and a structure that is weak to moderate fine and medium sub-angular blocky granular. Below 100cm, the mottles become prominently reddish yellow.

2.2.4 Kokofu series (Gleyic Lixisol)

Kokofu series is found below *Nzima series* and occupies lower slope sites with slope gradients of 1-3%. It is developed from colluvial material from upslope. The soil is deep, non gravelly and moderately well or imperfectly drained. The topsoil consists of dark brown friable silt loam. The underlying subsoil consists of yellowish brown silt clay loam, faintly mottled yellow

2.2.5 Temang series (Haplic Gleysol)

This soil is developed from alluvial material and occupies the valley bottoms of 0-1% slope and depressions that are subjected to water-logging during the rainy season. The soil is deep and poorly drained. The topsoil consists of brown, faintly mottled dark yellowish brown friable loam. The underlying subsoil is pale brown to light brownish grey friable sandy loam with dark yellowish brown mottles.

Fig. 3. Typical *profile of Nta* soil series (FAO/WRB: *Eutric-Gleysol*)

Horizon	Depth (cm)	Description
Apg	0-6	Dark grayish brown (10 YR 4/2) with clear smooth rusty mottles, moderate sub-angular blocky with a sandy loam texture
Bacg	6-17	Dark grayish (10 YR 3/2) with clear waxy yellow mottles, moderate granular and sandy loam in texture
Bcg	17-34	Grayish (10 YR 6/2) with clear smooth yellow mottles, moderate crumbly and sandy loam in texture
Bcg2	34-75	Grayish (10 YR6/1) with clear smooth rusty mottles, fine granular with a sandy clay loam texture
Cg1	75-89	Grayish (10 YR 6/2) with clear smooth rusty mottles, moderate crumbly and sandy clay loam
Cg2	89-140	Grayish (10 YR7/1) clear smooth rusty mottles, fine granular and a sandy clay loam texture

Table 3. Profile description of typical *Oda* soil series (FAO/WRB: *Eutric-Gleysol*)

Fig. 4. Typical profile of *Oda* soil series (FAO/WRB: *Eutric-Gleysol*)

3. Some research findings

3.1 General fertility status of lowland soils in West Africa

Across the sub-region, various studies (Issaka et al, 1999a, b, 1997, Buri et al, 1996, 1998, 2000, 2006) have shown that lowlands are generally low in soil fertility with wide variations in fertility levels across the different and varied agro-ecological zones. Soil fertility levels as compared to other regions of the world showed the sub region to be quite deficient in available phosphorus and relatively lower in the basic cations particularly calcium and potassium, thus reflecting lower levels of eCEC. Thus soils of West African lowlands in general are characteristically low in basic plant nutrients.

3.2 General fertility status of lowland soils in Ghana
3.2.1 Soil reaction

In Ghana, soil pH within the drier Savannah agro-ecological zones, particularly both the *Volta* and Lima series are strongly acid (mostly < 5.0). Topsoil pH ranges from strongly acid to neutral for *Lapliki* series. However, pH of lowlands within the Forest agro-ecological zones is relatively uniform. Soil pH is slightly higher and generally greater than 5.5. Even though some of the soils within this zone are relatively acid, this is on a limited scale. Exchangeable acidity is also relatively higher within the savannah agro-ecology (mean = 1.0 cmol (+) kg⁻¹) which can adversely affect basic cation balances particularly Ca and Mg leading to adverse effect on rice growth. However, generally and under reduced conditions, pH levels may not pose any serious problem for rice production, since hydromorphic or reduced conditions under rice cultivation tend to favour and enhance pH increases.

3.2.2 Total carbon and nitrogen

Within the forest agro-ecological zones, lowland soils have low to moderate levels of organic carbon. Organic Carbon levels could be as low as 4 g kg⁻¹ at some sites and rising to 37g kg⁻¹ at some locations. Mean levels are around 12.0 g kg⁻¹. However, within the savannah agro-ecology, organic carbon levels are comparatively lower with general mean levels around 6.0 g kg⁻¹ with a 50% coefficient of variability across locations. *Volta* series shows relatively higher organic matter content (13-22 g kg⁻¹) than *Lima* series (8-19g kg⁻¹) within the topsoil. In the same vein total Nitrogen levels show a similar trend to that of Carbon, being slightly higher for the forest than the savanna agro-ecological zones. Total Nitrogen has a mean value of 1.1g kg⁻¹ within the forest agro-ecology with very little variability. The savannah zones show much lower levels of total Nitrogen with much lower variability compared to the forest ecology. Mean levels across locations is lower than 0.7g kg⁻¹.

3.2.3 Available phosphorus (P)

Available P is generally very low for all the soil types and across all agro-ecological zones (Buri et al, 2008a, b). Available P is the single most limiting nutrient. It varies very greatly across locations within the forest agro-ecological zones. Mean levels within the Forest is about 5 mg kg⁻¹ but varied very greatly (CV > 90%). Within the Savanna zones, mean available P levels for lowlands is even lower and also varies significantly (CV > 60%). Mean level is about 1.5 mg kg⁻¹. Under hydromorphic conditions, P utilization and availability is enhanced. This makes current available P levels very inadequate and therefore very limiting to the utilization of these lowlands for rice cultivation due to its significant inlake.

3.2.4 Exchangeable bases

Exchangeable cations (K, Ca, Mg) levels within the forest ecology are generally moderate to medium across most locations, even though they also vary significantly (CV > 60%). Exchangeable Potassium (K) has a mean value of 0.4cmol (+) kg⁻¹. Exchangeable Ca and Mg have mean values of about 7.5 cmol (+) kg⁻¹ and 4 cmol (+) kg⁻¹ respectively. Exchangeable Na levels are lower {mean = 0.32 cmol (+) kg⁻¹} but show much higher variability (CV > 80%). Effective Cation Exchange Capacity (eCEC) values within the forest agro-ecology are relatively moderate, a reflection of the moderate levels of exchangeable cations. Most lowlands within the forest are therefore relatively adequate in Ca, Mg and Na, but with some areas showing potential K deficiencies.

Exchangeable cation levels within the Savannah agro-ecological zones are, however, generally low when compared to those of the forest agro-ecological zone. Topsoil exchangeable calcium is moderate and relatively higher for *Volta* series {2.2-5.8 cmol (+) kg⁻¹}

than the Lima series {1.76-2.24 cmol (+) kg^{-1}}. Mean levels of exchangeable K {0.22 cmol (+) kg^{-1}}, Mg {0.9 cmol (+) kg^{-1}} and Na {0.11 cmol (+) kg^{-1}} are also quite low with coefficient of variability levels of over 74%, 90%, and 77% respectively. Effective cation exchange capacity levels are therefore relatively low across locations, indicative of the need to consider improving upon the levels of these nutrients under any effective and sustainable cropping program (Table 4).

3.2.5 Soil texture

Within the drier Savannah agro-ecological zones, lowland soils are relatively low in clay content. Most locations show less than 10% clay content but again with higher variability (CV > 60%). The soils, however, show appreciable levels of silt (mean > 60%) and are therefore mostly Silt loam in texture, with isolated areas being sandy loam. They occur abundantly and cover a greater part of the lowlands. The soils are generally deep but water retention capacity may be low due to low clay contents. Typical examples are as described earlier. Within the forest ecology, the soils are also relatively low in clay (mean = 13%). They also contain relatively higher levels of silt (mean > 50%). Some are deep while others are very shallow. Textures vary from sandy loam through silt loam to loam. The water retention capacity of these soils is better when compared to those of the savannah zones.

Parameter	Savannah agro-ecology (Ghana)	Forest agro-ecology (Ghana)	West Africa lowlands	Paddy fields of S. E. Asia
Sample (No.)	90	122	247	410
pH (water)	4.6	5.7	5.3	6.0
Total Carbon (g kg^{-1})	6.1	12.0	12.3	14.1
Total Nitrogen(g kg^{-1})	0.65	1.10	1.08	1.30
Available Phosphorus (mg kg^{-1})	1.5	4.9	8.4	17.6
Exch. Calcium {cmol (+) kg^{-1}}	2.1	7.5	2.8	10.4
Exch. Magnesium {cmol (+) kg^{-1}}	1.0	4.1	1.3	5.5
Exch. Potassium {cmol (+) kg^{-1}}	0.2	0.4	0.3	0.4
Exch. Sodium {cmol (+) kg^{-1}}	0.1	0.3	0.3	-
Effective CEC {cmol (+) kg^{-1}}	4.4	12.7	5.8	17.8
Clay (g kg^{-1})	66	127	230	280

Table 4. Soil nutrient levels of lowlands in Ghana in comparison with West Africa and paddy field of South East Asia

Research further shows that productivity of these soils can be improved and sustained when water and nutrient management structures are put in place. Buri et al, (2004) observed that, soils of lowlands in Ghana respond significantly to the application of soil amendments like farm organic materials and mineral fertilizer. The authors observed that, applying 7.0 t ha^{-1} of poultry droppings gave similar rice grain yields as the full dose of recommended mineral fertilizer of 90kg N, 60kg P$_2$O$_5$and 60kg K$_2$O per ha or a 50% dose of recommended mineral fertilizer rate and 3.5t ha^{-1} of poultry droppings combined. Furthermore application of cattle dung or rice husk resulted in significant increases in rice grain yield over the control. In a similar study, Issaka et al, (2008) also reported significant increases in rice grain yield with improvement in water management and land preparation. Rice grain yield increased significantly in the order: farmers practice < bunded only < bunded and puddled < bunded, puddled and leveled. The introduction of improved soil, water and nutrient management, to some selected farmers resulted in significant increases in total rice grain production. Mean rice grain yields among such farmers compared to the national mean is presented in Table 4. There was an increase in yield ranging from 154% to 235% over the period 2003 to 2009. Such significant yield increases in rice grain could lead to improved and sustainable management of these lowlands, as enough revenue will be generated which can be re-invested in effective and proper soil management.

*National mean rice grain yield (t ha^{-1})

Year of Production	Grain yields under "Sawah" system (t ha^{-1})	National mean rice grain yield (t ha^{-1})	% increase as a result of "Sawah" adoption
2003	5.1	2.0	155
2004	5.6	2.0	180
2005	5.0	2.0	150
2006	5.7	2.0	185
2007	5.7	1.7	235
2008	6.0	2.3	161
2009	6.1	2.4	154
Mean	**5.6**	**2.1**	**167**

*source – Ministry of Food and Agriculture, Ghana

Table 5. Mean rice grain yields under the "Sawah" system compared to the national mean in Ghana

4. Conclusions

On the basis of generally observed nutrient levels and heterogeneous nature, lowland soils in Ghana are deficient in most basic nutrient elements which vary considerably from location to location. Site specific nutrient management options are therefore recommended. However, nutrient deficiency limitations can be corrected through improved organic matter management, additions of mineral fertilizers, integrated soil fertility management methods and adoption of sustainable and improved rice production technologies (e.g. "Sawah" system). The "Sawah" system which is intrinsic and conservative can help improve and/or maintain nutrient levels within these environments which will enhance sustainability. Considering the variable nature of lowlands, further research in soil, water and nutrient management (particularly organic matter) may be necessary for the development of specific suitable and sustainable technologies for the various ecologies. This will encourage and promote easy adoption for sustainability and increased productivity.

5. Recommended management practices

Lowland soils in Ghana are deficient in most basic nutrient elements which vary considerably across locations. These are soils with heterogeneous characteristics and therefore require different/varied management options. The major constraints to the use of these sites for rice cultivation include luck of proper land preparation methods, ineffective water management and low soil fertility. Generally, greater emphasis is laid on the development of improved planting material (varieties) to the neglect of the micro-environment in which the improved planting material will grow. Consequently, the high yielding varieties perform poorly because soil fertility cannot be maintained. This has led to farmers not realizing the full potential of improved rice varieties. There is the need for the integration of genetic and natural resource management. Therefore for the effective utilization of these lowlands, the development of technologies that will result in a balance between bio-technology (varietal improvement) and eco-technology (environmental improvement) should be promoted.

The provision of water management structures will greatly improve the utilization and nutrient management options of these lowlands. The development of technologies that will be easy-to-adopt and using affordable materials for water harvesting will make more farmers adopt water harvesting for use on their rice fields. This therefore calls for the development of technologies that will enhance water harvesting from small streams and springs that occur abundantly in these lowlands. Due to farmers inability to use the recommended amounts of mineral fertilizers, integrated nutrient management options are very necessary. The combined use of farm organic materials, mineral fertilizers and effective cropping systems will help improve soil fertility. The recycling of farm organic matter will be very useful. Farm organic materials may be used directly on rice fields or may be treated (composted, ashed, charred). The constant burning off of farm organic matter should be discouraged. Instead, farmers should be educated on how to do partial burning under special conditions.

The adoption of improved rice production technologies such as the "Sawah" systems will significantly lead to improved water and nutrient management. The concept and term

"Sawah" refers to man-made improved rice fields with demarcated, bunded, leveled, and puddle rice fields with water inlets and outlets which can be connected to various irrigation facilities such as canals, ponds, weirs, springs, dug-outs or pumps. Field demarcation based on soil, water and topography need to be considered seriously for the sustainable use of these lowlands. Site specific nutrient management options are therefore recommended. However, while nutrient deficiency limitations can be corrected through improved organic matter management, additions of mineral fertilizers and/or integrated nutrient management options, water management under current traditional systems is very poor. This tends to negatively affect both soil nutrient retention and availability for plant use. As a step towards improving water management, simple and cost effective management structures are necessary for harvesting surface water for temporal storage and use. Land preparation methods for rice cultivation should be improved to include the construction of bunds and leveling in addition to ploughing. The uses of heavy land preparation machinery such as tractors are not suitable for most lowlands due to their sizes, topography, wetness and nature of soils. The use of lighter and affordable land preparation machinery such as the power tiller (two wheel tractor) should be preferred.

6. References

Adu S. V. (1995a). Soils of the Nasia Basin, Northern Region, Ghana. Memoir No. 11 CSIR-Soil Research Institute, Kwadaso – Kumasi. Advent Press.

Adu S. V. (1995b). Soils of the Bole-Bamboi Area, Northern Region, Ghana. Memoir No. 14 CSIR-Soil Research Institute, Kwadaso – Kumasi. Advent Press.

Adu S. V. (1969). Soils of the Navrongo-Bawku Area, Upper Region, Ghana. Memoir No. 5. Soil Research Institute (CSIR), Kumasi

Asubonteng K. O, Andah W. E. I., Kubota D., Hayashi K., Masunaga T., & Wakatsuki T. (2001). Characterization and Evaluation of Inland Valleys of the Sub-humid Tropics for Sustainable Agricultural Production: Case study of Ghana. In Proceedings of the International workshop on Integrated Watershed Management of the Inland Valley – Ecotechnology Approach, Novotel, Accra, Ghana, February 6-8th, 2001

Buri M. M., Oppong J., Tetteh F. M. & Aidoo E. (2008). Soil Fertility Survey of Selected Lowlands within the Mankran Watershed in the Ashanti region of Ghana. CSIR – Soil Research Institute Technical Report No. 274

Buri M. M., Aidoo E. & Fujii H. (2008). Soil Fertility Survey of Selected Lowlands within the Jolo Kwaha Watershed near Tamale in the Northern region of Ghana. CSIR – Soil Research Institute Technical Report No. 279

Buri M. M., Issaka R. N. & Wakatsuki T. (2007). Determining Optimum Rates of Mineral Fertilizers for Economic Rice grain Yields under the "Sawah" system in Ghana. West African Journal of Applied Ecology, Vol. 12. 19-31

Buri M. M., Issaka R. N. & Wakatsuki T. (2006). Selected lowland soils in Ghana: Nutrient levels and distribution as influenced by agro-ecology. Proceedings of the International Conference and the 17th and 18th Annual General Meetings of the Soil Science Society of Ghana. Pp 149-162

Buri M. M., Issaka R. N., Wakatsuki T. & Otto E. (2004). Soil Organic Amendments and Mineral fertilizers: Options for Sustainable Lowland Rice Production in the Forest Agro-ecology of Ghana. Agriculture and Food Science Journal of Ghana. Vol. 3, 237-248

Buri M. M., Masunaga T. & Wakatsuki T. (2000). Sulfur and Zinc levels as limiting factors to rice production in West Africa lowlands. Geoderma 94 (2000), 23-42.

Buri M. M., Fusako I., Daisuki K., Masunaga T. & Wakatsuki T. (1998). Soils of Flood plains of West Africa. General Fertility Status. Soil Sci. and Plt Nutr. 45(1). 37-50

Buri M. M. & Wakatsuki T. (1996). Soils of flood plains of West Africa: Geographical and regional distribution of some fertility parameters. Proceedings of the International Symposium on "Maximizing Sustainable Rice Yields through Improved Soil and Environmental Management." held in Khon Kaen, Thailand, November 11-17, 1996. Pp 445-455

Dedzoe C. D., Senayah J. K. Adjei-Gyapong T., Dwomo O., Tetteh F. M. & Asiamah R. D. (2001a). Soil characterization and suitability assessment for lowland rice production in the Northern Region. Phase 1- SILLUM VALLEY. SRI Technical Report No. 203. Kwadaso, Kumasi

Dedzoe C. D., Senayah J. K., Adjei-Gyapong T., Dwomo O., Tetteh F. M. & Asiamah R. D. 2001b. Soil characterization and suitability assessment for lowland rice production in the Northern Region. Phase III – KULDA-YARONG VALLEY. SRI Technical Report No. 213.

Dedzoe C. D, Seneyah J. K & Adjei-Gyapong T. (2002). Soil characterization and Suitability Assessment for lowland rice production in the Northern region of Ghana. SRI Technical Report No. 213. FAO, ISRIC & ISSS (1998). World Reference Base for Soil Resources. World Soil Resources Report No. 84

Issaka, R. N., Ishida, M., Kubota, D. & Wakatsuki, T. (1997): Geographical Distribution of Soil Fertility Parameters of West Africa Inland Valleys. Geoderma 75: 99-116.

Issaka R. N., Masunaga T., Kosaki T., & Wakatsuki T. (1996). Soils of Inland Valleys of West Africa. General Fertility Parameters. Soil Sci. and Plt Nutr.. 42. 71-80.

Issaka, R. N., Masunaga, T. & Wakatsuki, T. (1996): Soils of Inland Valleys of West Africa: Geographical Distribution of selected soil fertility parameters. Soil Sci. Plant Nutr., 42. 197-201

Senayah J. K. Adjei-Gyapong T., Dedzoe C. D., Dwomo O., Tetteh F. M. & Asiamah R. D. 2001. Soil characterization and suitability assessment for lowland rice production in the Northern Region. Phase 2. KULAWURI VALLEY. SRI Technical Report No. 207. Kwadaso, Kumasi

Senayah J. K, Issaka R. N. & Dedzoe C. D. (2009). Characteristics of Major Lowland Rice-growing Soils in the Guinea Savanna Voltaian Basin of Ghana. Ghana Journal of Agricultural Sciences: Agricultural and Food Science Journal of Ghana.

Wakatsuki T, Buri M. M. & Fashola O.O (2004a). Rice Green Revolution and Restoration of Degraded Inland Valley Watersheds in West Africa through Participatory Strategy for Soil and Water Conservation. ERECON Institute of Environmental Rehabilitation and Conservation. 241-246

Wakatsuki T, Buri M. M. & Fashola O.O (2004b). Ecological Engineering for Sustainable Rice Production of Degraded Watersheds in West Africa. World Resource Research 2004, IRRI

Part 3

Nutrient Management and Crop Production

8

Impact of Salinity Stress on Date Palm (*Phoenix dactylifera* L) – A Review

Mohamed S. Alhammadi[1] and Shyam S. Kurup[2]
[1]*Research & Development Division, Abu Dhabi Food Control Authority*
[2]*Faculty of Food and Agriculture, United Arab Emirates University*
United Arab Emirates

1. Introduction

The growth and productivity of palms are primarily affected by salinity stress in the arid regions apart from drought and heat. The world scenario on salinity problems is critical, over 20 percent of the cultivated area and half of the irrigated land in the world is encountering salinity stress of different magnitudes. The importance of date palm culture for its high nutritive, economic and social values is well recognized, especially in arid and semi-arid areas where it plays an important role influencing microclimate in a way that enhances the production of other agricultural crops. Worldwide production, utilization and industrialization of dates are increasing continuously (Botes and Zaid 2002).

Different productivity parameters such as germination of seeds, seedling growth, vigor, reproductive flushing and fruiting are adversely affected reflecting on optimum production and productivity. The basic cause of salinity hazards is enhanced ion toxicity causing impaired sequestering of sodium ions into the vacuoles. The increased ionic concentration of the soil solution decreases the osmotic potential of the soil creating severe water stress derailing the uptake process. This can create imbalance in the absorption process of other minerals. In order to achieve salt-tolerance, the foremost task is either to prevent or alleviate the damage, or to re-establish homeostatic conditions in the new stressful environment (Parida and Das, 2005).

The date palm growth and production in the arid regions is adversely affected by salinity problems apart from other crops like vegetables and fruits (Figures 1 & 2). Date palm is considered as the subsistence crop of the Middle East and North Africa. Despite date palms outstanding agronomic and socio-economic significance, attempts to use date palm biodiversity to screen against salinity tolerance have been limited and therefore of urgent priority. Although potential salt tolerant cultivars are available, there is no systematic approach to characterize such genotypes employing molecular diagnostic techniques. The understanding of the molecular, physiological, biochemical basis and soil factors will be helpful in developing selection strategies for improving salinity tolerance. Therefore pooling the information through the present review on salt responses in relation to various factors is crucial in developing strategies and improving salt tolerant mechanism in date palms. The review article covers the different aspects under the various growth stages of date palm, like seedling, vegetative phase and reproductive phase. The review also focuses on the use of remote sensing technology in date palm responses to salinity.

Plate 1. Deterioration of date palm trees due to high level of salinity.

Plate 2. Date palm field neglected due to the salinity.

The aim of this paper is to pool the various aspects of responses of date palm to salinity tolerance and to review the existing knowledge in relation to biodiversity useful to researchers and developmental agencies, engaged in date palm. The review also facilitates interdisciplinary studies to assess the ecological significance of salt stress in relation to date palm cultivation.

2. Background

Date palm (*Phoenix dactylifera* L) accounts for more than 1500 cultivars around the world (FAO, 2002). In addition to its commercial and nutritional value, the date palm tree has a minimum water demand, tolerates harsh weather, and tolerates high levels of salinity (Diallo, 2005); in fact, it is more salt tolerant than any other fruit crops (FAO, 1982). There are very few trees that can tolerate the desert environment, which is characterized by high temperatures, low soil moistures and high salinity levels. Furr (1975) reported that it is obvious that date palm is more salt tolerant than barley and may be the most salt tolerant of all crop plants since barley known as one of the most salt-tolerant field crops is usually grown in the cool season; in contrast date palms grow faster in hot weather when salinity has the most adverse influence on plants.

On the other hand, increasing of soil salinity is starting to show negative impact on the date palm agro-ecosystem in arid region, especially in the Middle East (Dakheel, 2005). Accurate information about the growth of date palm in saline environment and the variability in salt tolerance among cultivars is largely unknown. A serious attention is needed to maintain the diversity and growth of such plant in the arid regions.

3. Salinity stress and soil response

Salt-affected soil is a worldwide problem; however, it is more common in arid and semi-arid regions because of high evaporation rates and lack of fresh water resources that is required to leach salts. As water evaporates from the soil surface, the salts move upward to the soil surface but stay within or on the soil (Miller & Donahue, 1990). In many irrigated lands, salts occur in the irrigated water leading to accumulation of salts in the soil. The relationship between water and soil salinities and date palm production is illustrated in figure 3.

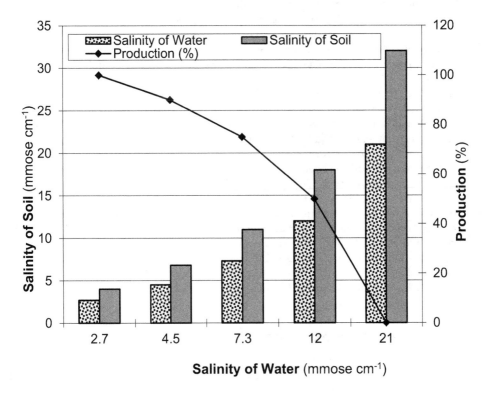

Fig. 3. Relationship between salinity of irrigated water, salinity of soil and production of date palm trees (Zaid and Liebenberg 2005).

Plant development will be affected when the soil has high levels of soluble salts, and the salt may become concentrated enough to be toxic to plants. Presence of salts on the soil is usually associated with osmotic and ionic negative effects, which will then lower the biological activity. Furthermore, salts have significant effect on fertility of the soil as well as its physical, chemical and biological characteristics (Srour et. al, 2010).

The approximation of global salt-affected area is 1 billion hectares, which correspond to about 7% of the earth's continental extent or about 20 times the size of a country such as France. About 77 million hectares have been salinized as a result of human activities, with 58% of these concentrated in irrigated lands. Generally speaking, over 40% of irrigated lands in the world are subject to different degrees of salinity (Dakheel, 2005). As a consequence of increasing population pressure, more arid land will be put into agricultural production in future, which means more salinization danger associated with irrigations.

Soil is a complex system involving interaction between cations on the soil solution with other cations on the soil particles exchange site. The role of the cations on the soil solution is highly dependent on the pH and the negative charge of soil colloids. The common inorganic solutes present in the soil solutions are Na^+, Mg^{2+}, Ca^{2+}, K^+ , Cl^-, SO_4^{2-}, HCO_3^-, NO_3^- and CO_3^{2-}, in addition to boron, selenium, molybdenum, and arsenic that may found in small amounts. Plants will show signs of injury and yield reduction when soluble salts exceed certain level of concentrations (Essington, 2003).

Salt-affected soils are classified into three classes, saline, sodic, saline-sodic soils. They differ based on their chemistry, morphology, and pH. The United States Salinity Laboratory classified salt-affected soils into three classes based on the basis of two criteria, the total soluble salt content and the exchangeable sodium percentage (ESP) or sodium adsorption ratio (SAR).

1. Saline Soils: these soils have a saturated paste electrical conductivity (EC_e) of 4 dS m^{-1} or more and SAR of <12 (ESP <15%). The pH of these soils, which were formerly referred to as white alkali soils, is less than 8.5. Because of a high salt content, these soils reduce water uptake by plants, and increase ion toxicity to the plant tissues. Furthermore, ion imbalances may occur in some soils.

2. Sodic Soils: these soils have an EC_e of less than 4 dS m^{-1}, and SAR of 12 or more (ESP ≥ 15%), and their pH exceeds 8.5. Exchangeable sodium percentage of 15 means that Na^+ occupies more than 15 % of the soil's cation exchange capacity. Formerly, these soils were called black alkali because of dispersed black organic matter coatings on peds and the soil surface. The most important problems for these soils are poor structure due to breakdown of structural units, ion toxicity (mainly Na^+ and Cl^-), and ion imbalances, especially deficiencies in Ca^{+2}, Mg^{+2}, and K^+.

3. Saline-Sodic Soils: these soils have EC_e ≥ 4 dS m^{-1}, SAR ≥ 12, and their pH is less than 8.5. They have similar problems to saline soils, especially reduced water uptake due to high soil osmotic potential.

4. Plant responses to salinity stress

When plant exposed to low-moderate salinity, it may metabolize normally and does not show symptoms of injury. However, more energy is required to maintain normal metabolism demand (Gale and Zeroni, 1985), which may cause reduction in growth and yield (Subbarao and Johansen, 2010). In most crops, lose of production can be significant even before the appearance of foliar injury (Francois and Maas, 2010).

In general, salinity can reduce plant growth through osmotic effects, toxicity of ions, nutrient uptake imbalance, or a combination of these factors (Karim and Dakheel, 2006; Maathuis, 2006). Dynamics observed in juvenile plant water uptake and tree growth observed in of juvenile date palms (*Phoenix dactylifera* L., cv. Medjool) for salinity and boron occurring independently and together were summarized by decreased water uptake but not ion accumulation for NaCl and CaCl salts and by boron that was accumulated in leaves and subsequently was associated with reduced tree size. It is suggested that while mechanisms for plant response to salinity are dominated by lowered soil water potential (osmotic stress); boron becomes toxic as it accumulates to a threshold level in plant tissue (Trippler et. al, 2007).

4.1 Morphological responses

The excess of salinity in the soil water can cause significant morphological changes in the plant growth responses. Morphological parameters like plant height, leaf production, and collar girth of different varieties, in the vegetative phase of growth subjected to high salinity irrigation showed differential responses. There is a threshold level of salinity that each palm can tolerate under the progression of salinity (Kurup et. al, 2009). Effects of salinisation of soil on emergence, growth, and physiological attributes of seedlings of the date palm showed negative relationship between percentage seed germination and salt concentration when a mixture of chlorides and sulfates of Na, K, Ca, and Mg maintained at 4.3, 6.0, 8.2, 10.5, 12.8, and 14.6 dS m^{-1}. Seedlings did not emerge when soil salinity exceeded 12.8 dS m^{-1}. Seedlings survived and grew up to a soil salinity of 12.8 dS m^{-1} and evidently this species is salt tolerant at the seedling stage as well (Ramoliya and Pandey, 2003). Furr and Ream (1968) studied the effect of salts ranging between 520 to 24,000 ppm on growth and salt uptake of 'Deglet Noor' and 'Medjool' varieties of date palm. The result of their study was that the average growth rate of leaves was depressed as the salinity increased and that the decline in the growth was more related to salinity of irrigation water than to salt content of the plants. Khudairi (1958) evaluated seed germination of 'Zahedi' cultivar grown in three NaCl solutions (0.5, 1.0, and 2.0%). He found that NaCl suppressed seed germinations during the early stages and the maximum germination did not exceed 50%. He also evaluated seeds of the same cultivar on several NaCl concentrations ranging between 0.1% and 2.5%. He observed that NaCl concentration below 0.8% did not affect seed germination, while the germination sustained in solutions up to 2%.

Aljuburi (1992) studied the growth of four cultivars of date palms, 'Lulu,' 'Khalas,' 'Boman,' and 'Barhee,' using four salinity concentrations (0, 0.6, 1.2, and 1.8%). He found that the cultivar 'Lulu' was more affected by salinity compared to the other cultivars. The research done by Hewitt (1963) tested the effect of different salts and salt concentrations on the germination of 'Deglet Noor' seeds using various combinations of NaCl, calcium chloride (CaCl$_2$), sodium sulfate (Na$_2$SO$_4$), NaCl + CaCl$_2$, and NaCl + Na$_2$SO$_4$ with salt's concentrations from 10,000 to more than 30,000 ppm. He indicated that the growth of 'Deglet Noor' decreased slightly at 10,000 ppm, decreased drastically at 20,000 ppm, and was prevented at 30,000 ppm except for three seedlings in NaCl + Na$_2$SO$_4$ 34,000 ppm treatment.

The behavior of mature date palm was studied by Furr and Armstrong (1962). They examined the growth of 'Halawy' and 'Medjool' 17-year-old cultivars using salinities ranging between 2,500–15,300 ppm. They found little or no effect on growth rate of leaves, yield, size or quality of fruit, or on chloride content of the leaf pinnae.

Response of tissue culture plantlets of date palm Khalas variety to marginal saline water irrigation at varied frequencies with and without the use of mulch influenced the growth factors considerably (Al-Wali et. al, 2011).

4.2 Physiological and biochemical basis of salt stress

The physiological basis of salt tolerance in date palm was found as a strict control on Na^+ and Cl^- concentration in leaves and keeping up the K^+ content (Alrasbi et. al, 2010). It can be recommended that date palm plants of certain varieties can be irrigated with saline water during vegetative growth. Youssef and Awad (2008) conducted a study to enhance photosynthetic gas exchange in date palm seedlings under salinity stress (subjected to seawater treatments at 1-, 15-, and 30- mS cm^{-1}) using a 5-Aminolevulinic Acid-based fertilizer. They found that date palm seedlings accumulated significant amounts of Na+ in the foliage with increasing salinity, about a threefold increase in the accumulated Na+ between the control and 30 mS cm^{-1} salinity treatment. Electrolyte leakage indicated a significant reduction in membrane integrity as salinity increased. A strong linear correlation was observed between the chlorophyll (chl) a/b ratio and assimilation rate throughout salinity treatments. Salinity did not induce any change in the carboxylation efficiency of the rubisco enzyme (Vc,max), or in the rate of electrons supplied by the electron transport system for ribulose 1,5-bisphosphate (RuBP) regeneration.

The physiological basis of salt tolerance in date palm was found to be based on Na and Cl concentration in leaves and keeping up the K content. Therefore it can be recommended that date palm (seedlings of varieties 'Khalas', 'Khunaizy' and 'Abunarinjah') can be irrigated with saline water during vegetative growth. However, a significant decline in growth is expected when the EC of irrigation water exceeds 9 dS m^{-1} that may reach up to 50% with water EC 18 dS m^{-1} (in sandy soil with very good drainage). (Alrasbi et. al, 2010).

4.3 Molecular basis of salt stress mechanism

The metabolic adjustments to salt stress at the cellular level are the main focus to molecular characterization and identification of a large number of genes induced by salt.

Salt tolerance is a multigenic trait and a number of genes categorized into different functional groups are responsible for encoding salt-stress proteins (Parida and Das, 2005).

In order to identify salinity tolerant date palms, evaluation of four varieties was conducted using fourteen random primers to detect the DNA polymorphism. Randomly amplified polymorphic DNA (RAPD) technique was employed to characterize these varieties. Primer OPD-02 distinguished Bugal white, which proved to be salinity tolerant, with a DNA fragment of about 1200 bp (Kurup et. al, 2009). Sedra et. al, (1998) used RAPD marker system as a tool for the identification of date palm cultivars and examined 43 cultivars from Morocco. RAPD markers showed considerable difficulties when characterizing cultivars, and these difficulties include mainly low polymorphisms, irreproducibility and lack of evident organization. Microsatellite markers have been applied to assess the genetic relationships of 45 date palm cultivars for salinity tolerance collected from Sudan and Morocco (Elshibli and Korpelainen, 2008).

Somaclonal variations in tissue culture-derived date palm plants using isoenzyme analysis and activities of peroxidase (PER), polyphenol oxidase (POD) and glutamate oxaloacetate (GOT) and randomly amplified polymorphic DNA (RAPD) fingerprints were analyzed for salinity. The frequency of somaclonal variations was found to be age dependent. Similar isoenzyme patterns for PER and GOT were detected in all analyzed plants (Saker et. al, 1999).

5. Remote sensing techniques

The internal status of plants grown in saline environment is not clearly observable particularly when dealing with halophytes such as date palm. Response of halophytes to salinity needs longer time to be measurable using usual experimental methods. Satellite and digital imagery play a considerable function in remote sensing, offering great information about the area studied. It has the ability to observe vegetation changes at different periods, which assists in the study of vegetation change quickly and precisely (Alhammadi & Glenn, 2008; Howari, 2003; Jackson, 1986). Several attempts were carried out using various satellite sensors to study the date palm responses to salinity (Harris 2001; Alhammadi & Glenn 2008).

One of the recognized techniques of satellite imaging is to map the vegetation through change detection. This technique is commonly used to observe vegetation changes at different time periods, providing quick and precise information of the plants condition. Since plant is a dynamic component, any vegetation change occur can be detected using a comparison of several images from different periods.

The technique simply depends on the implementation of vegetation indices (VI) obtained from multispectral imagery, such as Soil Adjusted Vegetation Index (SAVI) and Normalized Difference Vegetation Index (NDVI). Vegetation indices combine the low reflectance in the visible region of the electromagnetic spectrum (red) with the high reflectance in the near infrared band (NIR). When a correlation between VI and the vegetation biomass is found, any decline in the plant pigment or condition can be detected by the difference of the index images (Tucker 1979).

It has been recognized that in the saline environment, the plant excludes the salts from the cytoplasm accumulating them outside the cells. Therefore, water moves from the cell to outside to accommodate the ion changes causing dehydration of the cell (Volmar et al., 1998). The NIR reflectance of the electromagnetic spectrum is highly affected by the cellular structure of the leaves, while the short wave infrared region is highly affected by the water content in the leaves (Gausman et al. 1978; Gausman 1985; Tucker 1980). In healthy active vegetation, chlorophyll is a strong absorber in the red region, while the cellular structure is a strong reflector in the NIR region. On the other hand, unhealthy vegetation has a lower photosynthetic activity causing increase reflectance in the red region and decrease reflectance in the NIR region (Weiss et al. 2001). Based on these findings, several studies were able to correlate spectral characteristics and vegetation health (Alhammadi and Glenn 2008; Peñuelas et al. 1997; Wiegand et al. 1992).

However, date palm tree was not intensively studied with remote sensing due to the special characteristics of canopy structure and the variations of the tree spacing (Harris 2001, Alhammdi and Glenn 2008, Alhammadi 2010). Alhammadi and Glenn (2008) used two types of satellite sensors, Thematic Mapper (TM) and Enhanced Thematic Mapper Plus (ETM+) to study the health of date palm trees grown in salt-affected environment. They implemented change detection technique from two dates 1987 and 2000 using SAVI, which work more with areas of low vegetation cover. They found that a serious decrease in date palm trees health was associated with increased salinity levels. Although, the official census data show a general increase in agriculture lands, the condition and the production of date palm decreased. They recommended using high resolution sensors for future studies due to the canopy structure of the palm trees and extent of background soil. In another study, Alhammadi (2010) investigated the ability of using high resolution satellite sensor (QuickBird) with pixel size of 60 cm to evaluate growth rate of eighteen cultivars of date palm trees grown in three salinity levels (5, 10, 15 dS m^{-1}) during years 2003 and 2008. The results showed that QuickBird sensor was able to detect significant variations in the growth of date palm cultivars at the three salinity levels in year 2008. Furthermore, date palm trees health were highest in year 2003 and started to decrease by time. Further investigations are needed to develop a consistent correlation between the date palm growth parameters on the ground and vegetation indices of satellite images.

6. Conclusion

Date palm growth and development with regard to salinity tolerance and use of remote sensing diagnostic tools are discussed in the review paper. The paper provides information on two major aspects classified as salinity stress in relation to soil and plant response with emphasis morphological, physiological, biochemical, and molecular basis of salt tolerance in date palm in seedling, vegetative and reproductive phases. In general, salinity can reduce plant growth through osmotic effects, toxicity of ions, nutrient uptake imbalance, or a combination of these factors. Results of studies indicated that there are differences in salt tolerance between date palm cultivars, which appear to be related to the salt exclusion mechanisms by the root parts (Greenway and Munns 1980), resulting in reduced Na^+ translocation to the shoots. Khnaizi, Lulu, Nabtat Safi, and Razez cultivars showed greatest growth parameters and Na:K ratios that indicates higher sodium discriminations from plant parts than other cultivars. Efforts should be made to compare the relative sensitivity of various cultivars to salt, uptake and transport of NaCl and their interactions with nutrients. In addition, it is required to identify differences in salinity tolerance between date palm cultivars, and thus start new breeding programs to improve salinity tolerance. In order to bring arid and semiarid regions into production, future researches should focus on using halophytes as an alternative crop and seawater for irrigation. Remote sensing as a tool to detect salinity stress should be effectively applied in future date palm development programs.

7. References

Alhammadi, M.S. 2010. Using QuickBird Satellite Images to Study the Salinity Effect on Date Palm Field. International Congress Geotunis 2010. 29 November – 3 December 2010. Tunis.

Alhammadi, M.S. and Glenn, E.P. 2008. Detecting Date Palm Trees Health and Vegetation Greenness Change on the Eastern Coast of the United Arab Emirates Using SAVI. International Journal of Remote Sensing. 29: 1745-1765.

Aljuburi, H.J. 1992. Effect of sodium chloride on seedling growth of four date palm varieties. Annals of Arid Zone. 31:4:259-262.

Alrasbi, S.A.R N., Hussain, H. Schmeisky, 2010. Evaluation of the Growth of Date Palm Seedlings Irrigated with Saline Water in the Sultanate of Oman, ISHS Acta Horticulturae. 882: IV, International Date Palm Conference.

Al-Wali, B.R., Kurup, S.S., Alhammadi, M.S. 2011. Response of Tissue Culture Plantlets of Date Palm (Phoenix dactylifera L) var. Khalas to Moisture Stress under the UAE Condition. M.Sc. Thesis, UAE University, Al-Ain.

Botes, A. and Zaid, A. 2002. The economic importance of date production and international trade. In: Zaid, A. Arias-Jimenez, E. J. Date palm cultivation. FAO plant production and protection paper, 156 rev.1. FAO, Rome

Dakheel, A., 2005. Date Palm Tree and Biosaline Agriculture in the United Arab Emirates. In: The Date Palm: From Traditional Resource to Green Wealth. pp. 247-263. UAE Center of Studies and Strategy Researches, Abu Dhabi, UAE.

Diallo, H. 2005. The role of date palm in combat desertification. In: The Date Palm: From Traditional Resource to Green Wealth. pp. 13-19. UAE Center of Studies and Strategy Researches. Abu Dhabi, UAE.

Elshibli S., and Korpelainen H. 2008 Microsatellite markers reveal high genetic diversity in date palm (Phoenix dactylifera L.) germplasm from Sudan. Genet 134:251-260.

Essington, M.E. 2003. Soil and Water Chemistry: An Integrative Approach. CRC Press. Florida, USA.

FAO. 1982. Plant production and protection paper. Date production and protection. Food and Agriculture Organization of the United Nation. Rome, Italy.

FAO. 2002. Date Palm Cultivation. FAO plant production and protection papers-156 Rev.1. Food and Agriculture Organization of the United Nation. Rome, Italy.

Francois E. F. and E. V. Maas, 2010. Crop response and management of salt-affected soils. In: Pessarakli, M. 2010. Handbook of Plant and Crop Stress. p.p. 169-201. Ed. 3. CRC Press. Florida, USA.

Furr, J.R. 1975. Water and salinity problems of Abadan Island date gardens. Date Growers' Inst. Rept. 52:14-17.

Furr, J.R. and W.W. Armstrong. 1962. A Test of Mature Halawy and Medjool date palms for salt tolerance. Date Growers' Inst. Rept. 39:11-13.

Furr, J.R. and C.L. Ream. 1968. Salinity effects on growth and salt uptake of seedlings of the date, Phoenix dactylifera L. Proc. Amer. Soc. Hort. Sci. 92:268-273.

Gale, J. and M. Zeroni. 1985. The cost of plant of different strategies of adaptation to stress and the alleviation of stress by increasing assimilation. Plant Soil. 89:57-67.

Gausman, H. W., 1985, Plant leaf optical properties in visible and nearinfrared light. p. 78. Graduate Studies, Texas Tech University (No. 29). Lubbock, Texas: Texas Tech Press.

Gausman H. W., Escobar, D. E., Everutt, J. H. A., Richardson, J., and Rodriguez, R. R., 1978, Distinguishing succulent plants from crop and woody plants. Phot. Eng. Rem. Sens, 44, 487-491.

Greenway, H., and Munns, R. 1980. Mechanisms of salt tolerance in nonhalophytes. Ann. Rev. Plant Physiol. 31:149-190.

Harris R. 2001. Remote Sensing of Agriculture Change in Oman. International Journal of Remote Sensing. 24: 4835-4852.

Hewitt, A.A. 1963. Effect of different salts and salt concentration on the germination and subsequent growth of Deglet Noor date seeds. Date Growers' Inst. Rept. 40:4-6.

Howari, F. M, 2003. The Use of Remote Sensing Data to Extract Information from Agricultural Land with Emphasis on Soil Salinity. Australian Journal of Soil Research. 41: 1243-1253.

Jackson, R. D., 1986, Remote sensing of Biotic and Abiotic Plant Stress. Annual Review of Phytopathology. 24: 265-287.

Karim, F.M. and Dakheel, A.J. 2006. Salt-tolerant Plants of the United Arab Emirates. International Center for Biosaline Agriculture (ICBA). Dubai, UAE.

Khudairi, A.K. 1958. Studies on the germination of date-palm seeds. The effect of sodium chloride. Physiol. Plantarum. 11:16-22.

Kurup, S.S., Hedar, Y.S., Al Dhaheri, M.A., El-Hewiety, A.Y., Aly, M.M., and Alhadrami, G. 2009. Morpho-Physiological Evaluation and RAPD Markers – Assisted Characterization of Date Palm (Phoenix dactylifera L) Varieties for Salinity Tolerance. Journal of Food, Agriculture & Environment. 7: (3-4) 503-507.

Maathuis, F.J. 2006. The Role of Monovalent Cation Transporters in Plant Responses to Salinity. Journal of Experimental Botany. 57, (5):1137-1147.

Miller, R.W. and Donahue, R.L. 1990. Soils: An introduction to Soils and Plant Growth. 6th Edition. Prentice Hall. Florida, USA.

Parida,A.K and Das,A.B.2005. Salt tolerance and salinity effects on plants: a review Ecotoxicology and Environmental safety, 60(3):324-349.Pessarakli, M. and Szabolcs, I. 2010. Soil Salinity and Sodicity as Particular Plant/Crop Stress Factors. In: Pessarakli, M. 2010. Handbook of Plant and Crop Stress. p.p. 3-15. 3rd Edition. CRC Press. Florida, USA.

Peñuelas, J., Isla, R., Filella, I., and Araus, J. L., 1997, Visible and near-infrared reflectance assessment of salinity effects on barley. Crop Science, 37, 198-202.

Ramoliya, P.J. and A. N. Pandey, 2003.Soil salinity and water status affect growth of Phoenix dactylifera seedlings New Zealand Journal of Crop and Horticultural Science, 31, (4): 345-353.

Saker, M.M.,S.A. Bekheet, H.S. Taha, A.S. Fahmy and H.A. Moursy.1999. Detection of Somaclonal Variations in Tissue Culture-Derived Date Palm Plants Using Isoenzyme Analysis and RAPD Fingerprints. Biologia Plantarum. 43(3): 347-351.

Sedra MH, Lashermes P., Trouslot P et al. (1998) Identification and genetic diversity analysis of date palm (Phoenix dactylifera L.) varieties from morocco using RAPD markers. Euphy 103:75-82.

Srour, R.K., McDonald, L.M., and Evangelou, V.P. 2010. Influence of Sodium on Soils in Humid Regions. In: Pessarakli, M. 2010. Handbook of Plant and Crop Stress. p.p. 55-88. Ed. 3. CRC Press. Florida, USA.

Subbarao, G.V. and Johansen C. 2010. Strategies and scope for improving salinity tolerance in crop plants. In: Pessarakli, M. 2010. Handbook of Plant and Crop Stress. p.p. 1069-1087. Ed. 3. CRC Press. Florida, USA.

Trippler, E., Gal, A.B., and Shani, U. 2007. Consequence of Salinity and Excess Boron on Growth, Evapotranspiration and Ion Uptake in Date Palm (*Phoenix dactylifera* L., cv. Medjool). Plant and Soil, 297, (1-2): 147-155.

Tucker, C. J., 1979, Red and photographic infrared linear combinations for monitoring vegetation. Remote sensing of the environment, 8, 127-150.

Tucker, C. J., 1980, Remote sensing of leaf water content in the near infrared. Remote Sensing of Environment, 10, 23-32.

Volmar, K. M., Hu, Y., and Steppuhn, H., 1998, Physiological responses of plants to salinity: a review. Can. J. Plant Sci., 78, 19-27.

Weiss, E., Marsh, S. E., and Pfirman, E. S., 2001, Application of NOAA-AVHRR NDVI time-series data to assess change in Saudi Arabia's rangelands. Int. J. Remote Sensing, 22, 1005-1027.

Wiegand, C. L., Everitt, J. H., and Richardson, A. J., 1992, Comparison of multispectral video and SPOT-1 HRV observations for cotton affected by soil salinity. International Journal of Remote Sensing, 13, 1511-1525.

Youssef,T and M. A. Awad.2008.Mechanisms of Enhancing Photosynthetic Gas Exchange in Date Palm Seedlings (*Phoenix dactylifera* L.) under Salinity Stress by a 5-Aminolevulinic Acid-based Fertilizer. J. Plant Growth Regulation, 27(1):1-9.

Zaid, A. and Liebenberg P.J. 2005. Date palm irrigation. In: Date palm cultivation. pp. 164-179. FAO. Roma. Italy.

9

Impact of Mineral Fertilizer Integration with Farmyard Manure on Crop Yield, Nutrient Use Efficiency, and Soil Fertility in a Long-Term Trial

Melkamu Jate

Research Centre Hanninghof, Yara International, Hanninghof, Duelmen
Germany

1. Introduction

Plant nutrient should be well managed to increase productivity of crop production with responsibility to protect environment. The major purposes of plant nutrient management include: (1) to budget and supply nutrients for crop production; (2) to properly utilize manure or organic byproducts; (3) to minimize agricultural nonpoint source pollution of surface and ground water resources; (4) to protect air quality by reducing nitrogen loss; and (5) to maintain or improve the physical, chemical, and biological condition of soil (NRCS and NHCP, 2006). Availability of sufficient amount of nutrients at the right place and time is an essential factor to maximize crop yield per area. Plant nutrient management is defined as application of right form, quantity, and ratios of nutrient at a right location and growth stages of crop to increase yield per area with a minimum nutrient loss. It is described as fertilizer best management practices, integrated plant nutrient management, code of best agricultural practices, site-specific nutrient management, and other similar expressions (Roberts, 2010).

Nutrient recycling by application of organic fertilizer is needed to replace nutrient removed by crop yield from fields in order to restore crop production potential of a soil. But application of organic fertilizer alone insufficiently increases crop yield per area because nutrient content of organic fertilizer is unbalanced and if it is applied in a large quantity to balance nutrient supply the loss will increase. Therefore integrated plant nutrient management (IPNM) can minimize the problem. IPNM is application of mineral fertilizer in combination with locally available organic fertilizer to maintain soil fertility and to balance nutrient supply in order to increase crop yield per area. It is one of the best practices of plant nutrient management to take into consideration mineral fertilizer integration with organic sources of the plant nutrients to optimize social, economic, and environmental benefits of crop production. The main objectives of the IPNM are: (i) to maintain or enhance soil productivity, (ii) to improve stock of plant nutrient in the soil; (iii) to limit nutrient loss to the environment by improvement of nutrient use efficiency (FAO, 1998).

One of the most important challenges to continuously satisfy growing food demand is maximization of crop production on limited areas of agricultural land. Sustainable production of crop requires adequate supply of plant-available nutrients to support crop growth and that the nutrients removed in the harvested material or in the exported product of livestock systems must be replaced so that soil fertility is not depleted over time: and at the same time, excess nutrient accumulation must be avoided to reduce the risk of nutrients moving out of the root zone to the air and water (Aulakh and Grant, 2008). Management of plant nutrient can potentially address the challenge through soil fertility improvement and reduction of nutrient loss. Soil fertility is the capacity of a soil to retain, cycle and supply essential nutrients for plant growth over extended periods of time (Alley and Vanlauwe, 2009). Nutrient and organic matter content of the soil determines status of its fertility. Soil organic matter is an important index of soil fertility (Rahman and Parkinson, 2007). It improves soil fertility with the following functions: source of inorganic nutrient for crops and microbial biomass, exchange of ions, chelating agent and buffer, aggregating soil particles, support root development, and water conservation (Allison, 1973). In addition, it reduces level of atmospheric CO_2 that contributes positively to climate change (USDA and NRCS, 2003).

The concept of IPNM primarily optimizes the use of nutrients in organic fertilizer to maximize crop yield per area and to improve nutrient use efficiency (NUE) synergistically. Improvement of NUE implies reduction of nutrient losses so that it indicates environmental impact of nutrient management. NUE in agriculture can be considered from three major perspectives: (i) as 'agronomic efficiency' that concerns optimization of nutrient uptake by crop; (ii) as 'economic efficiency' that implies improvement of nutrient input increases profitability of crop production; and (iii) as 'environmental efficiency', i.e. minimization of losses of nutrient to the environment (Robert, 2005). There are different methods to express NUE. This study uses the 'difference method', i.e. total nutrient removal of fertilized crop minus total nutrient removal of unfertilized crop per fertilizer rate (Varvel and Peterson, 1990). It is appropriate to use the method in the long-term trials because nutrient is residually accumulated each year.

This paper analyzes a long-term agronomic field trial comparing different schemes of integration of mineral fertilizer with Farm Yard Manure (FYM) in the production of potato, rye, and oat from 1958 – 2008. The objective of the trial is to measure the effects of IPNM on economic, social, and environmental dimensions of crop production sustainability. Moreover it investigates whether organic or mineral fertilizer alone can sustain nutrient demand of crop at the right balance to achieve the highest yield. It evaluates benefits of supplementing organic with mineral fertilizer and balanced mineral fertilizer application in crop production compared to application of organic fertilizer alone. Nutrient management affects ability to maximize crop yield per area, which is a primary factor to produce sufficiently high quality food, feed, and fiber in economically viable systems of production. There are three major components of plant nutrient managements, which are known as nutrient recycling or organic plant nutrient management (OPNM), IPNM, and balanced plant nutrient management (BPNM). In this study the three components of plant nutrient managements are considered as FYM alone (OPNM), FYM + mineral fertilizer (IPNM), and mineral N+PK+Mg fertilizers alone (BPNM). The analysis deals with the evaluation of social and economic (crop yield), environmental (NUE), and soil fertility (organic matter and nutrient index) benefits of the IPNM and the BPNM compared to the OPNM.

2. Material and methods

2.1 Location and history

The Hanninghof long-term trial is located near Duelmen in Western Germany. The experiment started in 1958 with potato cultivation. The potato was followed by winter rye in 1959 and oat in1960. Since then each crop was cultivated 17 times in rotation. Long-term experiments (LTE) are classified as: classical (longer than 50 years); medium length (20 – 50 years old); and young (less than 20 years) long-term trials (Steiner and Herdt, 1993). The Hanninghof long-term trial completed its medium phase in year 2008 and it entered its classical stage in 2009. It is listed along the most prominent classical LTE in the world.

2.2 Soil and climate
The soil is a sandy with the following initial soil parameters: carbon total 1%, N total 0.1%, pH 6, P2O5 12 mg (100 g) $^{-1}$ and K2O 5 mg (100 g) $^{-1}$. Annual average rain fall (1961 – 2008) was 885 mm and average air temperature (1962 – 2008) was 10 °C. Spring and summer averages were 201mm and 243 mm rainfall and 9 °C and 18°C air temperature, respectively.

2.3 Layout
The trial is a two factorial experiment with the factors mineral fertilizer with and without FYM. The layout is a split-plot design with a randomized complete block design. The cultivated area of the trial is 0.3ha (72 × 42m). The field is split into two parts, one receiving FYM every three years and one receiving no additional organic material. Each of the two parts is subdivided into 32 plots i.e. 64 plots in total. The gross area of each plot is 4.5×10.5m with a harvested net area of 4×10m.

2.4 Treatment
A total of 16 treatments were established as shown in Table 1. Each treatment is replicated four times and randomly assigned to 64 plots. In 1960, a treatment with N only (#8 and #16) was introduced. Since the trial was already ongoing for two years a new control for treatments # 8 and # 16 was established. Because they were not different from the old control treatments (#2 and #10) the new control treatments (#7 and #15) are omitted from analysis of the result (Table1).

Mineral fertilizer with FYM		Mineral fertilizer without FYM	
#	Treatments	#	Treatments
1	FYM + N + P	9	N + P
2	FYM	10	Control (without mineral fertilizers)
3	FYM + N + K	11	N + K
4	FYM + N + P + K	12	N + P + K
5	FYM + P + K	13	P + K
6	FYM + N + P + K + Mg	14	N + P + K + Mg
7	FYM	15	Control (without mineral fertilizers)
8	FYM + N	16	N

Table 1. Description of treatments.

Mineral fertilizer N, P, K, and Mg application rates for each crop are given in Table 2. The mineral fertilizer rates were the same for the two parts of the trial with and without FYM.

Crop	Years	Mineral fertilizer application rate (kg ha^{-1})			
		N	P2O5	K2O	MgO
Potatoes	Initial 1958	100	90	160	50
	Since1979	140	90	160	50
Winter rye	Initial 1959	60	90	120	50
	Since 1980	140	90	120	50
Oat	Initial 1960	100	90	120	50

Table 2. Mineral fertilizer application rate for potatoes, winter rye, and oat from 1958 – 2008.

P, K, and Mg mineral fertilizer were applied once at planting for all crops. N mineral fertilizer was applied once at planting for potatoes but split applied for winter rye and oat (Table 3).

Crop	Years	N application according to growth stages (kg ha^{-1})				
		Planting	Early vegetative	2 weeks later	Stem elongation (BBCH 30/31)	Booting (BBCH 49)
Winter rye	Initial 1959	-	60	-	40	40
	Since 1995	-	30	30	40	40
Oat	Initial 1960	60	-	-	-	40

Table 3. Mineral N fertilizer application time for winter rye and oat from 1959 – 2008.

FYM was applied as pig manure at a rate of 25 t ha^{-1} once every three years in spring 10 days before potato planting. The 25 t ha^{-1} FYM was applied once in rotation to supply nutrient requirement of potato, winter rye, and oat production. It was a typical manure application rate in 1958. Nutrient content of FYM is given in Table 4 (YARA and KTBL, 2005).

Unit	Amount of N, P, K, and Mg in pig manure			
	N total	P2O5	K2O	MgO
kg t^{-1}	7	6.7	7.2	2.2
kg ha^{-1}	175	167.5	180	55

Table 4. Nutrient contents of 25 t ha^{-1} pig manure (FYM).

Since 1958, lime (CaO) was applied to the whole field at a rate of 1000kg ha^{-1} every three years to stabilize soil pH.

2.5 Measurements

Crop fresh and dry matter yields were recorded. N, P, and K concentrations of tuber, grain, and straw were analyzed. The straw of winter rye and oat were removed from the field.

Soil organic matter content was measured as C total and N total at depth of 0 – 30 centimetres (cm).

Soil P2O5, K2O, and pH levels were measured at 0 – 30 cm.

Mineralized N as NH_4^+ & NO_3^{-2} and mineralized sulphur as SO_4^{-2} were measured at three depths: 0 – 30 cm, 30 – 60 cm, and 60 – 90 cm.

2.6 Statistical analysis and calculations

The differences between average dry matter yields of treatments were analyzed statistically. During 1958 – 2008 each crop was grown 17 times in rotation. The average yield of each crop was calculated as an average of 17 years for each of the 4 replicates of a treatment in which each crop was grown. The average replicates were considered in comparison of means of potato tuber yield, winter rye grain yield, and oat grain yield.

The difference method was used to calculate nutrient use efficiencies. NUE (%) = Nutrient removal with fertilized crop minus nutrient removal with unfertilized crop divided by fertilizer rate × 100 (Varvel and Peterson, 1990). The N fertilizer use efficiency, for example, was calculated as the total N removal of the crop (tuber, grain, and straw) yield fertilized with N minus total N removal of crop yield without fertilizer (control treatment) divided by total N fertilizer rate times 100. The calculation was done in a similar way for P and K fertilizer use efficiencies.

Soil fertility levels were indicated by relative increase of C-total, and   P2O5 and K2O content.

3. Results

This paper mainly focuses on the results of a combination of FYM with mineral fertilizers as an example of the IPNM in comparison to the results of application of FYM alone as an example of the OPNM.

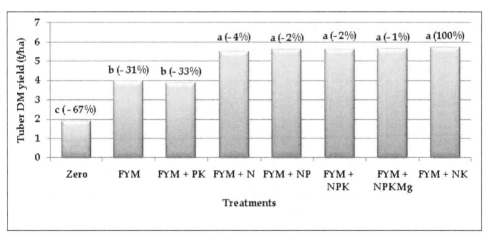

Fig. 1. Average potato yield (1958 –2006, n=17) at different combination of mineral fertilizer with FYM.

3.1 Effects of application of FYM alone, the combination of FYM with mineral fertilizer, and mineral fertilizer alone on crop dry matter (DM) yield
3.1.1 Average tuber yield of potato
Application of FYM plus mineral fertilizer was increased potato yield. The highest yield of potato tuber was measured at 5.74 t ha $^{-1}$ in the treatment of the combination of FYM with mineral NK fertilizer. The average yield of 5.74 t ha $^{-1}$ is quite low and can be explained by

the low yielding variety of potato at early decades of the trial and low water supply because of the sandy texture of the soil at the site. Integration of mineral P fertilizer with FYM did not achieve the highest yield because potato tuber yield is more responsive to K than P fertilizer. Application of FYM (i.e. organic fertilizer) alone decreased potato yield by 31% in comparison to integration of mineral NK fertilizer with FYM (Figure 1). Application of FYM alone and omitting application of mineral N+Mg, PK+Mg, and Mg fertilizers reduced potato yield by 15%, 57%, 41%, and 12%, respectively in comparison to the application of N+PK+Mg fertilizers without FYM (Figure 4). Balanced mineral fertilizer application (i.e. the N+PK+Mg treatment without FYM) achieved more additional yield than application of FYM alone even though less amount of N, P, K, & Mg were applied as mineral fertilizer in comparison to nutrient content of FYM at the year of potato cultivation of each of the rotations (Table 4).

3.1.2 Average grain yield of winter rye

The crop yield was increased with application of FYM plus mineral fertilizer. Application of mineral NP fertilizer with FYM achieved the highest average grain yield of 5.1 t ha^{-1} but application of FYM alone reduced crop yield by 56% in comparison to the highest yield (Figure 2). Organic fertilizer (FYM treatment) alone and omitting application of mineral N+Mg, PK+Mg, and Mg fertilizers reduced winter rye yield by 51%, 61%, 9%, and 3%, respectively compared to the yield of the treatment with N+PK+Mg (balanced mineral fertilizer) application (Figure 4). Effect of organic fertilizer (FYM) on winter rye yield was reduced, because most of the nutrient in the FYM was consumed by cultivation of potato before winter rye in a rotation (Figure 4).

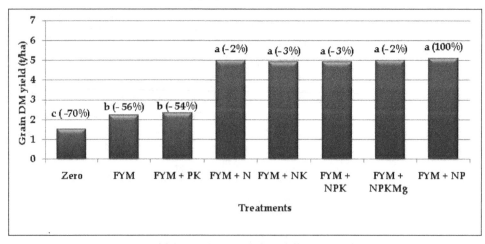

Fig. 2. Average winter rye yield (1959 –2007, n=17) at different combination of mineral fertilizer with FYM.

3.1.3 Average grain yield of oat

The highest average yield of oat grain was measured at 3.97 t ha^{-1} in the combination of mineral NP with organic fertilizer. With application of FYM alone the yield was reduced by

56% compared to the highest yield, but the yield was increased with application of FYM
plus mineral fertilizer (Figure 3). Organic fertilizer (FYM) alone and omitting application of
mineral N+Mg, PK+Mg, and Mg fertilizers reduced oat yield by 52%, 60%, 28%, and 5%,
respectively in comparison to the balanced mineral fertilizer application (Figure 4).
Unbalanced mineral fertilizer application is the result of either omitting, insufficient, or
over application of one or more nutrient. Omitting application of mineral N, PK, and Mg
fertilizers reduced crop yield, because specific function of a nutrient cannot be replaced with
specific functions of other nutrients.

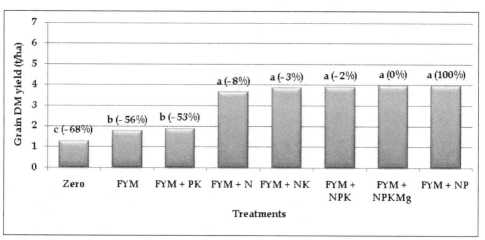

Fig. 3. Average oat yield (1960 –2008, n=17) at different combination of mineral fertilizer
with FYM.

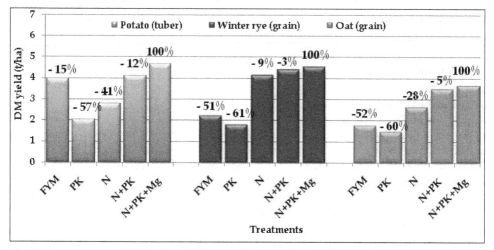

Fig. 4. Average crop yield (1958 –2008) at application of FYM alone and mineral fertilizer
without FYM.

3.2 Effects of application of FYM, combination of FYM with mineral fertilizer, and mineral fertilizer alone on nutrient use efficiency of crop

The effect of nutrient management on N, P, and K fertilizers use efficiency of crops during 50 years of the experiment is explained as follows.

3.2.1 The N fertilizer use efficiency (NFUE)

The combination of mineral N and P fertilizers with organic fertilizer (FYM) achieved the highest NFUE in comparison to all the treatments with FYM (Table 5). Application of FYM alone decreased NFUE of crop by 27% in comparison to the highest NFUE (Table 5). The combination of mineral with organic fertilizer resulted in a higher NFUE than application of FYM alone. Application of FYM alone and omitting application of mineral PK+Mg or Mg fertilizer reduced NFUE in comparison to the balanced mineral fertilizer application (Figure 5). Neither application of organic fertilizer (FYM) alone nor unbalanced mineral fertilizer application combines benefits of high yield and improvement of NFUE. Improvement of NFUE is required to minimize risk of environmental degradation, because increase of recovery of N by high yield of crop is responsible to reduce loss of N. High crop yield and improvement of NFUE are achieved with the approaches of IPNM and balanced mineral fertilizer application (i.e. the BPNM).

Treatments	N, P, and K fertilizers use efficiency					
	NFUE (%)	Relative to the highest	PFUE (%)	Relative to the highest	KFUE (%)	Relative to the highest
FYM	26.5	-27%	21.7	-63%	54.3	-57%
FYM+PK	29.6	-19%	9.5	-84%	20.5	-84%
FYM+N	33.8	-7%	54.4	-6%	112.1	-12%
FYM+NPKMg	34.8	- 4%	23.8	-59%	48.7	-62%
FYM+NK	35.5	- 3%	57.9	100%	47.8	-62%
FYM+NPK	35.7	- 2%	23.5	-59%	48.8	-62%
FYM+NP	36.4	100%	23.9	-59%	126.9	100%

Table 5. Nutrient use efficiency of crop.

3.2.2 The P Fertilizer Use Efficiency (PFUE)

Application of FYM without mineral fertilizer reduced PFUE of crop by 63% in comparison to the highest PFUE (Table 5). The PFUE of crop was increased in integration of mineral N or NK fertilizers with FYM (Table 5). The highest PFUE was achieved with application of FYM plus mineral NK fertilizer, because nutrient supply for crop demand is balanced. Application of organic fertilizer (FYM) alone and omitting application of mineral N+Mg or Mg fertilizer reduced PFUE in comparison to the balanced mineral fertilizer application (Figure 5). The PFUE of crops in integration of mineral PK with FYM and application of mineral PK without FYM is very low, because mineral P and K fertilizers application without mineral N fertilizer achieved very low yield and poor recoveries of P in tuber, grain, and straw of crops. Yield and the PFUE of crops were increased with the approaches of IPNM and balanced mineral fertilizer application (i.e. the BPNM).

Impact of Mineral Fertilizer Integration with Farmyard Manure on Crop Yield, Nutrient Use Efficiency,
and Soil Fertility in a Long-Term Trial

171

3.2.3 The K fertilizer use efficiency (KFUE)

The combination of mineral N and P fertilizers with organic fertilizer (FYM) resulted the highest KFUE of crop (Table 5). Application of organic fertilizer alone decreased KFUE of crop by 57% in comparison to the highest KFUE (Table 5). Integration of mineral N and NP fertilizers with organic fertilizer increased KFUE of crops in comparison to application of organic fertilizer alone, because it balances nutrient need of crops. Application of mineral NPK and NPK+Mg without FYM did not achieve higher KFUE than application of organic fertilizer alone, because the quantity of K fertilizer applied as mineral was 122% higher than the amount of K applied as FYM alone at each of the complete rotation (Tables 2 and 4). Integration of mineral NP fertilizer with FYM combines the advantages of the highest crop yield and improvement of the KFUE.

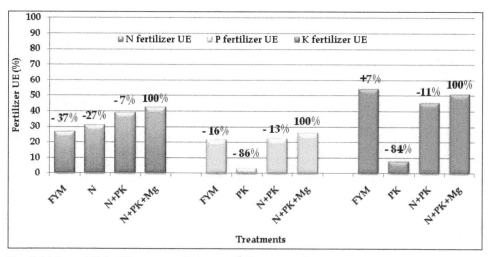

Fig. 5. N, P, and K fertilizers use efficiency of all crops and years.

3.3 Effects of application of FYM alone, the combination of FYM with mineral fertilizer, and mineral fertilizer alone on soil fertility

The effects of application of organic fertilizer (FYM) alone, integration of mineral with organic fertilizer, and balanced mineral fertilizer application on soil fertility during 50 years of the experiment are measured with the levels of soil organic matter content and the development of the P, and K content of the soil. Organic matter and nutrient content are considered as the major indicators of soil fertility.

3.3.1 Level of soil C total

The soil organic matter content was measured as soil C total. Organic matter improves soil fertility through its positive impact on chemical, physical, and biological properties of a soil. In general the level of soil C total is low because crop residues (straw of winter rye and oat) have been removed from the field. Even though the crop residues were removed from the field a significant amount of carbon has been accumulated by root biomass. The level of C total in top soil was increased with the application of farm yard manure (FYM) plus mineral fertilizer and the balanced mineral fertilizer application (the N+PK+Mg treatment).

However organic matter content of the soil was depleted by application of organic fertilizer alone (the FYM treatment) and by unbalanced application of mineral fertilizer (the PK treatment without FYM).The highest accumulation of C total was achieved with application of FYM plus mineral NP fertilizer (Table 6). Integration of mineral N and P fertilizers with FYM resulted in an increase of the soil organic matter content and in addition to this benefit it achieved the highest yield and nutrient use efficiency in comparison to application of FYM without mineral fertilizer.

3.3.2 Level of soil P2O5

The soil P2O5 level indicates the potential of a soil to supply P to a crop. The soil P index in Germany is classified as 'very low' (< 5), 'low' (6 – 9), 'medium' (10 – 20), 'high' (21 – 34), and 'very high' (> 35) mg P2O5 per 100 g soil at 0 – 30 cm depth (YARA and KTBL, 2005). Integration of mineral fertilizers with FYM; and the application of mineral PK and N+PK+Mg fertilizer without FYM increased the status of soil P from medium to the `high` index (Table 7). Application of organic fertilizer (FYM) without mineral fertilizer kept soil P index at the 'medium' level.

The soil P content was depleted from medium to the `low` index by application of mineral N fertilizer without FYM (Table 7). It was depleted by unbalanced mineral fertilizer application (mineral N fertilizer without FYM), because P has been removed through tuber, grain, and straw yields of crops without replacement. Integration of mineral NP fertilizer with FYM and balanced mineral fertilizer application increased soil P content in addition to the high yield of crop, better nutrient use efficiency, and improvement of soil organic matter content in comparison to the organic fertilizer alone and unbalanced mineral fertilizer application.

3.3.3 Level of soil K2O

The soil K2O level indicates the potential of a soil to supply K to a crop. The soil K index in N-Western Germany is ranked as 'very low'(< 2), 'low' (3 – 5), 'medium' (6 –12), 'high' (13 – 19), and 'very high' (> 20) mg per 100 g soil at depth of 0 –30 cm (Landwirtschaftskammer Nordrhein-Westfalen, 2011). The status of soil K was changed from the low to the `high` index with integration of mineral NPK fertilizer plus FYM, mineral PK fertilizer pus FYM, and application of mineral PK fertilizer without FYM (Table 7). This positive transformation of soil K did not result the highest crop yield because nutrient supply was unbalanced. Integration of mineral NP, NPKMg, and NK fertilizers with FYM ; the application of mineral fertilizers without FYM; and organic fertilizer (FYM) increased status of soil K from low to the `medium` index (Table 7).

The status of soil K was depleted from low to the ` very low` index by application of mineral N fertilizer without FYM (Table 7). Depletion of soil K content is the result of continuous K removal in crop yield without any replacement or with insufficient replacement. Integration of mineral NP fertilizer with FYM and balanced mineral fertilizer application increased soil K content in addition to maximization of crop yield, high nutrient use efficiency, and improvement of soil fertility. However application of FYM alone and unbalanced mineral fertilizer application did not result the highest crop yield and nutrient use efficiency. In general crop yield and nutrient use efficiency were reduced with application of FYM without mineral fertilizer (Figures 1, 2, & 3) and unbalanced mineral fertilizer application

(Figure 4) due to poor practices of nutrient management. Nutrient has to be effectively available to support the highest crop growth in order to maximize crop yield per area. Therefore nutrient should be efficiently managed to address the nutrient demand of crop effectively.

Treatments	C total in different years				Relative to initial	
	1958 (Initial)		2008 (Final)		2008 (Final)	
	%	kg ha $^{-1}$	%	kg ha $^{-1}$		kg ha $^{-1}$
FYM	1	42000	0.98	41160	- 2%	-840
PK	1	42000	0.96	40320	- 4%	-1680
N+PK	1	42000	0.99	41580	- 1%	- 420
N+PK+Mg	1	42000	1.01	42420	+ 1%	+420
N	1	42000	1.02	42840	+ 2%	+840
FYM +N	1	42000	1.04	43680	+ 4%	+1680
FYM +PK	1	42000	1.06	44520	+ 6%	+2520
FYM +NPKMg	1	42000	1.07	44940	+ 7%	+2940
FYM +NPK	1	42000	1.11	46620	+ 11%	+4620
FYM+NK	1	42000	1.11	46620	+ 11%	+4620
FYM+NP	1	42000	1.14	47880	+ 14%	+5880

Table 6. Soil C total at 0 to 30 cm depth.

Treatments	Milligrams of P2O5 per 100 gram soil		P index	Milligrams of K2O per 100 gram soil		K index
	1958	2008	2008	1958	2008	2008
FYM	12	20	Medium	5	7	Medium
N	12	7	Low	5	2	Very low
FYM+N	12	16	Medium	5	4	Low
N+PK	12	19	Medium	5	10	Medium
FYM+NK	12	21	High	5	12	Medium
PK	12	23	High	5	13	High
N+PK+Mg	12	23	High	5	11	Medium
FYM+NP	12	29	High	5	7	Medium
FYM+NPK	12	29	High	5	13	High
FYM+NPKMg	12	31	High	5	12	Medium
FYM+PK	12	33	High	5	15	High

Table 7. Soil P and K indexes at 0 to 30 cm depth in different years.

4. Discussion

Sources of plant nutrients are mainly mineral or organic origin. The organic source includes crop residues, animal manure, nitrogen fixation, green manure, and organic wastes. The organic materials need to get decomposed and available at a right amount, ratio, and time to effectively support growth and development of crop. Response of crop growth to organic source of nutrient depends on management and environmental conditions affecting decomposition rate of organic materials. Therefore management of organic materials is an important component of both organic and integrated plant nutrient managements. The management is mainly related to source of organic materials and , rate, method, and time of application.

Recycling of organic materials at farm level does often not fully compensate nutrient removal by crop yield, because agriculture is not a closed system. Development of export of agricultural raw materials into the world markets has enormously increased the distance nutrient travelled from fields and in the vast majority of current farms a return flow of nutrients in waste products is no longer feasible and the nutrient cycle has thus become a nutrient flow process based on mining, with substantial on-site losses in each cycle and accumulation in urban areas (Noordwijk, 1999). Hence cost of transportation limits the use of organic fertilizer over a long distance. It is also difficult to predict the availability of nutrients and to apply the right quantity and ratio of nutrients from organic fertilizer to meet nutritional requirement of the crop at the right time. Therefore locally available organic fertilizer need to be supplemented by mineral fertilizer to replace nutrient loss in a long distance of export and to minimize management difficulties in order to sustain productivity of crop production and to restore soil fertility.

4.1 Effect of nutrient management on crop yield

It is desirable to use both mineral and organic sources of plant nutrients in an integrated principle (Emeritus and Roy, 1993). Integration of mineral with organic fertilizer improves availability and corrects the balance of nutrient to achieve healthy growth and development of crop. It increased potato tuber and cereal grain yields, because nutrient availability is improved. Combination of FYM with mineral NK fertilizers increased potato yield by 45% in comparison to application of FYM alone (Figures 1). Maize grain yield was similarly increased by 52% with integration of mineral N, P, and K fertilizers with cattle manure (Abunyewa, 2007). Also application of mineral N in mixture with FYM at N supply of 50% urea and 50% FYM increased wheat grain yield by 66% in comparison to application of FYM alone at the same rate of mineral N fertilizer (Zahir and Mian, 2006). The highest potato and cereal yields with the application of FYM plus mineral NK fertilizer and FYM plus mineral NP fertilizer, respectively confirm the economic benefit of the IPNM (Figures 1, 2, and 3).

The balanced mineral fertilizer application as the BPNM resulted a higher crop yield than with FYM alone (the OPNM) and an unbalanced mineral fertilizer application (Figure 4). Potato yield in the treatment of FYM alone was 15% lower than the potato yield of the treatment with balanced mineral fertilizer application , even though quantities of N, P, K, and Mg applied as FYM was higher than the N, P, K, and Mg applied as mineral fertilizer during potato cultivation (Figure 4 and Tables 4 & 2). Application of organic fertilizer alone does not fully satisfy nutritional requirement of crop because it is relatively difficult to balance nutrient availability to maximize crop growth and yield per area. Application of FYM alone decreased potato yield by 31% in comparison to integration of mineral NK fertilizer with FYM and it also reduced winter rye and oat yield by 56% in comparison to application of mineral NP fertilizer plus FYM (Figures 1, 2, and 3). In general with the best nutrient management practices, the IPNM and the BPNM, the highest crop yield per area were achieved in comparison to inferior management practices like the OPNM and the unbalanced mineral fertilizer application.

Maximization of crop production per area is the basis to achieve sufficient and affordable food to effectively meet the demand of a growing world population. Promotion of optimal and efficient plant nutrition is required on a large scale to achieve the 700 million tonnes of additional cereals that will be required by 2020. About 80% of the additional demand will have to come from already cultivated areas (Roy et al. 2006). Harvesting a maximum yield per area is the primary criteria to secure supply of food considering the limited potential to extend the crop land area globally (Bruinsma, 2003). Integration of mineral with organic fertilizer is one of the effective practices of plant nutrient management to increase crop yield per area. Therefore the highest yields of crop with FYM+NK (potato) and FYM+NP (cereal) treatments also supports the social benefit of the IPNM (Figures 1, 2, and 3).

4.2 Effect of nutrient management on nutrient use efficiency

Nutrient in organic materials should be used efficiently in order to increase crop yield per area and to reduce nutrient loss. Integration of mineral fertilizer with FYM increased nutrient use efficiency (NUE) of crops. However with application of organic fertilizer (FYM) alone the NFUE, the PFUE, and the KFUE of crops were reduced by 27% , 63%, and 57%, respectively in comparison to the highest nutrient use efficiency at the integration of mineral fertilizers with FYM (Table 5). The highest NFUE and KFUE were achieved with the integration of mineral NP fertilizer with FYM and the highest PFUE was achieved with the integration of mineral NK fertilizer with FYM (Table 5). Integration of NP with FYM increased NFUE and KFUE and application of FYM plus mineral NK fertilizer increased PFUE of crops; because application of one nutrient increases use efficiency of other nutrient through synergistic effect, i.e. the total essential functions of two or more nutrients is higher than the sum of the essential functions of a single nutrient.

The NFUE and the PFUE of crops were decreased by 37% and 16%, respectively with application of FYM alone in comparison to the balanced mineral fertilizer application (Figure 5). This indicates that nutrient in the FYM is not sufficiently available for uptake by crop at the right growth stages. Balanced mineral fertilizer application (the N+P+K+Mg treatment) resulted the highest yield of crops with the highest nutrient use efficiency: thereby, it balances economic, social, and environmental conditions for sustainability of crop production (Jate, 2010). Loss of N and P to the environment is reduced through improvement of NFUE and PFUE of crops by integration of NP and NK mineral fertilizers with FYM and balanced mineral fertilizer application. Therefore negative effects of N and P loss on environment is minimized with the approaches of the IPNM and the BPNM as the best practices of plant nutrient management.

4.3 Effect of nutrient management on soil fertility

Improvement of soil fertility is one of the basic criterions to maximize crop yield and to minimize nutrient loss per area. Integration of mineral with organic fertilizer increases soil fertility through improvement of physical, biological and chemical properties of soil. Improvement of soil organic matter improves soil physical properties and it increases nutrient availability that these improvements should ultimately lead to increase of crop growth and yield (Onemli, 2004). Application of FYM plus mineral fertilizer improved soil fertility through improvement of organic matter and nutrient content of the soil (Tables 6 and 7). Fertilizer application significantly increased the concentrations of N, P, K and organic carbon in the plough layer of soil (Ishaq et al. 2002).

Integration of mineral N and P fertilizers with FYM achieved 16% more organic matter (carbon) content of the soil in comparison to application of FYM alone (Table 6). It resulted the highest crop yield per area with application of FYM plus mineral NP fertilizer (Figures 2 and 3). Similarly the integration of compound NPK (20:10:10) fertilizer at 150 kg ha^{-1} with FYM (cocoa pod ash) at 10 t ha^{-1} increased maize grain yield and soil carbon content by 24% and 16%, respectively in comparison to FYM without mineral fertilizer (Ayeni, 2010). Accumulation of soil organic matter content improves growth condition for crop production through improvement of soil fertility and it also reduces CO_2 emission through sequestration of carbon in the root biomass. Therefore both soil fertility and environmental benefits are achieved with the practice of IPNM.

Fertilizer application increases crop yield per area through direct improvement of nutrient concentration in the soil. Integration of mineral NP and NK fertilizers with FYM and balanced mineral fertilizer application sustained high soil P (Table 7) and medium soil K (Table 8) status in addition to improvement of crop yield, nutrient use efficiency, and soil organic matter. Neither organic fertilizer alone nor unbalanced mineral fertilizer application ensures sustainability of crop production. Integration of mineral with organic fertilizer and balanced mineral fertilizer application sustain crop production through improvement of soil fertility and nutrient use efficiency. Therefore the IPNM and the BPNM are the best practices of plant nutrient management to increase crop production per area, to restore soil fertility, and to minimize negative effect of nutrient loss on environment.

5. Conclusion

Availability of sufficient quantity and effective form of nutrient just on a time at a right ratio and even distribution at root zone and canopy surface are the major parameters of nutrient management responsible to optimize nutrient uptake and crop yield. Application of organic fertilizer (FYM) plus mineral fertilizers as integrated plant nutrient management (IPNM) increases crop production per unit area through improvement of nutrient availability. It also improves nutrient use efficiency and soil fertility. The approach of IPNM combines economic, social, environmental, and soil fertility benefits of the best practice of nutrient management. These benefits are the basic criteria to sustain high crop yield per area in order to secure physical availability and socio-economic accessibility of food at a family, a local, a country, a regional, and global levels.

The IPNM is one of the best practices of plant nutrient management that determines crop production potential of a soil in addition to its positive effect on nutrient use efficiency to minimize negative impacts of agriculture on quality of environment. It contributes to environment protection through reduction of N and P losses and sequestration of carbon in the soil organic matter. It reduces imbalances of nutrients; it limits uncertainty of nutrient availability; it minimizes nutrient loss; it enhances soil organic matter content; and it avoids degradation of soil fertility. These major positive effects are very important to maximize crop yield per area, to improve nutrient use efficiency, and to ensure sustainability of soil fertility.

The highest crop yield with the approach of the IPNM, in comparison to application of organic fertilizer alone (the OPNM), is the result of improvement of nutrient availability, balance, and uptake. The practice of the IPNM can balance rate and ratio of nutrient availability at the right growth stages of crop to increase yield per area with environmental responsibility. Increase of crop production per area is needed to sustain physical availability and socio-economic ability to access food for humans and feed for animal consumptions. Nutrients of organic sources are taken up by plant roots after they are broken down (mineralized) into ionic forms. For the crop uptake there is no ionic difference between N, P, K, Ca, Mg, S, etc available from mineral and organic fertilizers. Therefore integration of mineral with organic fertilizer (the IPNM) is securer than the organic fertilizer alone (the OPNM): because more crop yield and less nutrient loss per hectare has been achieved with the IPNM compared to the OPNM.

6. References

Allison, F.E. (1973). *Soil organic matter and its role in crop production.* P. 277 – 360, Development in soil sciences 3, Elsevier Scientific Publishing Company, ISBN 0 – 444 - 41017-1, Amsterdam, London, New York

Abunyewa, A.A.; Osei, C.; Asiedu, E.K.; and Safo, E.Y. (2007). Integrated manure and fertilizer use, maize production and sustainable soil fertility in sub humid zone of west Africa. *Journal of Agronomy. 6(2): 302 – 309,* ISSN 1812-5379

Alley, M.M. and Vanlauwe, B. (July 2009). *The Role of Fertilizers in Integrated Plant Nutrient Management.* International Fertilizer Industry Association, and Tropical Soil Biology and Fertility Institute of the International Centre for Tropical Agriculture, Retrieved from: www. fertilizer.org/ifacontent/.../2/file/2009_ifa_role_plant_nutrients.pdf

Aulakh, M.S. and Grant, C.A. (2008). *Integrated nutrient management for sustainable crop production.* P.29, CRC Press, Taylor & Francis Group. ISBN 978 – 1-56022 -304 – 7, London

Ayeni, L.S. (2010). Effect of Combined Cocoa Pod Ash and NPK Fertilizer on Soil Properties, Nutrient Uptake and Yield of Maize (Zea mays). *Journal of American Science 6 (3): 79 – 84,* ISSN 1545 – 1003

Bruinsma, J. (2003). *World Agriculture: Towards 2015/2030, an FAO perspective.* Earthscan Publications Ltd, Retrieved from: http://www.fao.org/fileadmin/user_upload/esag/docs/y4252e.pdf

Emeritus, D. and Roy, R.N. (December 1993). Summary report, conclusion, and recommendations. In: *Integrated plant nutrition systems: Report of an expert consultation,* FAO, 04. 07. 2011 Available from: www. ftp://ftp.fao.org/agl/agll/docs/ipnseng.pdf FAO, (1998). *Guide to efficient plant nutrition management.* Land and water development division of FAO of the United Nations. Retrieved from www.ftp://ftp.fao.org/agl/agl/docs/gepnm.pdf

Ishaq, M. Ibrahim, M, and Lal, R. (2002). Tillage effects on soil properties at different levels of fertilizer application in Punjab, Pakistan. *Soil Tillage Research 68:* 93–99.

Jate, M. (2010). Long-term effect of balanced mineral fertilizer application on potato, winter rye, and oat yields; nutrient use efficiency; and soil fertility. *Archives of Agronomy and Soil Science 56 (4): 421 – 432.*

Landwirtschafts-kammer Nordrhein-Westfalen. (2011). Düngung mit Phosphat, Kali, Magnseium. In: *Ratgeber 2011.* 04. 07. 2011. Available from: http://www.landwirtschaftskammer.de/landwirtschaft/ackerbau/pdf/phosphat-kalium-magnesium-pdf.pdf

Noordwijk, M. V. (1999). Nutrient cycling in Ecosystems versus nutrient budgets of agricultural systems. In: *International Centre for Research in Agroforestry*, 04. 07. 2011, Available from: http://www.worldagroforestrycentre.org/sea/Publications/files/bookchapter/B C0104-04/BC0104-04-1.PDF NRCS and NHCP. (August 2006). Nutrient management (Ac.) code 590. In: *Natural resources conservation service conservation practice standard*. 05. 07. 2011. Available from: ftp://ftp-fc.sc.egov.usda.gov/NHQ/practice-standards/standards/590.pdf

Onemmli, F.(2004). The effects of soil organic matter on seedling emergence in sunflower (*Helianthus annuus* L.). *Plant and soil environment, 50*: 494–499

Rahman, S. and Parkinson, R.J. (2007). Productivity and soil fertility relationships in rice production systems, Bagladesh. *Agricultural.Systems. 92: 318 – 333*

Robert, M. (2005). Nutrient Use Efficiency: using nutrient budgets. In: *Western nutrient management conference volume 6 (1-7)*, 07. 04. 2011, Available from: http://isnap.oregonstate.edu/WERA_103/2005_Proccedings/Mikkelsen%20N%20 Use%20Efficiency%20pg2.pdf

Roberts, T.L. (2010). Nutrient best management practices: Western perspectives on global nutrient stewardship, *19th World Congress of soil science, soil solutions for a changing world*, pp172 – 175, Brisbane, Australia. August 1-6, 2010

Roy, R.N., Finck ,A., Blair, G.J. and Tandon, H.L.S.(2006). *Plant nutrition for food security. A guide for integrated nutrient management.* Land and water development division of FAO of the United Nations. Retrieved from www.fao.org/icatalog/search/dett.asp?aries_id=107288

Steiner, R.A. and Herdt, R.W. (1993). In McRae, K.B. and Ryan, D.A.J. (1996). Design and planning of long-term experiments. *Canadian Journal of Plant Science 76*: 595–602. USDA and NRCS (2003). Managing soil organic matter, the key to air and water quality. In: *Soil quality technical note No. 5*, 04. 07. 2011., Available from: www.soils.usda.gov/sqi/concepts/soil_organic_matter/.../sq_tn_5.pdf

Varvel, G.E. and Peterson, T.A. (1990). Nitrogen fertilizer recovery by corn in monoculture and rotation systems. *Agronomy Journal. 82*:935–938.

YARA and KTBL. (2005). *Faustzahlen für die Landwirtschaft.* p. 238, KTBL-Schriftenvertrieb im Landwirtschaftsverlag, ISBN 3-7843-2194-1, Darmstadt

Zahir, S. and Mian, I.A. (2006). Effect of integrated use of farm yard manure and urea on yield and nitrogen uptake of wheat. *Journal of agricultural and biological science.* 1 (1): 61– 65. ISSN 1990-6145

10

Vegetable Waste Compost Used as Substrate in Soilless Culture

Pilar Mazuela[1], Miguel Urrestarazu[2] and Elizabeth Bastias[1]
[1]Universidad de Tarapacá
[2]Universidad de Almería
[1]Chile
[2]Spain

1. Introduction

One of the main environmental impacts of forced systems in horticulture – such as plastic covers and soilless culture - is the generation of organic plant residues and substrate waste. For example, the surface area of greenhouse cultivated crops in the province of Almeria, in southeastern Spain, exceeds 30,000 ha. These generate approximately 1,000,000 tons of solid plant waste per year. Greenhouse industry residues cause serious environmental and visual pollution, making it necessary to look for new ways to eliminate these plant residues. This mass not only acts as a host for pests, microorganisms, rats and insects; it also has other harmful environmental effects such as pollution of the soil by toxic elements, effluent runoff, and the emission of bad smells. Conway (1996) indicated that an important factor for sustainable agriculture in areas using protected systems is the need to eliminate the harvest residues of these crops. Controlled composting appears to be an effective method of eliminating residues by recycling them. For example, Ozores-Hampton *et al.* (1999) reported that in Florida 1.5 million tons of compost could be produced per year.

The wastes genereted by intensive agriculture systems are very varied and frequently cannot be reused directly. Cara and Ribera (1998) indicated that greenhouses generate 29.1 tons of vegetable waste per ha and 6-10 tons of substrate remains per year in the province of Almería (Spain). A less indiscriminate form of management of these residues, however, could turn them into usable products. This would also reduce their environmental impact. Callejón et al., 2010 indicated that the assessment of the environmental impact of a potential waste treatment plants showed that it would be better to recycle and compost waste than to try to obtain energy from it through combustion. This compost can be used as a soil conditioner or to improve the structure in degraded soils or those with low organic matter content. Another alternative is to reuse these residues, incorporating them as ecologically friendly substrates in soilless cultivation in the form of compost.

Using waste materials, most of them locally produced, as soilless growing media has been the subject of an important number of studies, especially as an alternative to peat for ornamental potted plants (e.g., Ingelmo et al., 1997; Offord et al., 1998; Lao and Jiménez, 2004a,b), and less frequently for vegetable production (Shinohara et al., 1999; Ball et al., 2000) and even for tomato transplant production (Ozores-Hampton et al., 1999). However, it has been suggested that certain types of compost alone are unsuitable as growing media due to unacceptably high salt and pH content (Spiers and Fietje, 2000), in particular when immature, unstable compost is used (Ozores-Hampton et al., 1999). Another disadvantage of the use of compost as substrate is that it is a very heterogeneous material and therefore needs to be amended so that it can be used as substrate (Urrestarazu et al., 2000; Urrestarazu et al., 2001; Urrestarazu et al., 2003; Sanchez-Monedero et al., 2004; Carrión et al., 2005; Mazuela et al., 2005; Mazuela et al., 2010)). Once physical-chemical properties were adjusted for soilless culture, yield trials proved the suitability of compost as an acceptable soilless growing media and as a viable and ecologically friendly alternative substrate.

Fig. 1. Leaching compost assays

Fig. 2. New and re-used compost from horticultural waste crops

Fig. 3. Texture and coarseness index in new compost

Fig. 4. Melon production using compost as substrate in southeastern Spain

2. Vegetable waste use as an alternative and friendly substrate

Many people are keen on the research and development of ecologically friendly substrates. Recently it has been demonstrated that these substrates are a perfectly viable alternative to other more traditional methods such as rockwool, perlite or hydroponic systems. However, in order to be competitive for vegetable production in the Mediterranean region, they must be used for at least one year. In recent years there has been an increase in soilless crop cultivation in southeast Spain (Almeria, Murcia and Granada) with a current surface area estimated at 5,000 ha, using substrates such as rockwool, perlite, sand, coconut fibre and other minor types (Urrestarazu and Salas, 2002).

Fig. 5. Cherry production using compost as substrate in northern Chile

Fig. 6. Tomato production using compost as substrate in greenhouse

Alternative substrates have the advantage of being locally produced, renewable and contaminant and disease-free (Salas et al., 2000), and may be less expensive than any other traditional growing media used in soilless crop production. Abad et al. (2002) reported that coconut coir waste may be used for ornamental crops, and the decision on whether to use it as a peat substitute will depend primarily on economic and technical factors, and secondly on environmental issues.

Fig. 7. Effect of high salinity on a tomato crop

Selected characteristics of some alternative substrates are given in Table 1.

	C[1]	AS[2]	CF[3]	PF[4]	RV[5]
Bulk density (g cm[-3])	0.38	0.40	0.059	0.061	< 0.40
Real density (g cm[-3])		1.40-1.45	1.51	1.46	1.45-2.65
Coarseness index (%)	62.2	84.2-86.3	34	97.5	
Total pore space (% vol)	80	71-72	96.1	95.8	> 85
Total water-holding capacity (mL L[-1])	388	188-194	523	187	600-1000
Shrinkage (%)	11.1	10-12.3	14	10	<30
Organic matter content (% dry wt)	58.5	99	93.8	99	> 80
pH	7.8-8.0	5.1-5.2	5.71	5.76	5.2-6.3
Electrical conductivity (dS m[-1])	22.2-34.3	2.44-2.70	3.52	0.63	0.75-1.99

Source: [1]Mazuela et al., 2005; [2]Urrestarazu et al. 2005a; [3]Abad et al., 1997 ; [4]Urrestarazu et al., 2006; [5]Abad et al., 1993

Table 1. Selected physical, physical-chemical and chemical properties of alternative substrates: compost (C), almond shells (AS), coir fiber (CF), pine fiber (PF) and reference value (RV)

Fig. 8. Melon transplant production

2.1 Plant pathogen elimination by composting

Many studies indicate that the elimination of plant pathogens is possible through composting, although for different time periods at different temperatures. High temperatures eliminate phytopathogens such as *Pythium irregulare, Pythium ultimun* (Suárez-Estrella et al., 2007), *Rhizoctonia solani* (Hoitink et al., 1976; Christensen et al., 2001; Suárez-Estrella et al., 2007), *Fusarium oxysporum f.sp. melonis* (Suárez-Estrella et al., 2003, 2004), *Xanthomonas campestris pv. Vesicatoria, Erwinia carotovora* and *Pseudomonas syringae pv. syringae* (Elorrieta et al., 2003; Suárez-Estrella et al., 2007), and several viruses such as tomato spotted wilt virus (TSWV) and pepper mild mottle virus (PMMV) (Suárez-Estrella et al., 2002). Suárez-Estrella et al. (2003) suggested that composting is therefore a useful method to recycle horticultural waste, when it is ensured that all pathogenic bacteria are eliminated in the process.

2.2 Trace metals

The levels of trace metals in different composts are known to be much higher than in most agricultural soils (He et al., 1992), and depend on the origin of the compost. Pinamonti et al. (1997) reported that the use of compost from sewage sludge and poplar bark did not cause any significant increase in heavy metal levels in soil or plants in the short/medium term; by contrast, their experiments clearly demonstrated that the compost from municipal solid waste increased concentrations of Zn, Cu, Ni, Pb and Cr in soil, and in the case of Pb and Cd also in the vegetation and the fruits. In Spain, as in other European countries such as The Netherlands and Italy, the concentrations of some heavy metals are regulated in order to guarantee the safe use of compost.

Heavy metal levels or potentially toxic microelements (Table 2) were below the tolerated limits in compost according to both the European (BOE, 1998) and American (US, 1997) regulations and the limits established by authors such as Abad et al. (1993) for soilless production of vegetables and ornamentals.

| Metal | Average contents in compost | | Regulated limits | | | |
	USA[1]	Spain[2]	Vegetables[3]	Ornamentals[3]	Amendment[4]	USA[5]
Zn	503	95.9-179	1000	1500	1100	2800
Cu	154	37.2-98.5	100	500	450	1500
Cr	34.8	5.02-11.2	150	200	400	1200
Pb	215	6.18-9.1	600	1000	300	300
Ni	24.8	3.6-6.45	50	100	120	420
Co	-	1.29-2.07	50	50	-	-
Cd	2.9	0.11-0.25	5	5	10	39

Source: [1]Epstein et al., 1992; [2]Mazuela et al., 2005; [3]Abad et al.,1993; [4]BOE, 1998, [5]US Composting Council, 1997

Table 2. Concentrations of heavy metals (mg kg[-1]) of compost obtained from horticultural crop residues in the United States and Spain and regulated limits

2.3 Leaching of horticultural greenhouse crop waste for substrate

Mazuela and Urrestarazu (2009) determined the effect of leaching of compost for substrate preparation with two composting processes; C1 compost formed by mixing pepper, bean and cucumber waste; and C2, compost formed by melon plant waste. In both cases, sawdust was added (1:4 ratio v/v), as a C/N relation conditioner. The composting process was described by Suárez-Estrella et al. (2003), who indicated that piles of 2 m^3 were turned over and aerated periodically after the first 14 days of composting. Electrical conductivity (EC), anion content (NO_3^-, $H_2PO_4^-$, SO_4^{2-}, Cl^-) and cations (Ca^{2+}, K^+, Mg^{2+}, Na^+) were determined by the saturation extract method (Warncke, 1986).

Table 3 shows EC values above 21.38 and 11.84, in compost C1 and C2, respectively, likeness ratio indicated by McLachlan et al., 2004; Sanchez-Monedero et al., 2004; Mazuela et al., 2005. These values are higher than the recommended range of 0.75-1.99 reported by Abad et al. (1993) as optimum for soilless culture (Table 1). Amendment with leaching 1:6 volumes is sufficient to produce acceptable values of EC for horticultural purposes (Sanchez-Monedero et al., 2004; Mazuela et al., 2005).

Soluble salts represent dissolved inorganic ions in the solution and are typically measured in terms of electrical conductivity. EC readings and mineral element concentrations of composts decreased sharply with leaching and eventually reached acceptable levels despite the high initial value of this parameter in the compost. This drop in the EC was parallel to that found in the concentrations of soluble mineral elements, mainly SO_4^{2-}, K^+, Cl^-, Mg^{2+}, Ca^{2+} and Na^+, and showed significant differences for higher levels of elements at the end of the experiment independent of the initial values in the composts. Often, soluble salt measurements from different studies or laboratories cannot be cross-referenced or there is a lot of confusion when comparing the results. Dimambro et al. (2007) and Carrión et al. (2005) reported that total salts were higher in mixed waste composts, predominantly due to high concentrations of K^+, Ca^{2+}, SO_4^{2-}, and Na^+. Nitrates and phosphates in leaching had low levels in both composts without significant differences. This suggests that the low levels of available nitrogen and the chemical binding or adsorption of phosphorus found in the composts studied will reduce N and P concentrations in leachates.

sv:wv		EC	Anions				Cations			
			NO_3^-	$H_2PO_4^-$	SO_4^{2-}	Cl^-	K^+	Na^+	Ca^{2+}	Mg^{2+}
1:1	C1	21.38	0.97	0.14	191.83	72.53	172.20	22.55	46.00	49.00
	C2	11.84	8.12	0.98	88.91	17.44	49.80	8.85	25.50	20.56
		**	ns	ns	*	**	*	*	**	*
1:2	C1	10.78	0.25	0.00	128.97	29.15	96.00	10.50	25.50	28.60
	C2	5.51	5.26	1.25	46.64	4.71	22.20	7.25	16.60	7.94
		*	ns	ns	**	**	**	**	ns	*
1:6	C1	2.66	1.96	0.09	23.01	2.51	16.60	1.62	5.83	3.95
	C2	1.95	0.63	0.38	23.92	0.54	6.10	0.76	10.40	3.39
		*	ns	ns	ns	*	*	*	*	ns
1:8	C1	1.59	0.54	0.07	4.59	0.43	3.14	0.46	2.06	1.19
	C2	0.82	0.08	0.21	19.94	0.03	2.66	0.38	9.53	2.89
		**	ns	ns	**	ns	ns	ns	**	**
1:10	C1	1.47	0.23	0.08	1.27	0.13	2.10	0.16	1.27	0.86
	C2	0.36	0.00	0.18	19.13	0.03	2.00	0.28	9.13	2.83
		**	ns	ns	**	ns	ns	ns	**	**

Values are means of three replicates.
*, **, ***, ns are $P \leq 0.05$, $P \leq 0.01$, $P \leq 0.001$ and not significant or $P > 0.05$, respectively.
Source: Mazuela and Urrestarazu, 2009

Table 3. Electrical conductivity ($dS\ m^{-1}$), anion and cation contents ($me\ L^{-1}$) in leaching experiments of compost from two horticultural crop residue mixtures (C1, mixing pepper, bean and cucumber waste; C2, melon plant waste) using distilled water in different substrate volume: distilled water volume (sv:ws)

Thus to avoid environmental pollution, special emphasis must be paid to the management and treatment of effluents produced when leaching saline composts under commercial conditions. The preparatory operation needs about six times the water volume of the substrate and should be done inside a composting station, where the lixiviated fertigation is controlled. It is recommended to saturate the substrate with the standard nutrient solution before draining the bags (Villegas, 2004).

2.4 Characteristics of compost used as growing media and effects on yield and quality in horticultural crops

Physical properties are the most important characteristics in a new alternative substrate, because they do not change when the substrate is in the container. These characteristics determine the time and frequency of irrigation. Table 4 shows the particle-size distribution of composts and a coarseness index, expressed as the percentage weight of particles with Ø > 1 mm (Richards et al., 1986).

Particle Sizes (mm)								
<0.125	0.125-0.25	0.25-0.5	0.5-1	1-2	2-4	4-8	8-16	CI
2.59	6.00	11.88	17.36	36.31	18.14	5.14	2.58	62.2

Table 4. Texture and particle size distribution of compost originated from horticultural crop residues used as soilless growing media (% wt) and coarseness index (CI)

Texture was very similar to those recommended by Jensen and Collin (1985) for soilless vegetable culture. The coarseness index was about 62 %, similar to peat (63 %) and much higher than coconut coir waste (35 %) values reported by Noguera et al. (2000). hese values easily explain the high wettability of compost (Table 1) according to Bunt (1988). Bulk densities were within the limits of the optimal range. Total pore space showed lower than optimum levels. The total water-holding capacity of composts did not stay within the optimum values (Abad et al., 1993). Shrinkage, wettability and organic matter content stayed within the optimum range.

However, the deficient physical properties were not limiting for crop yield and quality (Table 5, Table 6) probably because crops were irrigated according to the physical analysis (Table 7) of the substrate, a method tailored to the water transport capabilities of each individual substrate (Drzal et al., 1999). The criteria of Smith (1987) and the necessary local adjustments (Salas and Urrestarazu, 2001) were adopted in the fertigation management.

	Tomato				Melon (Galia)			
	cv Josefina[1]		cv Daniela[2]		cv Yucatán[3]		cv Danubio[4]	
Substrate	kg m-2	n° m-2	kg m-2	n° m-2	kg m-2	n° m-2	kg m-2	n° m-2
CW	6.82	790	4.68	43	6.55	4.97	5.89	5.11
C	6.00	712	4.75	44	6.05	4.80	5.29	4.63

Source: [1]Urrestarazu et al., 2000; [2]Urrestarazu et al., 2003; [3]Mazuela et al., 2005; [4]Mazuela and Urrestarazu, 2009

Table 5. Effect of coconut coir waste (CW) and compost (C) on yield of melon crops.

	Yucatán[1]				Danubio[2]			
	F	TSS	pH	DWC	F	TSS	pH	DWC
CW	1.71	12.35	6.26	9.47	2.15	10.45	6.92	7.94
C	1.56	12.66	6.22	9.61	1.81	10.50	6.67	8.12

Source: [1]Mazuela et al., 2005; [2]Mazuela and Urrestarazu, 2009

Table 6. Effect of coconut coir waste (CW) and compost (C) on selected fruit parameters of melon crops.

	Yucatán[1]				Danubio[2]			
	Drainage			Uptake	Drainage			Uptake
	EC	pH	%	L m-2 crop-1	EC	pH	%	L m-2 crop-1
CW	4.76	5.71	21.41	2852	3.11	6.52	18.39	2049
C	4.00	7.02	31.72	2433	3.67	7.37	22.63	1943
P	ns	ns	ns	ns	ns	ns	ns	ns

EC: Electric conductivity (dS m-1)
Source: [1]Mazuela et al., 2005; [2]Mazuela and Urrestarazu, 2009

Table 7. Daily mean fertigation parameters and water uptake of melon crops in coconut coir waste (CW) and compost (C)

As part of a correct management procedure, previous acid rinsing and saturation with the standard nutrient solution are recommended in order to reduce the compost salinity and inadequate pH of the rhizosphere environment (Table 3). Once the physical-chemical properties were adjusted for soilless culture, yield trials proved the suitability of the compost as an acceptable soilless growing media and as a viable and ecologically-friendly alternative to rockwool and coconut coir waste (Table 8). In northern of Chile, Mazuela eta al., (2010) have similar results that shows in Table 9.

Fig. 9. Vegetable waste compost produced with grapes residues from CAPEL, Punitaqui, Chile

Substrate	Yield			Quality	
	kg m^{-2}	n° fruit m^{-2}	F (kg)	TSS (° Brix)	DWC (%)
Compost	6.05	4.80	1.6	12.7	9.6
Almond shells	6.52	4.56	2.2	12.5	9.8
Coir fiber	6.55	5.08	1.7	12.3	10.1
Pine fiber	6.19	5.12	1.6	11.8	9.9
Rockwool	6.57	5.00	1.7	11.8	7.4

F: Firmness; TSS: Total Soluble Solids; DWC: Dry Water Content
Source: Urrestarazu et al., 2006

Table 8. Yield and quality in melon crops of alternative substrates: compost, almond shells, coir fiber, pine fiber and rockwool

	Fertigation			Yield		Quality		
	EC	pH	%	kg m^{-2}	n° m^{-2}	TSS	FF	DWC
GH	2.86	7.51	20.59	3.79	342	8.43	1.11	8.47
AM	3.05	7.55	22.97	2.93	293	9.20	1.14	9.67

EC: Electric conductivity (dS m^{-1}); TSS: Total soluble solids (° Brix); FF: Firmness (kg); DWC: Dry weight content (%)

Table 9. Fertigation in drainage parameters, yield and quality in tomato (cherry) crop, in northern of Chile in Greenhouse (GH) and antiaphid mesh (AM)

2.5 Re-used substrate from waste materials

Recently, it has been demonstrated that the use of some ecologically friendly substrates are perfectly viable as alternatives to other more traditional media such as rockwool, perlite or hydroponic systems. Almond shell was found to be a viable culture substrate by Lao and Jiménez (2004a, b); these researchers used it as a peat substitute for an ornamental crop. Urrestarazu. (2008) reported that pure compost can be an acceptable substitute growing media for rockwool and coconut coir waste once it is leached and adjusted to physical-chemical proprieties. o limiting factors in comparison to rockwool were found for tomato and melon crops when alternative substrates were used as growing media.

Rockwool slab, perlite and coconut bag culture are used in southeastern Spain for two or three years for vegetable production (García, 2004; Villegas, 2004); consequently, in order to be competitive in the market of soilless crops and to have similar commercial opportunity, the unit with an alternative substrate must be usable for this time. Urrestarazu et al. (2008) showed that re-used alternatives substrates as compost or almond shells did not affect yield in melon and tomato crops (Table 10). Because environmental care and economic profit to the grower are of paramount importance, it was important to see if this re-use of alternative substrate would be viable in Mediterranean conditions.

Substrate	Melon Galia (cv Aitana)		Tomate (cv Pitenza)	
	kg m⁻²	n° fruits m⁻²	kg m⁻²	n° fruits m⁻²
New compost	5.1	4.5	7.7	99
Re-used compost	5.2	5.0	7.8	99

Source: Urrestarazu et al., 2008

Table 10. Yield of melon and tomato crops with new and reused compost

Fig. 10. Pepper seeds production in Quillota, Chile

Bulk density is a relevant substrate physical propriety, because this allows easier transportation of crop units in the greenhouse industry (Abad et al., 2004). The new substrates, before reutilization, were within the limit of optimal range (Urrestarazu et al., 2005b); in fact, this is the major disadvantage for transport in comparison to other more popular substrates such as rockwool and perlite (Mazuela et al., 2005; Urrestarazu et al., 2005b).

2.6 Fertigation management and reference values for nutrient dissolution in organic substrates

Part of a correct management procedure for the use of compost as substrate in growing media is to saturate the bags with nutrient solution before draining the containers, in order to reduce the compost salinity in the rhizosphere environment. Once the electrical conductivity level was adjusted for soilless culture, yield trials proved the suitability of compost as an acceptable growing medim and this ecologically friendly alternative did not affect production, yield or fruit quality of melon and tomato crops. Thus the use of compost in soilless culture is a viable alternative to resolve the environmental problem of vegetable waste.

In the Mediterranean region, the control of soilless vegetable cultivation is commonly through measurement of some fertigation parameter ranges in the drainage; pH, electrical conductivity and volume percentage (e.g., Villegas, 2004; García, 2004; Urrestarazu et al., 2005b). Table 11 shows reference values of nutrient dissolution for horticultural crops for each substrate. They are measured almost daily and are easier and cheaper than other analyses or/and fertigation methods based on nutrient solution content in the substrate (Sonneveld and Straver, 1994), which in practice are only used one or two times during the crop cycle. Smith (1987) and Urrestarazu et al. (2005a, 2008) suggested that under adjusted management of the fertigation according to the different proprieties of the substrate, and within certain limits, it is possible to maintain the main parameters used as control for the fertigation method. Since in southeastern Spain rockwool (García, 2004) and perlite (García et al., 1997) are commonly used for similar lengths of time, it is suggested that alternative substrates are economically viable, since their cost is the same.

Fig. 11. Pepper production in growing media in southeastern Spain

Crop	Substrate	mmol L⁻¹						
		NO_3^-	$H_2PO_4^-$	SO_4^{2-}	NH_4^+	K^+	Ca^{2+}	Mg^{2+}
Tomato	Rockwool[1]	10,5	1,5	2,5	0,5	7,0	3,75	1,0
	Perlite[2]	12,5	2,0	1,75		5,0	5,0	1,8
	Organic[3]	13,0	1,75	1,25	1,0	7,5	4,0	1,25
Pepper	Rockwool[4]	15,5	1,25	1,75	1,25	6,5	4,75	1,5
	Perlite[5]	13,5	1,5	1,35		5,5	4,5	1,5
	Organic[3]	13,0	2,0	2,0	1,0	6,0	4,25	2,0
Melon	Organic[3]	13,0	2,3	2,2	1,0	7,0	4,25	2,2
Cucumber	Rockwool[4]	16,0	1,25	1,375	1,25	8,0	4,0	1,375
	Organic[3]	15,0	1,75	1,25	1,0	7,75	4,00	1,25
Green Beans	Rockwool[4]	12,5	1,25	1,125	1,0	5,5	3,25	1,25
	Perlite[2]	13,5	1,75	1,65		6,0	3,25	1,75

Source: [1]Sonneveld, 1980; [2]García and Urrestarazu, 1999; [3]Urrestarazu and Mazuela, 2005; [4]Sonneveld and Straver, 1994; [5]Escobar, 1993

Table 11. Reference values for nutrient dissolution in rockwool, perlite and alternative (organic) substrates.

Fig. 12. Composting in arids zones, Arica, Chile

Fig. 13. Blueberry production in Arequipa, Peru

Fig. 14. Desert lands can be cultivates by soilless culture

3. Conclusion

Once of the main environmental impacts of forced systems in horticulture -such as plastic covered and soilless culture- is the generation of organic plant residues and substrate waste. Many people are keen on research and development of ecologically friendly substrates. The suitability of compost from horticultural residues as a growing medium in vegetable crop production is an acceptable substitute for rock wool and coconut coir waste. The unit of soilless crop: rockwool slab, perlite and coconut bag culture are used in South eastern Spain between two or three years by vegetable production, consequently in order to be competitive in the market of soilless crops and to have similar commercial opportunity the unit with the alternative substrate must be used during this time. Because the environmental care and economic profit to grower is of paramount importance, it was important to see if this reuse of alternative substrate would be viable in Mediterranean conditions.. Part of a correct management procedure for the use of compost as substrate in growing media is to saturate the bags with nutrient solution before draining the containers in order to reduce the compost salinity in the rhizosphere environment. Once the electrical conductivity level was adjusted for soilless culture, yield trials proved the suitability of the compost as an acceptable growing media and this ecologically friendly alternative does not affect production, yield and fruit quality of horticultural crops. In conclusion, the use of compost in soilless culture is a viable alternative to resolve the environmental problem of vegetable waste.

4. Acknowledgement

This work was financially supported by project FEDER AGL2010-18391 (Spain), proyect UTA 9720-10 and Convenio de Desempeño UTA-MECESUP2 (Arica Chile).

5. References

Abad, M., Martínez, P.F., Martínez, M.D. & Martínez, J.(1993). Evaluación agronómica de los sustratos de cultivo. *Actas de horticultura* 11, 141-154.

Abad, M.; Noguera, P. & Carrión, C. (2004). Los sustratos en los cultivos sin suelo. In: M., Urrestarazu (Eds.), *Manual del cultivo sin suelo*. Servicio de Publicaciones Universidad de Almería. Mundi-Prensa, Madrid, pp. 113-158.Abad, M.; Noguera, P.; Noguera, V.; Roig, A.; Cegarra, J. & Paredes, C. (1997). Reciclado de residuos orgánicos y su aprovechamiento como sustratos de cultivo. *Actas de Horticultura*, 19: 92-109.

Abad, M., Noguera, P., Puchades, R., Maqueira, A. & Noguera, V. (2002). Physico-chemical and chemical properties of some coconut coir dusts for use as a peat substitute for containerised ornamental plants. *Biores. Technol.*, 82:241-245.

Ball, A.S.; Shah, D. & Wheatley, C.F. (2000). Assessment of the potential of a novel newspaper/horse manure-based compost. *Biores. Technol.*, 73, 163-167.

Bunt, A.C. (1988). Media and mixes for container-grown plants, 2nd ed. Unwin Hyman Ltd., London. 309 pp.

BOE. (1998). Boletín Oficial del Estado. num. 131, Spain.

Callejón A.J.; Carreño A.; Sánchez-Hermosilla J. & Pérez J. (2010). Evaluación de impacto ambiental de centro de transformación y gestión de residuos sólidos agrícolas en la provincia de Almería (España). *Informes de la Construcción,* 62 (518): 79-93.

Cara, G. & Ribera, J. (1998). Residuos en la agricultura intensiva. El caso de Almería. *Encuentro medioambiental Almeriense: En busca de soluciones.* Almería, Spain. March 7-8. pp. 128-132.

Carrión, C., Abad, M., Maquieira, A., Puchades, R., Fornes, F. & Noguera, V. (2005). Leaching of composts from agricultural wastes to prepare nursery potting media. *Acta Hort.* (ISHS) 697:117-124

Christensen, K.K., Kron, E. & Carlsbaek, M., (2001). Development of a Nordic system for evaluating the sanitary quality of compost. Nordic Council of Ministers, Copenhagen.

Conway, K.E. (1996) An overview of the influence of sustainable agricultural system on plant diseases. *Crop protection, 15,* 223-228.

Dimanmbro, M.E.; Lillywhite, R.D. & Ranh, C.R. (2007). *Compost Sci. Util.,* 15(4): 243-252

Drzal, M.S., Fonteno, W.C. & Cassel, D.K. (1999). Pore fraction analysis. A new tool for substrate testing. *Acta Hort., 481,* 43-53.

Elorrieta, M.A., Suárez-Estrella, F., López, M.J., Vargas-García, M.C. & Moreno, J. (2003). Survival of phytopathogenic bacteria during waste composting. *Agriculture, Ecosystems and Environment* 96, 141–146.Epstein, E.; Chanay, R.L.; Henry, C. & Logan, T.J. (1992). Trace elements in municipal solid waste compost. *Biomass Bioenergy, 3* (3-4), 227-238.

Epstein, E.; Chanay, R.L.; Henry, C. & Logan, T.J. (1992). Trace elements in municipal solid waste compost. *Biomass Bioenergy, 3* (3-4), 227-238.

Escobar, I. (1993). Cultivo del pimiento en sustratos en las condiciones del sudeste español. En: E. Martínez y M. García (ed). *Cultivo sin suelo: hortalizas en clima mediterráneo.* Ediciones de Horticultura, Reus, Spain, pp.109-113.

García, A. (2004). Cultivo en lana de roca II. In: Tratado de cultivo sin suelo, 3rd Ed., ed. M. Urrestarazu, Mundi-Prensa, Madrid, pp. 622-636.

García, M. & Urrestarazu, M. (1999). Recirculación de la disolución nutritiva en los invernaderos de la Europa del Sur. Caja Rural de Granada. 171 p.García, A. 2004. Cultivo en lana de roca II. In: Tratado de cultivo sin suelo, 3rd Ed., ed. M. Urrestarazu, Mundi-Prensa, Madrid, pp. 622-636.

García, M., Guzmán, M., Urrestarazu, M., Salas, M.C. & Escobar, I. (1997). Evaluación de diferentes parámetros en cultivo de perlita para distintas especies hortícolas en invernadero. *Actas de Horticultura* 19, 519-525.

Jensen, M.H. & Collin, W.L. (1985). Hydroponic vegetable production. In *Horticultural Review;* Janick, J., Ed.,. The AVI Publishing Co: Westport, CT, USA, 7, 483-558.

He, X.; Traina S.J. & Logan, T.J. (1992). Chemical properties of municipal solid waste composts. *J. Environ. Qual.,* 21, 318-329.

Hoitink, H.A.J., Herr, L.J. & Schmitthenner, A.F., (1976). Survival of some plant pathogens during composting of hardwood tree bark. *Phytopathology* 66, 1369–1372.

Ingelmo, F., Canet, R., Ibañez, M.A., Pomares, F. & García, J. (1997). Use of MSW compost, driede sewage sludge an other wastes as partial substitutes for peat and soil. *Commun. Soil Sci. Plant Anal.*, 27:1859-1874.

Lao, M.T. & Jiménez, S. (2004)a. Evaluation of almond shell as a culture substrate for ornamental plants. I. Characterisation. Øyton. *International Journal of Experimental Botany* 53, 69-78.

Lao, M.T. & Jiménez, S. (2004)b. Evaluation of almond shell as a culture substrate for ornamental plants. II. *Ficus benjamina.* Øyton. *International Journal of Experimental Botany* 53, 79-84.Mazuela, P; Acuña, L; Alvàrez, M. & Fuentes, A. (2010). Producción y calidad de un tomate cherry en dos tipos de invernadero en cultivo sin suelo. *Idesia,* 28(2):97-100

Mazuela, P., Salas, M.C. & Urrestarazu, M. (2005). Vegetable waste compost as substrate for melon. *Commun. Soil Sci. Plant Anal.*, 36 (11-12): 1557-1572.Mazuela, P. & M. Urrestarazu. 2009. The Effect of Amendment of Vegetable Waste Compost Used as Substrate in Soilless Culture on Yield and Quality of Melon Crops. *Compost Science & Utilization* 17:103-107.

Mazuela, P.; Urrestarazu, M.; Salas, M.C. & Guillén, C. (2004). Comparison between different fertigation parameters and yield using pure compost and coir waste fibre in tomato (*Lycopersicon esculentum* cv Pitenza) crop by soilless culture. *Acta Horticulturae,* 659: 653-656.

McLachlan, K.L., Chong, C., Voroney, R.P., Liu, H.W. & Holbein, B.E. (2004). *Commun. Soil Sci. Plant Anal.*,, 12(2): 180-184

Noguera, P.; Abad, M.; Noguera, V.;. Puchades, R. & Maquieira, A. (2000). Coconut coir waste, a new and viable ecologically friendly peat substitute. *Acta Hort.,* 517, 279-286.

Offord, C.A., Muir, S. & Tyler, J.L. (1998). Growth of selected Australian plants in soilles media using coir as a substitute for peat. *Aust. J. Exp. Agric.,* 38:879-887.

Ozores-Hampton, M., Vavrina, C.S. & Obreza, T.A. 1999. Yard Trimmin-biosolids Compost: Possible Alternative to Spahgnum Peat Moss in Tomato Transplant Production. *Commun. Soil Sci. Plant Anal.*, 7:42-49.

Pinamonti, F.; Stringari, G.; Gasperi, F. & Zorzi, G. (1997) The use of compost: its effects on heavy metal levels in soil and plants. Resouces, *Conservation and Recycling,* 21, 129-143.

Richards, D.; Lane, M. & Beardsell, D.V. (1986). The influence of particle-size distribution in pinebark:sand:brown coal potting mixes on water supply, aeration and plant growth. *Scientia Hort.,* 29, 1-14.

Salas, M.C. & Urrestarazu, M. (2001). *Técnicas de fertirrigación en cultivo sin suelo,* Manuales de la Universidad de Almería, Servicios de Publicaciones de la Universidad de Almería, Spain. 280 pp.

Salas, M.C.; Urrestarazu, M.; Moreno, J. & Elorrieta, M.A. 2000. Sustrato alternativo para su uso en cultivo sin suelo. *Phytoma* 123, 52-55.

Sánchez-Monedero, M.A., Roig, A., Cegarra, J., Bernal, M.P., Noguera, P.; Abad, M. & Antón, A. (2004). *Commun. Soil Sci. Plant Anal.*, 12(2): 161-168

Shinohara, Y.; Hata, T.; Mauro, T.; Hohjo, M. & Ito, T. (1999). Chemical and physical properties of the coconut-fiber substrate and the growth and productivity of tomato (*Lycopersicon esculentum* Mill) plants. *Acta Hort.*, *481*, 145-149.

Smith, D.L. (1987). Rockwool in Horticulture. Grower Book: London. 153 pp.

Sonneveld, C. (1980). Growing cucumbers and tomatoes in Rockwool. Proceedings national congress on soilless culture. Wageningen, pp. 253-262

Sonneveld, C., & Straver, N. (1994). Nutrient Solutions for Vegetables and Flower Grow in Water or Substrates. Naaldiwijk, The Netherlands: Proefstation voor tuinbouw onder glas te Naaldiwjk.

Spiers, T.M. & Fietje, G. (2000). Green Waste Compost as a Component in Soilless Growing Media. *Compost Sci. Util.*, *8* (1), 19-23.

Suárez-Estrella, F., López, M.J., Elorrieta, M.A., Vargas-García, M.C. & Moreno, J., (2002). Survival of phytopathogen viruses during semipilot-scale composting. In: Insam, H., Riddech, N., Klammer, S. (Eds.), *Microbiology of Composting*. Springer-Verlag, Berlín, pp. 539-548.

Suárez-Estrella, F., Vargas-García, M.C., Elorrieta, M.A., López, M.J. & Moreno, J. (2003). Temperature effect on *Fusarium oxysporum* f.sp. *melonis* survival during horticultural waste composting. *J. Applied Microbiology* 94, 475-482.

Suárez-Estrella, F., Vargas-Garrcía, M.C., López, M.J. & Moreno, J. (2007). Effect of horticultural waste composting on infected plant residues with pathogenic bacteria and fungi: Integrated and localized sanitation. *Waste Management* 27, 886-892

Suárez-Estrella, F., Vargas-Garrcía, M.C., López, M.J. & Moreno, J., (2004). Survival of Fusarium oxysporum f.sp. melonis on plant waste. *Crop Protection* 23, 127-133.

Urrestarazu, M. & Mazuela, P. (2005). Introducción al manejo del fertirriego en cultivo sin suelo. Riegos y Drenajes XXI, 141:34-40.

Urrestarazu, M., Martínez, G.A., & Salas, M.C., (2005). Almond shell waste: possible local rockwool substitute in soilless crop culture. *Scientia Hort.*, 103, 453-460.

Urrestarazu, M., Mazuela, P. & Alarcón, A.L. (2006). Cultivo en sustratos alternativos. p 147-173. IN: A.L. Alarcón (ed.). *Cultivos sin suelo*. Ediciones de Horticultura, Reus, España

Urrestarazu, M., Mazuela, P., Del Castillo, J., Sabada, S. & Muro, J. 2005. Fibra de pino: un sustrato ecológico. *Horticultura Internacional*, 49: 28-33.

Urrestarazu, M., Mazuela, P.C. & Martínez, G.A. (2008). Effect of substrate reutilization on yield and properties of melon and tomato crops. *J. Plant Nutrition* 31, 2031-2043

Urrestarazu, M. & Salas, M.C. (2002). El papel de los cultivos sin suelo en la moderna agronomía. *Vida Rural*, 145:54-58.

Urrestarazu, M., Salas, M.C. & Mazuela, P. (2003). Methods of correction of vegetable waste compost used as substrate by soilless culture. *Act. Hort.* 609, 229-233.

Urrestarazu, M., Salas, M.C., Padilla, M.I., Moreno, J., Elorrieta, M.A., Carrasco, G. (2001). Evaluation of different composts from horticultural crop residues and their uses in greenhouse soilless cropping. *Acta Hort.* 549, 147-152.

Urrestarazu, M.; Salas, M.C.; Rodríguez, R.; Elorrieta, M.A. & Moreno, J. (2000). Evaluación agronómica del uso del compost de residuos hortícolas como sustrato alternativo en cultivo sin suelo en tomate. *Actas de Horticultura*, 32: 327-332.

US Composting Council. (1997). Test methods for the examination of composting and compost (Interim Draft). US Composting Council, Bethesda, Maryland.

Villegas, F.J. (2004). Cultivo en fibra de coco II. In: Tratado de cultivo sin suelo, 3rd Ed., ed. M. Urrestarazu, Mundi-Prensa. Madrid, pp 650-667.

Warncke, D.D., (1986). Analyzing greenhouse growth media by the saturation extraction method. HortScience 21, 223-225.

11

Efficacy of *Pseudomonas chlororaphis* subsp. *aurantiaca* SR1 for Improving Productivity of Several Crops

Susana B. Rosas, Nicolás A. Pastor, Lorena B. Guiñazú, Javier A. Andrés,
Evelin Carlier, Verónica Vogt, Jorge Bergesse and Marisa Rovera
Laboratorio de Interacción Microorganismo–Planta. Facultad de Ciencias Exactas, Físico-Químicas y Naturales, Universidad Nacional de Río Cuarto, Campus Universitario. Ruta 36, Km 601, (5800) Río Cuarto, Córdoba
Argentina

1. Introduction

The recognition of plant growth-promoting rhizobacteria (PGPR) as potentially useful for stimulating plant growth and increasing crop yield has evolved over the past years. Currently, researchers are able to successfully use them in field experiments. The use of PGPR offers an attractive way to supplement or replace chemical fertilizers and pesticides. Most of the isolates cause a significant increase in plant height, root length and dry matter production of plant shoot and root. Some PGPR, especially if they are inoculated on seeds before planting, are able to establish on roots. Also, PGPR can control plant diseases. These bacteria are a component of integrated management systems, which use reduced rates of agrochemicals. Such systems might be used for transplanted vegetables in order to produce more vigorous seedlings that would be tolerant to diseases for at least a few weeks after transplanting to the field (Kloepper *et al.*, 2004). Commercial applications of PGPR are being tested and are frequently successful. However, a better understanding of the microbial interactions that result in plant growth enhancements will greatly increase the success of field applications (Burr *et al.*, 1984).

In the last few years, the number of identified PGPR has been increasing, mainly because the role of the rhizosphere as an ecosystem has gained importance in the functioning of the biosphere. Several species of bacteria like *Pseudomonas, Azospirillum, Azotobacter, Klebsiella, Enterobacter, Alcaligenes, Arthrobacter, Burkholderia, Bacillus* and *Serratia* have been reported to enhance plant growth. Of these, the genera that are predominantly studied and increasingly marketed as biological control agents include *Bacillus, Streptomyces* and *Pseudomonas* (Glick, 1995; Joseph *et al.*, 2007; Kloepper *et al.*, 1989; Okon & Labandera-González, 1994).

2. Fluorescent *Pseudomonas*

The fluorescent pseudomonads produce a variety of biologically active natural products (Budzikiewicz, 1993; Leisinger & Margraff, 1979), many of which have an ecological function in these gram-negative bacteria. Some of these natural products contribute to the suppression of plant-pathogenic fungi (Dowling & O'Gara, 1994; Thomashow, 1996) whereas others are important virulence factors against certain plant-pathogenic *Pseudomonas* species (Bender, *et al.*, 1999). Because secondary metabolism significantly contributes to the molecular ecology of several *Pseudomonas* species and has also provided lead compounds for crop protection applications, several gene clusters that encode secondary metabolic pathways in this genus have been sequenced and characterized. For example, some *Pseudomonas* biosynthetic pathways have functionally combined modular, dissociated, or chalcone synthase-like polyketide synthases with adenylating enzymes (pyoluteorin, mupirocin, coronatine) or with components of fatty acid synthases (2,4-diacetylphloroglucinol) (Bender *et al.*, 1999; Guhien *et al.*, 2004). Phenazines are an important type of secondary metabolites. Phenazine antibiotics, nitrogen-containing aromatic compounds, are especially active against lower fungi and most gram-positive and gram-negative phytopathogenic bacteria (Chin-A-Woeng *et al.*, 2003). The most studied phenazinic substances are: phenazine 1-carboxilic acid, phenazine 1 carboxamide, aeruginosin A, pyocyanin, 2-hydroxyphenazine 1-carboxilic acid, 1-hydroxyphenazine (Price Whelan *et al.*, 2006; Raajmakers *et al.*, 1997). The phenazine antibiotics complex of *P. aurantiaca* B-162 consists of 1-oxyphenazine, phenazine, and phenazine-1,6 dicarboxylate (their common precursor) (Feklistova & Maksimova, 2005). Also, Feklistova & Maksimova (2008) reported the production of N-hexanoyl homoserine lactone (HHL) by *P. aurantiaca* B-16.

Mehnaz *et al.* (2009) described the characterization of a newly isolated strain, PB-St2, of *P. aurantiaca* and its main secondary metabolites. This strain showed antifungal activity against strains of *C. falcatum* and *Fusarium* spp., phytopathogens of agricultural significance. Two main antibiotics were isolated and identified to be phenazine-1-carboxylic acid and 2-hydroxyphenazine. In addition, strain PB-St2 produces HHL, a compound that indicates the presence of a quorum-sensing mechanism. Quorum sensing is the major mechanism by which many bacteria regulate production of antifungal factors. Strain PB-St2 produce hydrogen cyanide (HCN) and siderophores as PGPR traits. All of these data suggest that PB-St2 could be used as a potential effective biocontrol agent or biofertilizer to decrease the incidence of plant diseases and promote plant growth (Mehnaz *et al.*, 2009).

Liu *et al.* (2007) detected the production of two main antifungal substances produced by *Pseudomonas chlororaphis* GP72: phenazine-1-carboxylic acid and 2-hydroxyphenazine. This strain also produced the quorum-sensing signalling molecules N-butanoyl-L-homoserine lactone and N-hexanoyl-L-homoserine lactone. In addition, Ramarathnam & Dilantha Fernando (2006) reported that one isolate of *P. aurantiaca*, termed DF200, tested positive for the presence of pyrrolnitrin and phenazine biosynthetic genes.

Several *Pseudomonas* strains have already been marketed as commercial biocontrol products, such as Cedomon (BioAgri AB, Upsala, Sweden), a seed treatment based on a *Pseudomonas chlororaphis* strain providing protection against seedborne diseases in barley. Similarly, Mycolytin is an antifungal biopesticide containing *P. aurantiaca* M-518 (Omelyanets & Melnik, 1987). The genus *Pseudomonas* is well known for producing metabolites which stimulate plant growth and root colonization by beneficial microorganisms. The *Pseudomonas* also synthetize phytohormones and siderophores and solubilize phosphates (Mikuriya *et al.*, 2001; Kang *et al.*, 2006; Arshad & Frankernberger 1991; Rosas *et al.*, 2006). *Pseudomonas aurantiaca* S-1 can serve as a natural source of pesticides towards phytopathogens like *Fusarium oxysporum* P1 and *Pseudomonas syringae* pv. *glycinea* BIM B-

280. This strain was able to produce indole acetic acid and siderophores (Mandryk *et al.*, 2007).

Feklistova & Maximova (2009) extracted and identified gibberellins synthesized by *P. aurantiaca* B-162 and proposed a new method for selecting bacteria that produce plant growth-promoting substances. By NG-mutagenesis and a consequent selection on toxic analogues of the gibberellins, regulative analogue-resistant mutants of *P. aurantiaca* were obtained. These were capable of overproducing gibberellic acid (31 mg l^{-1}), that was 2,3 times in excess of that of the master strain.

Thus, research on *Pseudomonas* natural products provides an opportunity to study not only the ecological function of secondary metabolism but also the potential diversity found in secondary metabolic pathways.

3. *Pseudomonas chlororaphis* subsp. aurantiaca SR1

P. chlororaphis subsp. *aurantiaca* SR1 (GenBank accession number GU734089) was isolated from the rhizosphere of soybean in the area of Río Cuarto, Córdoba, Argentina. It was initially classified as *P. aurantiaca* by using the BIOLOG (Biolog Inc., Hayward, CA) system (Rosas *et al.*, 2001) and, more recently, by amplification and sequencing of a partial fragment from the 16S rDNA gene. The species *Pseudomonas aurantiaca* was recently reclassified as *P. chlororaphis* subsp. *aurantiaca* (Peix *et al.*, 2007).

Strain SR1 inhibits a wide range of phytopathogenic fungal species including *Macrophomina phaseolina*, *Rhizoctonia* spp. T11, *Fusarium* spp., *Alternaria* spp., *Pythium* spp., *Sclerotinia minor* and *Sclerotium rolfsii* (Rosas *et al.*, 2001). It produces siderophores, behaves as an endophyte and is capable of promoting plant growth through mechanisms that involve phytohormone-like substances (Rovera *et al.*, 2008). For instance, SR1 shows the ability to produce indole 3-acetic acid (IAA). In addition, strain SR1 is able to colonize the root-system of several crops, maintaining appropiate population densities in the rhizosphere area (Rosas *et al.*, 2005). SR1 was also shown to produce signal molecules such as acyl homoserine lactones (AHL's).

Recently, PCR assays were carried out to detect *phlD* and *phz*, genes involved in the biosynthesis of 2,4-diacetylphloroglucinol (DAPG) and phenazine-1-carboxylic acid (PCA), respectively, in strain SR1 through the use of primers and protocols described by Raaijmakers *et al.* (1997). Also, PCR assays involving the specific primers to detect *prnD* and *pltC*, genes encoding the production of pyrrolnitrin (PRN) and pyoluteorin (PLT), respectively, were performed as described by De Souza & Raaijmakers (2003). On the other hand, detection of *hcnAB* genes (involved in the biosynthesis of HCN synthetase) was performed by PCR using the primers PM2-F (5'-TGCGGCATGGGCGCATTGCTGCCTGG-3') and PM2-R (5'-CGCTCTTGATCTGCAATTGCAGGC-3') (Svercel *et al.*, 2007). As a result, fragments of the predicted size for PCA, PRN and HCN were amplified from the DNA of strain SR1.

4. Effect of inoculation with strain SR1 on agronomically important crops

4.1 Wheat (*Triticum aestivum* L.)

In these studies, we evaluated the effect of inoculating wheat seeds with strain SR1 on plant growth, under field conditions (Carlier *et al.*, 2008; Rosas *et al.*, 2009). Experiments were conducted with a complete randomized block design with seven blocks. Each block consisted of six plots (one per treatment and each of 7.20 m^2). Plots were separated in between by a distance of 1 m. Six rows (separated by 0.20 m) per block were sowed by using a plot seed drill. The sowing density was 120 kg ha^{-1} of seeds. The six treatments were: (1) uninoculated seeds in unfertilized soil (control); (2) uninoculated seeds in soil fertilized with 80 kg ha^{-1} of urea - 60 kg ha^{-1} of diammonium phosphate (100% dose); (3) uninoculated seeds in soil fertilized with 40 kg ha^{-1} of urea - 30 kg ha^{-1} of diammonium phosphate (50% dose); (4) seeds inoculated with SR1 in unfertilized soil; (5) seeds inoculated with SR1 in soil fertilized with the 100% dose; (6) seeds inoculated with SR1 in soil fertilized with the 50% dose. Seeds were inoculated with a formulation manufactured by *Laboratorios Biagro S.A.* containing strain SR1 at 10^9 CFU g^{-1} of peat. Briefly, 40 g inoculant, 20 g S2 adherent (*Laboratorios Biagro S.A.*), and 5 g cell protector S1 (*Laboratorios Biagro S.A.*) were mixed in 80 ml of water. Then, 12 g of this mixture was added to 1 kg wheat seeds to obtain a colony count of 10^5 CFU g^{-1} seeds.

Growth and yield parameters were recorded at the growth stages termed 1.5 (5 leaves), 3.0 (tillering), and 11.4 (ripe for harvest) (Feekes International Scale — Large 1954). At Feekes 1.5, the number of seedlings emerging per m^2 was evaluated. At Feekes 1.5 and 3.0, shoot length, root length, number of tillers, root volume (cm^3), shoot and root dry weight (72 h at 60 °C) were assessed. Yield parameters evaluated were: kg ha^{-1}, weight of 1,000 grains, number of spikes per plant, and number of grains per spike.

Inoculation had no effect on emergence of plants, as compared to control. On the other hand, the number of plants per m^2 was higher for inoculation treatments than for fertilization without inoculation. Increases in mean shoot length (14%) were observed for the inoculated/unfertilized treatment and for fertilization with a 50% dose (8%) during Feekes 1.5, compared to control plants. By comparison, a 60% increase in shoot length, relative to control plants, was observed during Feekes 3.0 in plants inoculated and fertilized with a 100% dose. Plants inoculated with strain SR1 showed increases between 47 and 78% in root length during Feekes 1.5 and between 65 and 75% during Feekes 3.0, compared to control. Also, root volume significantly increased during Feekes 1.5 after inoculation and fertilization with the 100% dose (Table 1).

Treatment	Root length (cm)		Root volume (ml)	
	1.5	3.0	1.5	3.0
Control	101b	204b	0.9c	2.7b
Uninoculated seeds in soil fertilized with the 100% dose	87b	357a	1.2b	4.2a
Uninoculated seeds in soil fertilized with the 50% dose	159a	235b	1.2b	2.4b
Seeds inoculated with SR1 in unfertilized soil	160a	339a	1.7a	4.4a
Seeds inoculated with SR1 in soil fertilized with the 100% dose	181a	346a	2.0a	3.9a
Seeds inoculated with SR1 in soil fertilized with the 50% dose	150a	359a	1.6a	3.6a

Values in each column with different letters are significantly different according to the LSD test ($P<0.05$)

Table 1. Root length and root volume of wheat plants during Feekes 1.5 and 3.0

All mean values of shoot dry weight from inoculation and/or fertilization treatments were higher than those of control plants during Feekes 1.5, but differences were not significant. The higher mean value was obtained after inoculating and fertilizing with the 50% dose, which increased shoot dry weight by 64 mg when compared to control. Throughout Feekes 1.5 and Feekes 3.0, root dry weight was significantly increased by inoculation with strain SR1 alone, as compared to control (Table 2). In addition, the number of tillers increased between 31 and 50 % at Feekes 3.0 in inoculated plants, with or without fertilization, compared to control plants.

Treatment	Dry weight (mg) at Feekes 1.5		Dry weight (mg) at Feekes 3.0	
	Shoot	Root	Shoot	Root
Control	198a	80d	790c	326c
Uninoculated seeds in soil fertilized with the 100% dose	217a	128bcd	1,510a	498ab
Uninoculated seeds in soil fertilized with the 50% dose	228a	109cd	1,080bc	377bc
Seeds inoculated with SR1 in unfertilized soil	223a	190a	1,310ab	536a
Seeds inoculated with SR1 in soil fertilized with the 100% dose	252a	183ab	1,310ab	420abc
Seeds inoculated with SR1 in soil fertilized with the 50% dose	262a	156c	1,370ab	512a

Values in each column with different letters are significantly different according to the LSD test (P<0.05)

Table 2. Shoot and root dry weight of wheat plants during Feekes 1.5 and 3.0

When considering the yield parameters, the value of kg ha^{-1} was significantly higher in plants inoculated with SR1 and fertilized with a 50% dose, as compared to control. Regarding number of grains per spike, values for inoculation treatments were always higher than for control. The highest value was observed after inoculation and fertilization with the 50% dose (40% more than the control) (Table 3).

Treatment	Yield (kg ha^{-1})	Number of grains per spike
Control	2,005b	32c
Uninoculated seeds in soil fertilized with the 100% dose	2,169ab	39b
Uninoculated seeds in soil fertilized with the 50% dose	2,264ab	40b
Seeds inoculated with SR1 in unfertilized soil	2,249ab	42b
Seeds inoculated with SR1 in soil fertilized with the 100% dose	1,776b	41b
Seeds inoculated with SR1 in soil fertilized with the 50% dose	2,641a	45a

Values in each column with different letters are significantly different according to the LSD test (P<0.05)

Table 3. Parameters of wheat yield

There are few reports on the contribution of inoculation of wheat seeds with *Pseudomonas* strains for improving plant growth and yield under field conditions. For instance, Shaharoona *et al.* (2007) tested several *Pseudomonas* spp. strains in the field to determine their efficacy to increase growth and yield of this crop plant. Their results revealed that all of the strains significantly increased plant height compared to uninoculated control. Strain *P. fluorescens* biotype F caused the maximum increase (16%). This strain also significantly increased the number of grains per spike (11.7% more than the uninoculated control). Another strain, *P. fluorescens* biotype G increased the number of tillers per m^2 by 9%, compared to uninoculated control plants. The maximum increase in 1,000-grain weight was recorded with *P. fluorescens* (ACC$_{50}$) (34% higher than the uninoculated control). They also reported that inoculation with strain *P. fluorescens* (ACC$_{50}$) increased grain yield by 39% when compared to the uninoculated control.

4.2 Maize (*Zea mays* L.)

The application of a SR1 formulation on maize seeds allowed us to evaluate its effectiveness as maize growth promoter in the field (Rosas *et al.*, 2009). For these experiments, plots were arranged in a completely randomized design, with four replicates of 156 m^2 for each treatment. Two treatments were included: 1. Seeds inoculated with strain SR1 and 2. Uninoculated seeds. Soil was fertilized with 100 kg ha^{-1} of diammonium phosphate at the sowing time and 100 kg ha^{-1} of urea during V7-8 stages for both treatments.

Length and dry weight of shoot were determined during the V2, V5, V13, R3 and R6 phenological stages. In addition, the following parameters were recorded during the first stages (V2 and V5): root length, root surface (Díaz Vargas *et al.*, 2001) and root volume (Carley & Watson, 1966). The weight of 1,000 grains and grain yield (kg ha^{-1}) were evaluated at the harvest time.

During V2 and V5, the beneficial effect of inoculation with strain SR1 was evidenced at the root system level. Root length increased 28% during V2 and 32% during V5 in inoculated plants. Similar results were obtained with root volume (42% and 36%, respectively) and root surface (39% and 34%, respectively). Shoot dry weight determinations indicated that inoculation with strain SR1 impacted favourably during the whole cycle of the crop. For instance, we observed a 22% increase in shoot dry weight during stage R3, as compared to control plants. Such beneficial effect was also observed for yield parameters. To illustrate, the weight of 1,000 grains and grain yield (kg ha^{-1}) were 11 and 20% higher in inoculated plants.

Egamberdiyeva *et al.* (2002) reported on the effect of a *Pseudomonas fluorescens* strain, termed PsIA12, and three *Pantoea agglomerans* strains (370320, 020315 and 050309) on the growth of maize in the field. Inoculation with these bacterial strains was found to significantly increase the root and shoot growth of maize grown in loamy sand at 16 °C. Also, K content was significantly increased in all treatments. More recently, Naveed *et al.* (2008) assessed the performance of an organic fertilizer and three *Pseudomonas* strains prepared as bio-fertilizers for improving growth and yield of maize in the field. Their results revealed that application of bio-fertilizers significantly improved the growth and yield of this crop. Indeed, plant height increased between 4 and 9% after inoculation with the bio-fertilizers and only 2% after treatment with the organic fertilizer, compared to control. Similarly, they observed that total biomass was enhanced between 21 and 39% by bio-fertilizers and 11.4% by the organic fertilizer, compared to control plants. The increases obtained for grain yield (t ha^{-1}) were

14.2% by the organic fertilizer and between 21 and 30% by treatment with the bio-fertilizers. Finally, bio-fertilization caused increases between 14 and 19% in 1,000 grain weight, as compared to control plants.

4.3 Soybean (*Glycine max* (L.) Merril)
Treatment of soybean seeds with strain SR1 was studied to determine the effect of inoculation on plant growth, under greenhouse conditions. At the present time, soybean is the most important oleaginous seed worldwide. In Argentina, soybean cultivation was introduced in the 1970's and it has been characterized by an incredible rate of adoption and growth. Indeed, Argentina is one the main exporters of soybean flour (27% of the world exports) as well as soybean oil (30% of the world exports) (Penna & Lema, 2002).
Soybean seeds were inoculated with a peat-based formulation prepared and packed by *Laboratorios Biagro S.A.* containing strain SR1 at 2.4×10^9 CFU g^{-1} peat. Then, plastic pots were filled with sterile soil and 4 inoculated seeds were placed into the soil surface in each pot. The four treatments were: (1) uninoculated seeds (control); (2) seeds inoculated with strain SR1 at 10^7 CFU g^{-1}; (3) seeds inoculated with SR1 at 10^8 CFU g^{-1}; (4) seeds inoculated with SR1 at 10^9 CFU g^{-1}. The inoculant containing 10^8 and 10^7 CFU g^{-1} was obtained by diluting the original formulation with sterile peat. Pots were incubated in a greenhouse. Shoot and root length as well as shoot and root dry weight (120 h at 60 °C) was recorded from each treatment after 25 days. Pots were arranged in a completely randomized design with five replicates per treatment.
SR1 at 10^{-9} CFU g^{-1} enhanced shoot length by 31%, as compared to control plants. There were no significant differences in root length. Although there were no significant differences among the three inoculation doses for shoot dry weight, the optimum inoculation dose proved to be 10^8 CFU g^{-1}. Compared to control plants, SR1 at 10^8 CFU g^{-1} increased shoot and root dry weight of inoculated soybean plants by 53 and 14%, respectively (Table 4).

Treatments	Shoot length (cm)	Root length (cm)	Shoot dry weight (mg)	Root dry weight (mg)
Control	34.2b	8.1a	430a	140a
Seeds inoculated with SR1 at 10^7 CFU g^{-1}	41.2b	6.3a	500a	160a
Seeds inoculated with SR1 at 10^8 CFU g^{-1}	41.7b	8.5a	660a	160a
Seeds inoculated with SR1 at 10^9 CFU g^{-1}	45.0a	6.6a	420a	110b

Values in each column with different letters are significantly different according to the Scheffé test (P<0.05)

Table 4. Soybean growth parameters

In addition, we evaluated strain SR1 for control of *Macrophomina phaseolina* (Tassi) Goid. in soybean, under greenhouse conditions. *Macrophomina* (the cause of charcoal rot, dry root rot and damping-off of many crop plants) is one of the most destructive plant pathogenic fungal genera. It prevails in the tropics and sub-tropics, inciting diseases in a wide range of hosts (Anjaiah, 2004). Significant yield losses of soybean are reported every year due to charcoal rot fungus *M. phaseolina* (Tassi) Goid. (Senthilkumar *et al.*, 2009). During biocontrol assays, soybean seeds were inoculated with strain SR1 prior to planting into infested soil. Growth parameters of soybean plants were recorded after 25 days. Compared to pathogen controls, strain SR1, inoculated at 10^7 CFU g^{-1}, increased shoot and root length by 277 and 290%, and shoot and root dry weight by 275 and 375%, respectively. Results suggest that strain SR1 provides effective control of M. *phaseolina* and that it might be applied as a biological control agent to protect soybean plants from this phytopathogen.

Wahyudi *et al.* (2011) isolated *Pseudomonas* sp. from the rhizosphere of soybean and tested them for promotion of seed growth. As a result, they found that two isolates (Crb-44 and 63) exhibited promoting activity for all of the measured parameters (length of primary root, shoot length and number of lateral roots). Also, they reported that other 15 *Pseudomonas* isolates showed promotion of soybean seed growth at varying degrees. After their experiments, they concluded that the *Pseudomonas* sp. isolates could be applied as inoculants of soybean plants because of their excellent growth promotion and biocontrol activities.

4.4 Alfalfa (*Medicago sativa* L.)

Finally, we also studied strain SR1 in co-inoculation with *Sinorhizobium meliloti* strain 3DOh13 to determine their effects on nodulation and growth of alfalfa plants (Rovera *et al.*, 2008). In our studies, both SR1 and *S. meliloti* strain 3DOh13 were cultured on tryptic soy broth (TSB) medium at 28 ± 1 °C. Their optical cell densities were 0.22 and 0.36 at 600 nm (OD_{600}), which corresponded to approximately 4.5 x 10^8 CFU ml^{-1} and 6.8 × 10^8 CFU ml^{-1} for SR1 and *S. meliloti* 3DOh13, respectively. The inoculant was prepared by mixing strain SR1 and *S. meliloti* 3DOh13 in a 1:1 ratio (v v^{-1}). The optical cell density at 600 nm (OD_{600}) was 0.25, which corresponded to approximately 6.6 × 10^8 CFU ml^{-1} of *S. meliloti* 3DOh13 and 6.3 × 10^8 CFU ml^{-1} of SR1. One gram of sterilized seeds was inoculated with the mixed bacterial suspension.

SR1, when inoculated alone, stimulated shoot and root length of alfalfa by 82 and 57%, respectively, compared to control plants. Co-inoculation of strain SR1 and *S. meliloti* 3DOh13 stimulated shoot and root length of alfalfa by 140 and 96%, respectively, as compared to control. Additionally, co-inoculation of alfalfa seeds with strain SR1 and *S. meliloti* 3DOh13 caused a significant increase in dry weight of shoot and root (Table 5). Finally, co-inoculation significantly enhanced nodulation and total N content, compared to inoculation with *S. meliloti* 3DOh13 alone or uninoculated control.

Treatment	Shoot lenght (cm)	Root lenght (cm)	Shoot dry weight (mg)	Root dry Weight (mg)
Control	3.4c	7.5c	4c	4c
N_2 Control	5.3c	9.7c	5c	5b
S. meliloti 3DOh13	7.0a	13.2b	26a	9b
P. aurantiaca SR1	6.2b	11.8b	19b	3c
Co-inoculation	8.2a	14.7a	29a	14a

Means with different letters in the same column differ significantly at P ≤ 0.05 (*Bonferroni test*)

Table 5. Effect of co-inoculation with *P. aurantiaca* SR1 and *S. meliloti* 3DOh13 on alfalfa growth

Ogata *et al.* (2008) sampled strains of *Pseudomonas* sp, *Rhizobium* sp, *Bradyrhizobium* sp, *Azotobacter* sp and actinomycetes from the rizosphere of the tara tree (*Caesalpinia spinosa*) and used them as seed inoculants for alfalfa, tara and bean (*Phaseolus lunatus* var. Sieva, *Phaseolus vulgaris* var. Camanejo and var. Caraota) to study the effects of these microorganisms on seed germination. A *Rhizobium* strain (rP2N3) significantly increased the germination percentage of alfalfa (154.9%) when compared to the control without inoculation. However, strain *Pseudomonas* spp. ps52b increased the germination percentage of alfalfa by 83.4%. These authors concluded that increase in germination percentage as well as production of IAA and phosphate solubilization by these bacteria outlines their potential as microbial inoculants for alfalfa, tara and bean.

5. Conclusions

Pseudomonas chlororaphis subsp. *aurantiaca* SR1 colonized the root system of the studied crops, persisted at appropriate population densities in the rhizosphere area and showed a significant plant growth-promoting effect that was reflected in the yield. In general, the promoting effect on growth parameters was observed throughout all of the phenological stages of the crops. A relevant finding was that wheat plants, after inoculation with SR1, presented higher yields with fertilization doses lower than those conventionally applied. This outlined the potential use of SR1 as a reasonable alternative for wheat production, with a minimization of negative environmental impacts.

In addition, maize yield values, expressed as weight of 1,000 grains, showed an 11% increase and were similar to others studies. Strain SR1 behaved as a nodulation-promoting rhizobacterium in alfalfa and soybean and was isolated as endophyte from both these crops. Strain SR1 mobilizes nutrients and produces IAA, PCA, PRN and HCN. These compounds might be contributing to the observed increase in growth parameters.

Our group's current goals are (1) to evaluate the influence of root exudates from different plant species on the expression of secondary metabolites in strain SR1 and (2) to determine SR1's effects on different plants under water and salinity stress conditions for being able to extend Argentina's cultivation areas.

Currently, a commercial formulation containing strain SR1, termed *Liquid PSA*, is registered with the Argentina's National Service for Agricultural Health (SENASA) for wheat growth promotion. *Liquid PSA* is produced by *Laboratorios Biagro S.A.*

6. Acknowledgments

This work was supported by grants from Secretaría de Ciencia y Técnica, Universidad Nacional de Río Cuarto (Córdoba, Argentina), Agencia Nacional de Promoción Científica y Tecnológica (Secretaría de Ciencia y Técnica de la Nación) and Consejo Nacional de Investigaciones Científicas y Técnicas (CONICET, Argentina).

7. References

Anjaiah, V. (2004). Biological Control Mechanisms of Fluorescent *Pseudomonas* Species involved in Control of Root Diseases of Vegetables/Fruits, In: *Disease Management of Fruits and Vegetables*, Mukerji, K.G., pp. 453-500. Kluwer Academic Publishers, Netherlands

Arshad, M. & Frankenberger, W.T. (1991). Microbial production of plant hormones. *Plant Soil* 133, 1–8

Bender, C.L., Alarcon-Chaidez, F., & Gross, D.C. (1999). *Pseudomonas syringae* phytotoxins: mode of action, regulation, and biosynthesis by peptide and polyketide synthetases. Microbiology & Molecular Biology Reviews 63, 266–292

Budzikiewicz, H. (1993). Secondary metabolites from fluorescent pseudomonads. FEMS Microbiology Reviews 10, 209–228

Burr, T.J., Caesar, A.M., & Schrolh, N. (1984). Beneficial plant bacteria. *Critical Reviews in Plant Sciences* 2, 1–20

Carley, H., & Watson, R.D. (1966). A new gravimetric method for estimating root surface area. *Soil Science* 102, 289–291

Carlier, E., Rovera, M., Rossi Jaume, A., & Rosas, S.B. (2008). Improvement of growth, under field conditions, of wheat inoculated with *Pseudomonas chlororaphis* subsp. *aurantiaca* SR1. *World Journal of Microbiology and Biotechnology* 24, 2653–2658

Chin-A-Woeng, T.F., Bloemberg, G.V., & Lugtenberg, B.J. (2003). Phenazines and their role in biocontrol by *Pseudomonas* bacteria. *New Phytologist* 157, 503–523

De Souza, J.T., & Raaijmakers, J.M. (2003). Polymorphisms whithin the *prn*D and *plt*C genes from pyrrolnitrin and pyoluteorin-producing *Pseudomonas* and *Burkholderia* spp. *FEMS Microbiology Ecology* 43, 21–34

Díaz Vargas, P., Ferrera-Cerrato, R., Almaraz-Suarez, J.J., & Alcantar Gonzalez, G. (2001). Inoculación de bacterias promotoras de crecimiento en lechuga. *Terra* 19, 329–335

Dowling, D.N., & O'Gara, F. (1994). Metabolites of *Pseudomonas* involved in the biocontrol of plant disease. *Trends in Biotechnology* 12, 133–144

Egamberdiyeva, D., Juraeva1, D., Gafurova, L., & Höflich, G. (2002). Promotion of plant growth of maize by plant growth promoting bacteria in different temperatures and soils, In: *Making conservation tillage conventional: building a future on 25 years of research*. Proc. of the 25th annual southern conservation tillage conference for sustainable agriculture, Van Santen, E.

Auburn, Al. Special Report No. 1. Alabama Agric. Expt. Stn. and Auburn University, Al. 36849. USA

Feklistova, I.N., & Maksimova, N.P. (2005). Synthesis of Phenazine Compounds by *Pseudomonas aurantiaca*, *Vest. Belorus. Un-ta*, Ser. 2. Chemistry. Biology. Geography 2, 66–69

Feklistova, I.N. & Maksimova, N.P. (2008). Obtaining *Pseudomonas aurantiaca* strains capable of overproduction of phenazine antibiotics. *Microbiology* 77, 176-180

Feklistova, I.N. & Maksimova, N.P. (2009). *Pseudomonas aurantiaca* bacteria gibberellins: biological activity, phytohormones producers creation and its application techniques. *Proceedings of the Belorusian State University Series of Physiological, Biochemical and Molecular Biology Sciences Scientific Journal* 4, 168-173

Glick, B.R. (1995). The enhancement of plant growth by free living bacteria. *Canadian Journal of Microbiology* 41, 109–114

Guihen, E., Glennon, J.D., Cullinane, M., & O'Gara, F. (2004). Rapid analysis of antimicrobial metabolites monoacetylphloroglucinol and 2,4-diacetylphloroglucinol using capillary zone electrophoresis. *Electrophoresis* 25, 1536–1542

Joseph, B., Patra, R.R., & Lawrence, R. (2007). Characterization of plant growth promoting Rhizobacteria associated with chickpea (*Cicer arietinum* L). *International Journal of Plant Production* 1, 141-152

Kang, B.R., Yang, K.Y., Cho, B.H., Han, T.H., Kim, I.S., Lee M.C., Anderson A.J., & Kim, Y.C. (2006). Production of indole-3-acetic acid in the plant-beneficial strain *Pseudomonas chlororaphis* O6 is negatively regulated by the global sensor kinase GacS. *Current Microbiology* 52, 473–476

Kloepper, J.W., Lifshitz, R., & Zablotowicz, R.M. (1989). Free-living bacterial inocula for enhancing crop productivity. *Trends in Biotechnology* 7, 39–43

Kloepper, J.W., Reddy, S.M., Rodreguez-Kabana, R., Kenney, D.S., Kokalis- Burelle, N., & Ochoa, N.M. (2004). Application for Rhizobacteria in Transplant Production and Yield Enhancement. *Acta Horticulturae* 631, 217-229

Mandryk, M., Kolomiets, E. & Dey, E. (2007). Characterization of Antimicrobial Compounds Produced by *Pseudomonas aurantiaca* S-1. *Polish Journal of Microbiology* 56, 245–250

Mehnaz, S., Weselowski, B., Aftab, F., Zahid, S., Lazarovits, G., & Iqbal, J. (2009). Isolation, characterization, and effect of *fluorescent pseudomonads* on micropropagated sugarcane. *Canadian Journal of Microbiology* 55, 1007–1011

Mikuriya, T., Fukushima, M., Yanagi, H., Nagamatsu, Y., & Yoshimoto, A. (2001). Production of plant hormone and antifungal antibiotics by *Pseudomonas fluorescens* S543 grown on ethanol. *Hiroshima Daigaku Seibutsu Seisangakubu Kiyo* 40, 33–43

Naveed, M., Khalid, M., Jones, D.L. Ahmad, R., & Zahir, Z.A. (2008) Relative efficacy of *Pseudomonas* spp., containing ACC-deaminase for improving growth and yield of maize (*Zea mays* L.) in the presence of organic fertilizer. *Pakistan Journal of Botany* 40, 1243-1251

Leisinger, T., & Margraff, R. (1979). Secondary metabolites of the fluorescent pseudomonads. *Microbiological Reviews* 43, 422–442

Liu, H., He, Y., Jiang, H., Peng, H., Huang, X., Zhang, X., Thomashow, L.S., & Xu, Y. (2007). Characterization of a Phenazine-Producing Strain *Pseudomonas chlororaphis* GP72 with Broad-Spectrum Antifungal Activity from Green Pepper Rhizosphere. *Current Microbiology* 54, 302-306

Ogata, K., Arellano, C., Zúñiga, D. (2008). Efecto de diferentes bacterias aisladas de rizósfera de *Caesalpina spinosa* en la germinación de algunas especies vegetales. Zonas Áridas 12, 137-153

Okon, Y., & Labandera-Gonzalez, C.A. (1994). Agronomic applications of *Azospirillum*, In: *Improving Plant Productivity with Rhizosphere Bacteria*, Ryder, M.H., Stephens, P.M., Bowen, G.D., pp 274–278. Commonwealth Scientific and Industrial Research Organization, Adelaide, Australia.

Omelyanets, T.G., & Melnik, G.P. (1987). Toxicological evaluation of the microbial preparation mycolytin. *Zdravookhranenie Turkmenistana* 6, 8

Peix, A., Valverde, A., Rivas, R., Igual, J.M., Ramírez-Bahena, M., & Mateos, P.F. (2007). Reclassification of *Pseudomonas aurantiaca* as a synonym of *Pseudomonas chlororaphis* and proposal of three subspecies, *P. chlororaphis* subsp. *chlororaphis* subsp. nov., *P. chlororaphis* subsp. *aureofaciens* subsp. nov., comb. nov. and *P. chlororaphis* subsp. *aurantiaca* subsp. nov., comb. nov. *International Journal of Systematic and Evolutionary Microbiology* 57, 1286-1290

Penna, J.A., & Lema, D. (2002). Adoption of herbicide resistant soybeans in Argentina: an economic analysis, Chapter 10, In: *Economic and Environmental impacts of Agbiotech*, Kalaitzandonakes, N. Kluwer Academic Publishers. http://www.inta.gov.ar/ies/docs/doctrab/adoption_dt_18.PDF. Last accesion, June 2011

Price-Whelan, A., Dietrich, L.E., & Newman, D.K. (2006). Rethinking 'secondary' metabolism: physiological roles for phenazine antibiotics. *Nature Chemical Biology* 2, 71–78

Raaijmakers, J.M., Weller, D.M., & Thomashow, L.S. (1997). Frequency of antibiotic-producing *Pseudomonas* spp. in natural environments. *Applied & Environmental Microbiology* 63, 881–887

Ramarathnam, R., & Fernando, D.W.G. (2006). Preliminary phenotypic and molecular screening for potential bacterial biocontrol agents of *Leptosphaeria maculans*, the blackleg pathogen of canola. Biocontrol Science & Technology 16, 567-582

Rosas, S.B., Altamirano, F., Schroder, E., & Correa, N. (2001). *In vitro* biocontrol activity of *Pseudomonas aurantiaca*. *Phyton-International Journal of Experimental Botany* 67, 203–209

Rosas, S., Rovera, M., Andrés, J.A., Pastor, N.A., Guiñazú, L.B., & Carlier, E. (2005). Characterization of *Pseudomonas aurantiaca* as biocontrol and PGPR agent. Endophytic properties, In: *Proceeding prospects and applications for plant associated microbes, 1st International conference on plant– microbe interactions: endophytes and biocontrol agents*, Sorvari, S., & Toldo, O., pp. 91–99. Lapland, Finland.

Rosas, S.B., Andrés, J.A., Rovera, M., & Correa, N.S. (2006). Phosphate solubilizing *Pseudomonas putida* can influence the rhizobia - legume symbiosis. Soil Biology & Biochemistry 38, 3502 - 3505

Rosas, S.B., Avanzini, G., Carlier, E., Pasluosta, C., Pastor, N., & Rovera, M. (2009). Root colonization and growth promotion of wheat and maize by *Pseudomonas aurantiaca* SR1. *Soil Biology and Biochemistry* 41, 1802–1806

Rovera, M., Carlier, E., Pasluosta, C., Avanzini, G., Andres, J., & Rosas, S. (2008). *Pseudomonas aurantiaca*: plant growth promoting traits, secondary metabolites and inoculation response, In: *Plant–bacteria interactions. Strategies and techniques to promote plant growth*, Ahmad, I., Pichtel, J., Hayat, S., pp 155–164. Wiley–VCH, Germany

Svercel, M., Duffy, B., & Défago, G. (2007). PCR amplification of hydrogen cyanide biosynthetic locus *hcnAB* in *Pseudomonas* spp. *Journal of Microbiological Methods* 70, 209-213

Thomashow, L. S. (1996). Biological control of plant root pathogens. *Current Opinion in Biotechnology* 7, 343–347

Senthilkumar, M., Swarnalakshmi K., Govindasamy, V., Keun Lee Y., & Annapurna, K. (2009). Biocontrol Potential of Soybean Bacterial Endophytes Against Charcoal Rot Fungus, *Rhizoctonia bataticola*. *Current Microbiology* 58, 288-293

Shaharoona, B., G.M., Jamro, Z.A., Zahir, M., Arshad, & Memon, K.S. (2007). Effectiveness of various *Pseudomonas* spp., and *Burkholderia caryophylli* containing ACC-deaminase for improving growth and yield (*Triticum aestivum* L.). *Journal of Microbiology & Biotechnology* 17, 1300-1307

Wahyudi, A.T., Astuti, & Giyanto, R.I. (2011). Screening of *Pseudomonas* sp isolated from rhizosphere of soybean plant as plant growth promoter and biocontrol agent. *American Journal of Agricultural and Biological Sciences* 6, 134-141

12

Effect of Socio-Economic Characteristics of Farmers on Their Adoption of Organic Farming Practices

O.M. Adesope[1], E.C. Matthews-Njoku[2], N.S. Oguzor[3] and V.C. Ugwuja[1]

[1]Department of Agricultural Economics and Extension
University of Port Harcourt
[2]Department of Agricultural Extension
Federal University of Technology, Owerri
[3]Federal College of Education (Technical), Omoku
Nigeria

1. Introduction

Organic farming is a form of agriculture which excludes the use of synthetic fertilizers, pesticides and plant growth regulators. The system also seeks to maintain the fertility demands of various crops to avoid excessive depletion of soil nutrients. Organic scientists and farmers in Africa, therefore deliberately integrates the age-long traditional organic system to enable a holistic development of the organic sector that would make use of the locally available resources, drawing from the pragmatic experiential knowledge of the farmers thereby making it highly relevant and acceptable to the majority of Africa. According to Parrot, Ssekyyew, Makunike, and Ntambi (2006), farmers have often resisted "Green revolution" viewing them not only as unsuitable but also risky and inaccessible. The concept of organic farming practices refers to the farm as an organism in which all the component parts (the soil minerals, plants, organic matter, micro-organisms, insects, animals including humans) interact to create a coherent and stable whole. Organic farming combines scientific knowledge of ecology and modern technology with traditional farming practices based on naturally occurring biological processes. Organic farming is based on ecological processes; knowledge of the agro-ecosystem is thus a pre-requisite to any organic farm. To this end, farmers with a traditional knowledge base are potentially better able to develop ecological processes. Traditional knowledge is not just a system for the present, but a source of institutional memory about what practices have worked best over time. Such knowledge has been described as a "reservoir of adaptations"; a whole set of practices that may be used again if the need arises (FAO, 2008).

Organic farming provides basis for maintaining environmental goods and services at the farm and landscape level. According to FAO, (2008), organic agriculture provides environmental goods and services. It promotes ecological resilience, improved biodiversity, healthy management of farms and the surrounding environment, and builds on community knowledge and strength. Hence, organic farming has been proved to be effective for enhanced adaptive capacity of farmers' socio-economic factors which necessitates social changes that are sometimes unacceptable to the local community. This might be successful in some communities but not in others; no doubt these factors include age, gender, farmers household size, educational status, farm size, level of personal income, membership of co-operatives, access to credit, number of extension contacts, policy problems and prospects, poverty, closeness to markets infrastructural facilities provided such as roads, electricity, housing, schools, clinics, potable water. Gender plays a role in organic farming practices, despite the fact that women participate more than men in agriculture in developing world, they remain more malnourished and less economically empowered because of the past, generally development assistance failed to reach women in rural areas.

Kang (2007) explains that today mainstreaming of gender of all levels in the agricultural sector is actively being done in order to address the issue of gender disparities. Mgbada (2000), Rahman and Usman (2004), and Ibrahim *et al* (2000), noted that rural women contribute two thirds of the labor force spent in agricultural production and marketing by producing up to 60-80% of food and other products in Nigeria and Africa at large. Adu, Famiyide, Adejoba, Ojo, Thomas, and Adebayo (2003), claimed that majority of rural women took active part in agricultural production, in addition to their domestic activities women should be encouraged when it comes to farming since they are responsible for small scale agriculture farm labor force and day-to-day family sustenance. This implies if women are well empowered in terms of organic farming practices, this will help to reduce scarcity of food and poverty level in the country. Poverty, does not only affect organic farming but the rural people in developing counties, and when the people are affected the production system is also affected. Ekong (2003), sees poverty as not only an expression of life condition, but a state of mind and preparation of self in the complex web of social relations. The high poverty level in Nigeria is ironic given the vast human materials and natural resources present in the country. Interestingly, this is why the country was placed among the poorest nations.

Benefits of Organic Farming Practices

Organic farming is beneficial in agriculture because it provides basis for healthy foods and healthy living. Organic agriculture reduces external inputs by controlling pests and diseases naturally, with both traditional and modern methods, increasing both agricultural yields and disease resistance. Organic farming practices enhance soil structures, conserve water and ensure sustained biodiversity. Through its holistic nature, organic farming integrates agro-biodiversity and soil conservation, and takes low intensity farming one step further by eliminating the use of chemical fertilizers, pesticides and genetically modified organism, which is not only an improvement for human health and agro biodiversity, but also for the associated off farm communities. According to Food and Agriculture Organization (FAO, 2008), organic agriculture promotes ecological resilience, improved biodiversity, healthy management, off-farms and the surrounding environment, and building on community knowledge and strength.

Despite the fact that much progress has been made, problems resulting from pests and diseases are still a major reason for lower yields but this problem could be addressed through plant breeding efforts under the special conditions of organic farming practices. Organic agriculture seeks to utilize those advances for consistent yield benefits (new varieties of crops, agricultural technologies, more efficient machinery) while discarding those methods that have impacted negatively on the society. Rod (1990), stated that sustainable agriculture systems rely more on crop rotation, crop residues, animal manures, legumes, green manures, off-farm organic waste, appropriate mechanized cultivation or minimal tillage to optimize soil biological control activities and thereby maintain soil fertility and crop fertility and crop productivity. According to Wikipedia (2007), organic farming technology relies on agronomic biological and mechanical method. This will help in environmental awareness and concern increase amongst the general population thereby enhancing inputs and sciences upon food because when technology and mechanical method has taken over it will take away some drudgery that which would have been a hindrance to increase in productivity.

Organic farming is beneficial when it comes to fauna and floral due to, in terms of its floral activities it helps to improve the soil fertility and this in turn increases the yield in crops while in fauna it is essential in the world of wildlife especially our livestock because most of our cattle's emits methane and we mostly produce them for its milk and meat, but with organic farming there is tendency for the grass which they feed on to grow all season of the year, this is to enable these livestock to still be in there physical health condition. Stolze (2000) reported that organic farming clearly performs better than conventional farming in respect to floral and fauna diversity, however it direct measure for wildlife and biotype conservation. Rigby and Caceres (2001), noted that in terms of soil, it is concluded that organic farming tends to conserve soil fertility and system stability. By conserving its soil fertility it will help the farmer to increase the humus content of the soil which will in turn improve the physical properties of the soil and support the life of micro-organism in the soil, which will help in increasing the pore spaces in the soil thereby improving drainage that aids to the stability of the soil structures and components.

Adoption of Organic Farming Practices

Rogers (1995), stated that the adoption of innovation is related to innovation decision process through which an individual passes from first knowledge of an innovation, to forming an attitudes towards the innovation, deciding to adopt or reject the innovation, implementing the new ideas, and confirming the innovation decision. This implies that organic farming can only be accepted by rural farmers when they have passed through the innovation to decision process and these farmers have picked interest concerning this because, when a farmer picks interest, he tends to seek for more information on his own and when this happens adoption can take place. Adoption models are generally based on the theory that farmers make decisions in order to maximize hand, farmers utility depends on optimizing the productivity and minimizing the cost of cultivation to attain maximum profits. Feder (1985), stated that farmers adopt or practice new technologies when they expect a more profitable outcome that is gained from the existing technology.

Optimizing this utility may also include considerations such as health benefits, environmental concerns, food security and risk (Napier, 2000). They further put explains that for an individual to go into adoption when it comes to organic farming he first wants to know if his health is at risk by finding out if the organic practice will be harmful to his crop and in turn becomes poisonous to human, and he will also consider that if he is using his farm for mixed cropping will organic farming practice enhance the growth of one plant and kill others or will the advantages be greater than the disadvantages before the farmers can finally adopt. The adoption of an innovation can be measured as the extent of its use, producing a continuous dependent variable, or simply the use of the innovation, with organic farming adoption defined as growing crops using organic method of cultivation. Adesina and Zinnah (2003) emphasized the impact of farmers perceptions on innovation related characteristics in measuring adoption. In this study organic farming practices, adoption was described as a mental process, farmers go through a stage or some stages of being aware or knowledgeable of organic farming related technology, to forming positive or negative perception towards organic farming and ultimately decide weather to adopt or not. This process can be influenced by a wide variety of factors, including household factors (socio-economic, resource base), community factors (access to extension, market, credit) and perception towards organic farming. Rogers (2003), reviews that diffusion theory provides a model for the diffusion-innovation process, which extension professionals as change agents can use as a media which will attract innovators and early adopters. According to Rogers (2003), the five important attributes of innovation related to an individual's attitude toward an innovation and whose stage in the innovation decision process summarized by Rogers are relative advantage, compatibility, complexity, observable and trial.

When all these attributes are put together with the guide of an extension agent it enhances or speeds up adoption and farmers will be willing to adopt despite the formal mindset the farmer had towards adoption. The perceived attributes of an innovation would vary according to individuals different personal characteristics (age, communication channels). Perception about these attributes of innovation will influence adoption behavior. Based on adoption, Rogers (2003), divided innovation adopters into five category; innovators, early adopters, early majority, late majority and laggards. Each category of adopters has different characteristics according to their socio-economic status, personality values, and communication behavior. This implies that how the extension service tends to communicate with the farmers because, when you communicate wrongly to the farmers there is the tendency of message misinterpretation and this will affect adoption. Rogers (2003), defined interpersonal delivery method as a face-to-face exchange between individuals. This explains that the extension services can speak to the farmer based on face-to-face contact were if the farmer has difficulty in any of the stages he can share with the extension agent and the correction will be made.

The main objective of the study was effect of socio-economic characteristics of farmers on their adoption of organic farming practices. Specifically, the study describes the socio-demographic characteristics of the respondents, determine the level of adoption of organic farming practices, determine the perceived benefits of using organic farming practices among respondents and determine the relationship between socio-economic characteristics of farmers and their adoption of organic farming practices.

2. Methodology

The study was carried out in Rivers State in Niger Delta, Nigeria. The sampling frame consisted of a list of farmers obtained from the Rivers State Agricultural Development Programme, the sole agency of Agricultural Extension activities in the State. A structured questionnaire was administered to 90 randomly selected farmers from the study area. Data analysis was by the use of frequency, mean, percentage, and Pearson correlation.

In order to measure the level of adoption (dependent variable), 14 items (organic farming practices) were presented to the respondents based on a 7-point scale of Not aware (NA) = 1, Aware (A) = 2, Interest (I) = 3, Evaluation (E) = 4, Trial (T) = 5, Adoption (A) = 6, Discontinuance (Discont) = 7. The independent variables (socio-economic characteristics) were analyzed thus; Gender (Male = 1, Female = 2); Age of farmers (years); Highest educational qualification (No formal education=1, Adult education=2, Primary=3,

Secondary=4, Tertiary=5). Household size (number of persons in a household); Extension visit (Yes =1, No = 2); Membership of cooperatives societies (Yes =1, No =2); Farming experience (years); Marital status (Single =1, Married = 2); Sources of information (friends/relatives/neighbor =1, Extension agent =2, Radio=3, Television=4, Newspaper=5, others =6)

3. Results and discussion

Table 1 shows that 34.4% of the farmers were male, while 65.6% were female. From the table it is obvious that majority of the respondents are females. Also, 14.4% of the farmers were between the ages of 31and 40 years old, 36.7% were between 41 and 50 years, 26.7% were between 51 and 60 years, while 22.2% were 61 years and above. The findings revealed that, a significant proportion of the farmers were between 41 and 50 years indicating that the farmers were mainly middle aged who are in their economically active stage and as such, can undergo the stress and this has implication for productivity of the farmers.

The result from the Table 1 shows that 26.7% of the farmers had no formal education, 33.3% had adult education, 22.2% of the farmers had primary education while 6.7% of the farmers attained primary education and 11.1% received tertiary education. The findings show that the a higher proportion of the respondents had adult education qualification (33.3%). Findings show that a reasonable percentage (48.9%) of the farmers had 6 to 10 members in their households. About 32.2% of the farmers were single, while 67.8% of the farmers were married. It was also found that 32.2% of the farmers were members of cooperative societies while 67.8% were not. Since majority were not members of cooperative societies, their access to farm resources like agro-inputs, credits and even extension contact might be lean and this would not encourage adoption.

Findings showed that 56.7% of the respondents had 6 to 10 years farming experience, 42.2% have been visited by extension agents, while 57.8% were not visited by extension agents. This is not too good because visit or contact with extension provides opportunity for transfer of skill, knowledge and information which facilitate adoption.

About 44.4% of the farmers know about organic farming practices through friends/relatives/ neighbors, 27.8% of the farmers know about the practice through extension visit, 7.8% of the farmers got to know about the practice through radio, 4.4% got the information through television, while 7.8% of the farmers got to know about organic farming through the newspaper.

Variables	Frequency	Percentage
Gender		
Male	31	34.4
Female	59	65.6
Age (Years)		
31-40	13	14.4
41-50	33	36.7
51-60	24	26.7
61 and above	20	22.2
Highest Educational Qualification		
No formal education	24	26.7
Adult education	30	33.3

Primary education	20	22.2
Secondary education	6	6.7
Tertiary	10	11.1
Household Size		
1-5	15	16.7
6-10	44	48.9
11-15	30	33.3
16 and above	1	1.1
Marital Status		
Single	29	32.2
Married	61	67.8
Membership of cooperative societies		
Yes	29	32.2
No	61	67.8
Years in Organic Farming Practices		
1-5	9	10.0
6-10	51	56.7
11-15	17	18.7
16 and above	13	14.1
Extension visit		
Yes	38	42.2
No	52	57.8
Sources of Information		
Friends/Relations/Neighbours	40	44.4
Extension Agents	25	27.8
Radio	7	7.8
Television	4	4.4
Newspaper	7	7.8
Others	7	7.8

Source: Field Survey, 2010.

Table 1. Socio-Economic characteristics of respondents (n=90)

4. Adoption of organic farming practices

The result in Table 2 indicated that about 68.9% of the farmers adopted the practice of crop rotation and mixed cropping as the major organic farming practices respectively, hoeing and hand weeding had 63.3%, and slash and burn were 58.9% while intercropping recorded 50.0%.

From the findings it is obvious that farmers adopted five (5) out of fourteen (14) listed organic farming practices, giving an adoption rate of 35.7%. This indicates that level of adoption of organic farming practices is generally low; this could be as a result of the strenuous nature of some of the practices like hand picking of insects, tillage with sticks. The farmers are not very convinced about its merits, inability and poor disposition of farmers as well as the inability of extension visit to facilitate their adoption.

ORGANIC FARMING PRACTICES	NA	A	I	E	T	A	DISCONT.
Crop Rotation	2(2.2)	14(15.6)	6(6.7)	3(3.3)	3(3.3)	62(68.9)	0(0)
Mixed Cropping	0(0)	13(14.4)	4(4.4)	2(2.2)	6(6.7)	62(68.9)	3(3.3)
Use of green manure	1(1.1)	14(15.6)	14(15.6)	14(15.6)	29(32.2)	18(20.0)	0(0)
Use of compost	3(3.3)	13(14.4)	20(22.2)	18(20.0)	19(21.1)	15(16.7)	2(2.2)
Hand picking of insects	12(13.3)	16(17.8)	29(32.2	19(21.1)	8(8.9)	3(3.3)	3(3.3)
Use of leaves as mulching materials	14(15.5)	18(20.0)	18(20.0)	10(11.1)	10(11.1)	19(21.1)	1(1.1)
Hoeing/weed removal by hand	4(4.4)	15(16.7)	5(5.6)	4(4.4)	4(4.4)	57(63.3)	1(1.1)
Slash and burn	9(10.0)	13(14.4)	4(4.4)	0(0)	9(10.0)	53(58.9)	2(2.2)
Tillage with sticks	11(12.2)	29(32.2)	13(14.4)	5(5.6)	13(14.4)	8(8.9)	11(12.2)
Use of air tight containers for storage	10(11.1)	20(22.2)	19(21.1)	13(14.4)	9(10.0)	16(17.8)	3(3.3)
Sun drying of farm produce	2(2.2)	16(17.8)	18(20.0)	14(15.6)	10(11.1)	30(33.3)	0(0)
Use of organic manures	3(3.3)	21(23.3)	18(20.0)	8(8.9)	19(21.1)	21(23.3)	0(0)
Inter cropping	6(6.7)	20(22.2)	4(4.4)	4(4.4)	9(10.0)	45(50.0)	2(2.2)
Tree/hedges planting	4(4.4)	24(26.7)	10(11.1)	6(6.7)	4(4.4)	7(7.8)	35(38.9)

NA; Not Aware, A; Awareness, I; interest, E; Evaluation, T; Trial, A; adoption, Discont; discontinuance
Figures in parenthesis are in percentages
Source: Field Survey, 2010.
Table 2. Level Of Adoption Of Organic Farming Practices

5. Perceived benefits of using organic farming practices

Fig. 1 shows that 41.1% of the respondents perceived that organic farming practices increased soil organic matter content, 22.2% perceived that they reduce input cost of farming, 26.7% perceived that they involved low risk in crop failure. Also, 81.1% indicated that it has a high social value of general acceptability. About 84% indicated that organic farming practices are compatible with their own cultural systems, 77.8% stated that they are inexpensive, 55.6% stated that organic farming practices are natural form of farming, while 22.2% indicated that they are environmentally friendly.

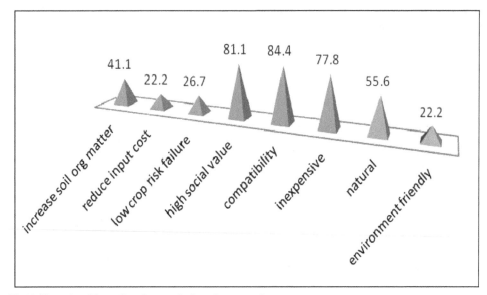

Fig. 1. Perceived benefitsof organic farming practices

6. Relationship between socio-economic characteristics of farmers and adoption of organic farming practices

Table 3 shows the results on relationship between socio economic characteristics of respondents and adoption of organic farming practices. From the findings only two variables correlated significantly with adoption of organic farming practices.

VARIABLES	r-VALUES
SEX (GENDER)	0.028
AGE	- 0.155
EDUCATIONAL QUALIFICATION	0.053
HOUSEHOLD SIZE	0.065
MARITAL STATUS	- 0.221*
MEMBERSHIP OF COOPERATIVES	0.084
FARMING EXPERIENCE	-0.277**
EXTENSION VISIT	- 0.001
SOURCES OF INFORMATION	0.020

* = Significant at 0.05% level; ** = Significant at the 0.01 level

Table 3. Correlation coefficient of socio-economic characteristics of farmers and adoption of organic farming practices

Farming experience is negatively correlated with adoption of organic farming practices and significant at 0.01 level. This implies that there exist an inverse relationship between farmers experience and adoption which means that those with less farming experience have higher adoption level. This is in consonance with the findings by Edeoghon's (2008), who reported that farmers usually are more involved in practices that they are more familiar with than other practices. The findings also agree with the literature that organic producers are newer entrants to farming (Padel, 2001). This can be attributed to the fact that farmers who have been long in the business are usually older, less educated and more resistant to change than new entrants.

Marital status of farmers was found to be negatively correlated with adoption of organic farming practices which means there is an inverse relationship, but is significant at 0.05 level which means if both variables are cordial there would be increase in adoption. Marital status of respondents was an important factor in adoption of organic farming practices. This is in line with Ekong (2000), Nwachukwu and Jibowo (2000), Bamneke (2003), who reported that majority of respondents involved in agricultural activities are married.

7. Conclusion

The study concludes that adoption of organic farming practices was low as farmers adopted only 5 of the 14 practices identified. These include crop rotation, mixed cropping, hoeing and hand weeding, slash and burn and intercropping. It was found that respondents perceived that organic farming practices increased soil organic matter content, reduce input cost of farming, involved low risk in crop failure. Also, it was found that organic farming practices have high social value of general acceptability, are compatible with their own cultural systems, are inexpensive, are natural form of farming, and are environmentally friendly.

8. Recommendations

There is need to intensify efforts to make agricultural extension services more functional so that farmers can get useful information to enhance adoption of organic farming practices. Farmers should be encouraged for form and belong to cooperatives to facilitate group dynamism. Adequate enlightenment programmes should be mounted on organic farming practices so that more farmers can adopt.

9. References

Adesina A and Zinnah, M. (2003), 'Technology charateristics, farmer's perception and adoption decision: A Tobit model applicationin Sierra Leone'. *Agricultural Economics* , 9, 197-331.

Adesina A A, Mabila D, Nakamleu G B & Endamana D 2000, ' Econometric analysis of the determinants of adoption of Alley Farming by farmers in the Forest Zone of southwest Cameroon.' *Agric. Ecosys. Environ*, 80, 255-265.

Adu, A. O., Famuyide, O.O, Adejoba, O.R., Ojo, M. O, Thomas, E.Y, and Adebayo, O. (2003): Gender uses in agro forestry development in three selected local government areas of Oyo state. A paper presented at the 9th Annual conference of the forestry Association of Nigeria in Calabar Cross river state. 6th-11th October, 2003. 128-134.

Edeoghon, C.O (2008). Awareness and use of sustainable Agricultural practices by Arable crop farmers in Ikpoba Local government Area of Edo State. Journal of sustainable Development in Agricultural and Environment: Vol. 3No. 2, Ph 55-63

Ekong E.W. (2000). Group and Non Group Women Farmers Access to Agricultural production, Responses in Akwa Ibom State, Nigeria. Ph.D. Thesis Department of Agricultural Ext and Rural Development, Ibadan: University of Ibadan.

FAO (2008). Organic Agriculture and climate change. Food and Agriculture Organization, Rome.

Feder, G., R.E.Just, and D. Zilberman (1985) "Adoption of Agricultural innovation in Developing Countries: A survey. *Economic Development and cultural change,* 33, 255-98

Kang Bhavdeep (2007). Women lead the way back to Organic Farming, IFOAM. www.ifoam.org/growing organic viewed 28th October, 2009

Mgbada, J. U., (2000): Production of stable food crops by Women in Enugu and Eboyi States: Lessons for enhancing poverty alleviation programme. In T.A. Olowu (ed) *Agricultural Extension and poverty alleviation in Nigeria. Proceedings of the 6th Annual National conference of the Agricultural Extension society of Nigeria.*

Napier T L, (2000), 'Adoption of conservation tillage production systems in three Midwest watersheds.' *J. soil cons.* 53: 123-134

Padel, S. (2001). Conservation to organic farming: a typical example of the diffusion of an innovation? *Sociologia ruralis* 41,40-61.

Parrott, N., Sockyew A., C., Makunike, C. and Ntambi, S.N. (2006). Organic farming in Africa. In: Helga, W. and M. Yussefi (eds). *The world of organic farming agriculture: statistics and emerging trends.* A joint Publication of International Federation of Organic Agricuture movement, Bonn, Germany and research Institute of organic agriculture (FiBl) Frick Switzerland.

Rahman, S.A and J.I Usman (2004): comparative Analysis of women participation in Agricultural production in Northern and southern Kaduna State. Mobilizing investors for sustainable Agricultural Research, Development and production in Nigeria. Proceedings of the 30th Annual conference of the Agricultural society of Nigeria held at the College of Agriculture Lafia, Nassarawa State, October 17th -21st p 103-113.

Rigby, D. and Caceres, D. (2001). Organic farming and Sustainablility of agricultural systems. *Agricultural systems,* 68: 21-40. Elsevier Ltd.

Rod (1990), in Matthews-Njoku,E.C., Adesope, O.M, Nakwasi, R.N and Agumagu A.C.; Benefits of Organic farming practices in Owerri West Local Government Area of Imo, state Nigeria. Proceedings of the 5th National conference of Organic Agriculture Project in Tertiary Institutions in Nigeria Federal University of technology, Owerri, Nigeria. November 15-19, 2009. p 318

Rogers E M 1995, *'Diffusion of Innovations (4th Edition)'.* New York: The Free Press.

Rogers E.M. (2003). *Diffusion of Innovations (5th ed.)* New York: The free press

Sanusi, R. A., C. A. Badejo and B. O. Yusuf (2006) *Measuring Household food insecurity in selected local government areas of Lagos and Ibadan, Nigeria.* Pakistan journal of Nut.

Stolze, M. (2000) Environmental impacts of organic farming in Europe. Organic farming in Europe: Economics policy. Stuttgart-Hohenheim 2000. Department of Economics, University of Hohenheim, Germany.

Wikipedia (2007). Importance of organic farming in Texas city

13

Analysis of Nutritional Interactions in Cropping Systems

Moses Imo
Chepkoilel University College, Moi University
Kenya

1. Introduction

Successful management of crop production systems requires design of cultural practices that enhance yield by ensuring that growth limiting factors are minimized or eliminated altogether. Together with moisture, light, competition, diseases and pests, availability and efficient utilization of soil nutrients is one of the major determinants of healthy plant growth and realization of optimum yield returns. Thus, assessment of nutrient availability, uptake and utilization by plants is critical for optimization of crop productivity. It is also important to note that these factors and other physiological processes characteristically involve complex relationships and interactions among many factors, some of which are of non-nutritional in nature. Traditionally, diagnoses of these nutritional responses and interactions have involved chemical analysis of soil and plant samples, while different approaches have been employed for the interpretation of the analytical results. These interpretations have been derived from empirical studies related to plant responses to specific nutrients in terms of visual symptoms, nutrient uptake and use, nutrient ratios, and soil, biochemical, physiological tests. Unfortunately, most of these techniques have focused on causes of nutrient deficiencies in an attempt to identify nutrients that are most limiting rather than in trying to optimize their availability, uptake and utilization by the crops. Also, interactions among different nutrients and other growth limiting factors such as moisture and light are seldom considered. Moreover, most of these interpretative approaches are only known to scientists in terms of their power and limitations, while land managers seldom use them to diagnose simple conditions in their farms.

Most of the diagnostic methods use chemical analysis of plants or plant parts to determine the qualitative or quantitative nutritional status since the plant itself is the subject of interest, thus its nutrient composition reflects most of the factors affecting its nutrition. These plant chemical analyses often use three plant parameters in interpretations of crop responses (growth and yield, nutrient concentration, and nutrient uptake also measured as nutrient content) to evaluate plant response to soil nutrient supply and fertilization. Unfortunately, overwhelming majority of studies on crop response to nutrient supply often express their results using only one or two of these parameters at a time, but not all of them combined. Our past research experience and review of other literature has clearly identified this glaring gap. Similar problems associated with excessive use of concentration data alone and improper use of terminology have been detected in plant nutrient studies several decades ago (e.g. Timmer and Stone 1978; Smith *et al.* 1981), which unfortunately is still common in current literature. There is now strong evidence to demonstrate that conclusions derived from such results may be misleading if any one of these plant parameters is excluded from the interpretation of plant nutrient analysis.

Current trends towards precision agriculture with intensive fertilization to achieve optimum nutrition is likely to lead to more complex nutritional disorders that will be more difficult to detect, thus the need for interpretative techniques that provide more comprehensive and precise plant nutritional status with the capacity to identify single and multiple nutrient deficiencies, sufficiency, imbalances, toxicity and interactions. The objective of this Chapter is to highlight the pitfalls associated with relying solely on concentration of nutrients only when interpreting results of plant responses to nutrient availability. The Chapter demonstrates that more insight into the mechanisms behind changes in plant nutritional status under changing soil nutrient conditions can be achieved by simultaneous examination of changes and interrelationships between all the three plant nutritional parameters i.e. growth or yield, nutrient concentration and nutrient uptake or content using a graphical approach called "Plant Nutrient Vector Analysis". Although this method has been used extensively in forestry research, little applications in evaluating nutritional status in agricultural crops have been. It is strongly suggested that agricultural scientists involved in intensive fertilizer management systems utilize the enormous potential of this method.

2. Background

2.1 Defining plant nutrient status

Plant nutrient status is often defined by either nutrient concentration or absolute content. By definition, the term nutrient *concentration* refers to the amount of a compound present in a unit amount of plant tissue and is expressed as a ratio (e.g. %, ppm, $mg\ g^{-1}$, or $mg\ cm^{-1}$ etc.), while nutrient *content* is the total amount of a compound present in a specific amount of plant tissue (e.g. whole plant, leaf, shoot, root etc.) and is expressed in any unit of mass such as mg, g, kg, and tons (Farhoomand and Peterson, 1968). Clearly, while concentration is *intensive* and *non-additive* (i.e. does not depend on the size of the sample but simply an indicator of plant quality), content on the other hand is *extensive* and *additive* (i.e. depends on the size of the plant or plant part and is an indicator of production and growth allocation). Further, it is good to note that concentration is simply a ratio between content and biomass (or area) in which content has been measured. Thus, the discussion on when to use concentration or content really is a matter to be determined by the objective of the analysis between variables; whether original or derived variables, a question that has been debated extensively in biological (e.g. Jackson and Somers, 1991) and statistical literature (e.g. Sokal and Rohlf, 1995) in the past.

In statistics for example, part of the drawbacks in the use of ratios (i.e.. concentration) is because these ratios are often derived from two independently measured variables, thus often resulting in inaccuracy and non-normal distribution (Sokal and Rohlf, 1981). Fortunately in the case of plant nutrient concentrations, they are not derived variables even though they are ratios since they are primary variable measured directly as a result of biochemical analyses. This is probably one of the reasons why most researchers prefer using concentration data directly since to obtain content requires multiplying concentration by the weight of the plant sample, which is rarely done. Statistically therefore, using content is likely to be more prone to error since it is derived from two independent variables i.e. concentration and biomass.

Koricheva (1999) has pointed out the major problems associated with the use of ratios and concentrations in interpreting changes in plant allelochemistry. The problem with the use of ratios is that they do not give any indications on the relationships between the two variables from which they are derived. This means that changes in a ratio can be caused by changes in the numerator, the denominator or both, and there is no way of distinguishing the mechanism responsible for the observed changes in the ratios. However, this would not be a problem if the ratio itself is the primary variable of interest, and the researcher is interested only on the implications of the shift in the ratio rather than in the mechanisms which brought about the change. If the biological process being investigated operates on the ratio of variables studied, then one must study the variables affecting the ratio in order to understand the mechanisms involved.

In terms of interpretations of plant responses to nutrient availability, the major problem often arises when researchers try to elucidate the mechanisms causing the observed changes in concentrations. Given the above definition, such changes can be brought about by either changes in plant growth (biomass accumulation), nutrient uptake (nutrient accumulation or content), or both. While changes in nutrient concentration due to changes in content can be considered as "*active*" since the plant altered nutrient uptake, synthesis or transport of the specific nutrient, changes in concentration due to changes in biomass can be considered as "*passive*" responses because they are simply by-products of plant growth and there is no specific effect on the metabolism of the compound analyzed.

It is important to distinguishing between these passive and active responses in order to enable establish the mechanisms that lead to observed changes in plant nutrient concentrations. This is important in enabling predictions of the effects of varying nutrient variability on plant physiological responses. When the focus of the study is on the mechanisms behind the changes in concentration, more can be achieved by studying the variables singly first and then examining their relationship to each other i.e. analysis of changes in content and plant biomass in addition to analysis of concentrations. Unfortunately, most research papers reporting plant nutrient data have not taken this message seriously probably because of the belief that concentrations do reflect changes in nutrient content but remove the effects of biomass, thus are superior indicators of plant nutrient status.

2.2 Problems with the use of plant nutrient concentrations alone

Although many researchers have often preferred using nutrient concentration (C) over nutrient content (U) by assuming that the former removes the effect of plant biomass (W) on nutrient content, and therefore good for standardizing data to allow for comparison between different plant individuals or plant parts by adjusting for differences due to biomass using the function $C = U/W$ i.e. concentration does not depend on the size of the plant (Jackson and Somers, 1991), there are many inherent problems with this approach. First, mathematically such a ratio can only remove the effect of W on U if and only if the relationship U/W is linear and passes through the origin (Raubenheimer and Simpson, 1992). Any nonlinearity of the function increases the probability of differences between the ratios calculated for the different parts of the curve, while non-zero intercept diminishes the statistical power to detect small differences between treatments. These assumptions are unlikely to hold for the U/W relationship, which is presumably non-linear and non-zero intercept may arise, for example, the rate of nutrient uptake is not equal to the rate of growth and biomass accumulation. As such, concentration is unlikely to remove the effect of plant weight from content, and analysis of covariance with content as the dependent variable and biomass as a covariate is a better option (Raubenheimer and Simpson, 1992).

Secondly, changes in nutrient concentration does not always reflect changes in nutrient content since changes in biomass are not due exclusively to changes in nutrient content since biomass production is a function complex interactions among many factors of plant growth including nutrients, light, water, temperature, cultural practices, weeds, pests and diseases (Jarrell and Beverly, 1981). Moreover, plant biomass (W) is the sum of the total contents of all plant constituents such as nutrients, carbon and other plant chemical compounds (i.e. W_1, W_2, W_3, W_n, where n is the number of plant constituents). The concentration of a specific nutrient C_1 can then be derived as $C_1 = U_1/(W_1+W_2+W_3 \ldots W_n)$. Therefore, any change in any of the plant constituents will result in change in concentration of nutrient C_1 even if its content remains the same. Decrease in concentration of a compound as a result in increase in content of other plant constituents is known as *dilution effect*, while increase in concentration of a compound due to decrease in content of another constituent is called *concentration effect*.

Although dilution and concentration effects will have relatively small impacts on the concentration of the major plant constituents such as carbon-based compounds, most nutrients are present in relatively very low concentrations making them potentially very sensitive to dilution and concentration effects. The dependence of concentration of a specific nutrient on the levels of other nutrients and plant constituents as well as environmental conditions has important implications on interpretation of concentration data with respect to active uptake e.g. in response to amelioration of nutrient deficiency, or passive uptake in response to accelerated or suppressed growth by other non-nutritional factors. Separation of these nutritional response mechanisms (active vs. passive) using only nutrient concentration data is seldom possible. To date, only a few studies have presented both nutrient concentration and content results demonstrating these distinct nutritional responses directly as for example using isotope techniques.

Thirdly, the use of nutrient concentration as a measure of nutrient uptake has no biological meaning since plants absorb molecules (content) of nutrients – concentration simply reflects distribution of these molecules within a given amount of plant biomass. Biologically, therefore, this distribution is not of any strategic importance but simply an inevitable consequence of plant growth (Timmer 1991). It has often been interpreted that higher nutrient concentration in young tissue is a strategy for resource allocation for higher productivity, while decline in nutrient concentration with leaf age may be largely a result of dilution due to accumulation of other metabolites as leaves expand. One alternative to using concentrations for assessing changes in allocation to nutrients in growing tissues is to plot nutrient content in the plant or plant part against plant biomass and to examine changes in the slope over time. Increase in slope indicates increased allocation to growth while decrease in slope indicates reduced allocation to growth

Fourthly, it is often assumed that plant resources are limited and thus plant preferential investment in certain organs of specific functions depending on prevailing environmental conditions. For example, increased root growth for increased drought resistance necessarily reduces investments in other plant organs such as shoot growth resulting in higher root/shoot ratio (Herms and Mattson 1992). Such trade-offs can only be examined in terms of the correlations between nutrient concentration and plant biomass that can demonstrate either positive or negative correlations. If a negative correlation is detected, then uptake of a given nutrient can be considered costly, while a positive correlation may be interpreted to be cheap or luxury uptake.

The problem however is whether two or more physiological processes jointly consume the same resource. Moreover, negative correlation between nutrient concentration and plant biomass may arise because biomass is a denominator in the ratio defining concentration. Thus, plotting nutrient concentration against biomass is the same as plotting a ratio against its denominator, which causes a negative relationship by default (Herms and Mattson 1992). Furthermore, the sign of correlation between nutrient concentration and plant biomass may depend on the timing of the compound synthesis during plant development. If uptake of a nutrient occurs early plant development and later metabolised or retranslocated to other plant parts, its concentration in the plant or old leaf will decrease with age, first due to dilution effect and later due to reduced nutrient content. Consequently, concentration of such nutrient would be negatively correlated with plant biomass. If however nutrient uptake equals growth demand and no retranslocation occurs, nutrient concentration will remain the same throughout the plant development stages, and will show no correlation with changes in plant biomass.

2.3 Reducing impacts of dilution and concentration effects

As discussed above, interpretation of concentration data is often confounded by dilution and concentration effects. Several methods have been developed to reduce the impacts of these dilution and concentration effects on the results of bioassays.

2.3.1 Concentration on a free-biomass basis

In cases where applied treatments are known to cause large changes in specific plant constituents (e.g. carbon), which may lead to dilution of other plant constituents (e.g. nitrogen), concentration of the other compounds (i.e. nitrogen) may be calculated on a carbon-free basis. For example, since elevated CO_2 usually causes large accumulation of carbon in plant tissue (Korner et al., 1995), then nutrient concentration can be calculated on carbon-free basis. According to Poorter et al. (1997), if the effect of CO_2 disappears when concentrations are compared on carbon-free basis, changes in concentration were due to dilution by the accumulated carbohydrates, while more pronounced differences when calculated on carbon-free basis means the CO_2 enrichment had real effect on the synthesis of the compound in question but this difference is obscured by larger changes in carbohydrate levels increased.

2.3.2 Concentration on a unit area basis

Expression of chemical concentrations on a unit area basis may also be an alternative to conventional expression on mass basis if the aim of the study is to assess the effects of non-seasonal environmental factors. Leaf area is usually more sensitive than leaf biomass to environmental changes hence less affected by seasonal variation and leaf age (Gholz 1978). Expressing nutrient concentration on area basis might be especially relevant for determination of large scale fertilizer prescriptions.

2.3.3 Concentration on fresh weight basis

Expressing concentration on fresh weight basis may also reduce impacts of dilution and concentration effects since fresh water-saturated plant material in mature organs is a more constant property than dry weight (Tamm, 1964). Unfortunately, this approach is based on the assumption that no difference in water content exists between the samples, which can only be tested by comparing water content among the treatments, sampling dates etc.

2.3.4 Comparison of nutrient concentration in similar plant developmental stages

As stated earlier, dilution effects are an integral part of plant development, thus nutrient concentrations of most plant constituents change during the plant life cycle if no environmental factors or cultural practices interfere with plant growth and nutrient uptake and concentration during the growing season. Therefore, assessing the effects of treatments at different dates during the season may result in confounding of environmental effects. This might be avoided if plants of the same the same development stage or age are compared between the treatments (Roumet et al. 1996). This approach may require changing the experimental design to facilitate comparisons between treatments through growth and developmental stages over time (Coleman et al., 1994). Thus, disappearance of treatment effects when plants of the same biomass are compared means that differences in nutrient concentrations were due to growth dilution, while similar effects on nutrient concentrations would suggest that the examined factor has an effect on plant nutrient status which is independent of developmental changes. Unfortunately, all these methods are inadequate for comprehensive analysis of plant responses to plant nutrient status. These are described in detail in the following sections.

3. Theoretical foundations

3.1 Nutrient uptake and growth relationships
3.1.1 Nutrient uptake

Uptake of mineral nutrients by higher plants occurs mainly through absorption of mineral elements from the soil environment by the roots, although leaves of some plants are also known to absorb limited amounts of mineral elements. Once absorbed, these elements accumulate in plant tissue. Although analyses of plant tissues often show accumulation of almost all elements found in the root environment, only a small number has been demonstrated to be essential for plant growth. Since nutrients accumulate in plant tissue during growth, and nutrient content of a plant gives an integrated estimate of both total uptake and use by a plant, studying the relationships between the two fundamental processes involved (nutrient accumulation and biomass production) can provide insight into the mechanisms involved. This can be achieved by studying the relationships between biomass, uptake and concentration. The following section discusses these fundamental relationships in terms of tissue nutrient composition, and its usefulness in elucidating mechanisms of plant nutrient interactions.

3.1.2 Plant nutrient composition

Chemical analysis of plants is frequently used to diagnose the nutritional status of plants since the plant itself is the object of interest, and its nutrient com position reflects many of the factors affecting its nutrition. Traditionally, plant nutrient com position is expressed either in relative term s (i.e. concentration [C], the amount of nutrient present per unit amount of biomass) or on total mass basis (i.e. absolute content [U], the total amount of nutrient present in a specific amount of plant tissue [W]) (see for example Imo 1999). Total content is obtained by multiplying concentration by dry mass of the sample, thus $U = C(W)$. Timmer (1991) has argued that using concentration alone does not reveal the mechanism on how nutrient content and dry mass are related, since changes in concentration may be caused by changes in either biomass or nutrient uptake or both, and there is no way of distinguishing between these mechanisms. Changes in concentration as a result of changes in content implies that the plant itself altered nutrient uptake and synthesis, while changes in concentration due to changes in biomass can be regarded as a growth response without any specific effects on metabolism of the nutrient.

Distinguishing between these processes is important to test hypotheses related to the effect of changing nutrient supply on plant growth and nutrient composition. One way of solving this problem is by first studying the effects of nutrient supply on each of the individual plant response variables (biomass, nutrient concentration and content), and then examining their interrelationships. Timmer (1991) presented a classic generalized interpretation of the relationship between these nutritional parameters diagrammatically (Fig. 1), which shows the possible relationships between increasing nutrient supply with plant growth, nutrient concentration and nutrient content. This diagram shows that growth responses to

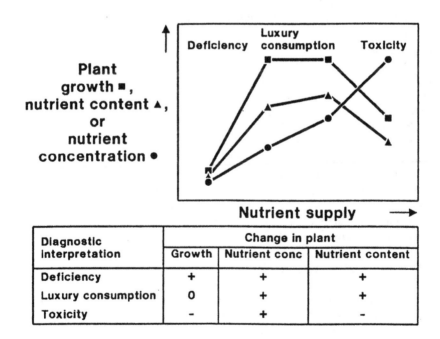

| Diagnostic | Change in plant | | |
Interpretation	Growth	Nutrient conc	Nutrient content
Deficiency	+	+	+
Luxury consumption	0	+	+
Toxicity	-	+	-

Fig. 1. Generalized interpretation of the relationships between plant growth, nutrient concentration and nutrient content with increasing nutrient supply. The lower box shows the expected direction of change (0, +, or -) in growth, nutrient concentration and nutrient content in the three phases. Adapted from Timmer (1991).

increasing supply of a limiting nutrient follows a curvilinear relationship that can be divided into three distinct nutritional phases: deficiency, luxury uptake, and toxicity (Timmer 1991). At each phase, changes in growth, nutrient concentration and uptake can be identified as either increasing (+), no change (0), or declining (-) as shown in Fig. 1. Notice that growth is restricted at low nutrient levels, increases sharply with increasing nutrient supply until sufficiency is reached, it levels off when nutrient supply is at luxury levels, and then declines gradually as nutrient concentrations become toxic (Fig. 1).

Nutrient content follows a similar curvilinear pattern as biomass, except that it continues to increase as a result of luxury uptake until toxicity occurs (Fig. 1). In contrast, nutrient concentration will continue to rise throughout the whole range of nutrient supply (Fig. 1), slowly in the deficiency range because of growth dilution, and more rapidly in the toxicity range because of accumulation. Thus, interpretation of concentration data alone is often confounded by dilution and concentration effects, and nutrient content data is often confounded by effects of total plant biomass.

One way of avoiding this problem is by examining all the three parameters simultaneously using vector diagnosis as discussed in the following section.

3.1.3 Nutrient interactions

While in general the term "interaction" is defined the mutual or reciprocal effects, it has been used in soil fertility evaluation in agriculture and forestry to mean both quantitative and qualitative responses to fertilizer nutrients in plant-soil systems, which involve both single or multi elements interactions as well other non-nutritional factors that may be occurring in these systems. These interactions often occur when the level of one factor of biological production influences the response of an intended product to another factor, resulting in mutual or synergistic (positive), reciprocal (compensatory) or antagonistic (negative) effects. For example, if the supply of one nutrient affects the absorption, distribution or function of other nutrients and thus modifies growth response then interactions can be said to occur. These interactions can occur in the soil as biochemical reactions, in the soil-plant interface due to uptake processes, or interactions within the plant itself due to varying requirements for nutrients by plants of varying phenotypes and life cycles. Inevitably, other non-nutritional factors that affect plant growth and development, and nutrient availability, uptake and utilization also influence the nature of nutrient interactions in soil-plant systems. Although management practices always strive to achieve positive interactions, methods for ascertaining achievement of the desired objectives are hardly explicit. A positive interaction occurs when the influence of the combined practices exceeds the sum of the influences of the individual practices. Such positive interactions have served as the science-based justification for development of a "balanced" plant nutrition program.

These interpretations often involve plant responses such as steady-state nutritional effects, synergistic or antagonistic dilution effects and induced toxicities are often quite complex and multi-factored. Detection of such interactions therefore requires specialized and comprehensive techniques and approaches for their quantification, especially in intensively managed cropping systems. Several methods and approaches have been proposed to detect these interactions in cropping systems including visual symptoms, critical nutrient approach, optimum nutrient ratios, factorial experiments, mathematical modeling and graphical nutrient analysis. Unfortunately, questions still a rise especially when dealing with nutritional elements of different biochemical nature, varying requirements and quantities by plants, multiple factors affecting their availability, prioritizing nutrient limitations, and making practical recommendations for optimum crop production. Separation of all these interaction effects in terms of site-specific fertilizer nutrient recommendation is seldom reported in research papers.

3.1.4 Steady state nutrition

Steady state nutrition refers to a condition whereby plants grow with constant internal nutrient concentration, free from stress (Ingestad and Lund 1986). This condition can be achieved by adding fertilizer nutrients at exponential rather than conventional (or constant) rates, which corresponds closer to the desired relative growth rates of the plants during their exponential phase of growth as has been demonstrated by Imo and Timmer (1992a; Imo and Timmer 1997). These studies also showed that although and nutrient accumulation of conventionally fertilized plants normally increases as the season progresses, internal nutrient concentration usually declines due to growth dilution, which suggests excess fertilization at the start but nutrient stress at the end of the growing season. In contrast, Plants growing at steady-state nutrition were growing relatively free of nutrient stress since it was characterized by stable internal nutrient concentration. Long term experiments are required to demonstrate achievement of this condition under field conditions, which would greatly improve nutrient use efficiency of applied fertilizers while reducing environmental pollution.

4. Vector analysis of plant nutritional responses

4.1 Diagnostic approach

Vector nutrient analysis is based on the biological dependence of plant growth on nutrient uptake. Fundamentally, the technique is a multivariate approach of examining changes in nutrient concentration (C) in relation to functional processes that cause these changes, namely nutrient uptake (U) and biomass accumulation (W), where $C = U/W$. This relationship is then compared graphically by plotting C on the y-axis and U on the x-axis (Fig. 2). Since concentration is a ratio between content and biomass, biomass is the inverse of the slope factor, and C and U values for each plant sample will follow a diagonal line (z-axis) corresponding to biomass (W) (Fig. 2). These diagonals also serve to separate biomass of various samples being compared. Changes in C, U and W can be plotted as absolute values (thus allowing standard error of the means to be shown on the diagram) or relative values (thus enabling multiple treatment, nutrients and inter-site comparisons by eliminating inherent differences in plant size and nutrient status).

In this graphical model, vectors are drawn to depict changes in the relationship between C, U and W. Plant samples for comparison can either be for different treatments (Timmer 1991) or changes over time (Imo and Timmer 1997). Interpretation of the relationships between C, U and W are based on changes in vector direction and magnitude observed as an increase [+], decrease [-], or no change [0] in biomass, nutrient content, and nutrient concentration relative to the reference plant status, and help quantify treatment effects (Fig. 2). The direction and magnitude of the vectors quantify treatment effects relative to the reference status, and facilitate diagnoses of nutritional effects of growth dilution, sufficiency, deficiency, luxury uptake, toxicity (excess uptake) and antagonism. Vector nutrient analyses of these nutritional effects are discussed here below.

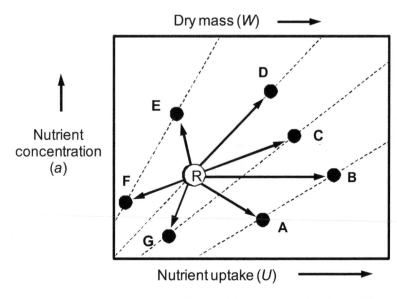

Fig. 2. Vector nutrient diagnosis of directional changes in relative dry mass (W), nutrient content (U) and concentration (C) of plants (or plant components) contrasting in growth and nutrient status. The reference status (R) is usually normalized to 100%. The dotted diagonal lines represent the biomass of samples being compared. Vector shifts (A to G) indicate increase [+], decrease [-], or no change [0] in dry mass and nutrient status relative to the reference status as summarized in the Box beneath. Vector magnitude reflects responsiveness of individual nutrients. From Imo (1999).

4.2 Diagnostic interpretations of nutritional responses, with examples
4.2.1 Growth dilution of nutrients

Growth dilution (Shift A, Fig. 2) occurs when nutrient concentration declines while growth and nutrient uptake increase (Armson 1977, Timmer 1991). Such dilution effects usually occur during periods of rapid plant growth when nutrient uptake cannot keep pace with the high rate of biomass accumulation (Ingestad and Ågren 1988; Jarrell and Beverly 1981). For

Vector shift	Change in W	Change in U	Change in a	Nutritional effect	Nutrient status	Possible diagnosis
A	+	+	-	Dilution	Non-limiting	Growth dilution
B	+	+	0	Accumulation	Non-limiting	Sufficiency, steady-state
C	+	+	+	Accumulation	Limiting	Deficiency
D	0	+	+	Accumulation	Non-limiting	Luxury consumption
E	-	-, +	+	Concentration	Excess	Excess accumulation
F	-	-	-	Antagonistic	Limiting	Induced deficiency
G	0, +	-	-	Depletion	Limiting	Retranslocation

example, Imo and Timmer (1992a) used single dose and constant top dressing fertilization regimes to induce growth dilution of nutrients in mesquite *(Prosopis chilensis)* seedlings under greenhouse conditions. because seedling growth rate was higher than rate of nutrient uptake during the growing season. Interpretation of the growth dilution of nutrient effects by vector nutrient analysis during seedling development is illustrated in Fig. 3 using the single dose treatment (Imo and Timmer 1997).

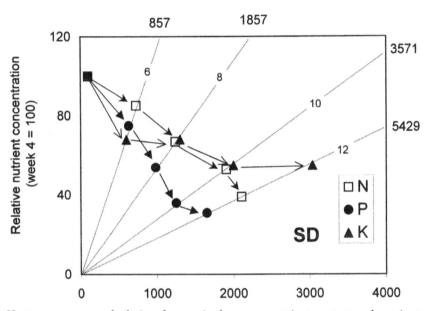

Fig. 3. Vector nomogram of relative changes in dry mass, nutrient content and nutrient concentration occurring at 2 week intervals of mesquite seedling shoots cultured under a single dose (SD) fertilization regime. Seedling status at week 4 was normalized to 100 for comparison with subsequent time intervals. Vectors reflect progressions in time from week 4 to 12. The downward-pointing vectors indicate growth dilution of nutrients (Shift A in Fig. 2) over time because nutrient uptake did not keep pace with growth demand by the seedlings (Imo and Timmer 1992 a; Imo and Timmer (1997).

In this diagram, the right-pointing vectors indicate dry mass and nutrient uptake increased while the downward-pointing vectors indicate decline in concentration over time, thus growth dilution (Shift A, Fig. 2). Growth dilution may also occur due to imbalanced nutrition, resulting in a decline in concentration of a non-limiting nutrient as a result of increased availability of a limiting factor (Armson, 1977; Timmer and Stone 1978), as demonstrated by Munson and Timmer (1989) using vector nutrient analysis (Fig. 4). In this study, addition of a limiting nutrient (N) induced a rapid increase in growth of black spruce seedlings resulting in a decline in concentration of a non-limiting element (K). From interpretation of the vector directions and length, they concluded that the result was primarily a response to N deficiency since the N vector is longer than the K vector

4.2.2 Nutrient deficiency

Nutrient deficiency response (Shift C, Fig. 2) is associated with increases in growth, nutrient uptake and concentration (see for example Imo and Timmer 1992a, Fig. 5), indicating that nutrient uptake rate is higher than rate of biomass accumulation. Such a response is characteristic of addition of a limiting nutrient (Timmer and Stone 1978). Imo and Timmer (1999) also examined the effects of 5-year old *Leuceana* hedgerows on growth and nutrient uptake of a maize intercrop over one cropping season in a humid highland of western Kenya. In this experiment, three between-alley spacing (2, 4 and 8m) and two within-alley spacing (1.0 and 0.5m) treatments plus a treeless sole crop control were compared with or without fertilization. Fig. 6 shows some of the results after evaluation using vector analysis. This diagram shows that the sole maize crop planted without any *Leuceana* trees experienced N deficiency since mulch application resulted in increased N availability due to and deficiency response (Fig. 6).

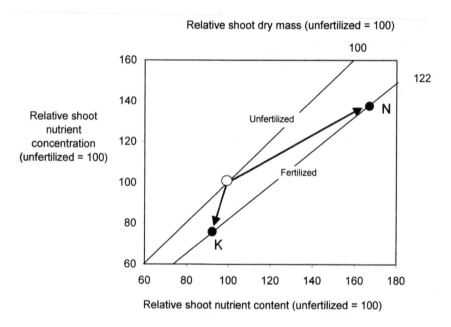

Fig. 4. Relative differences in nutrient content, concentration and shoot dry m ass between unfertilized and fertilized black spruce seedlings and planted and grown for 5 months in potted, intact blocks of forest hum us under nursery conditions with supplemental irrigation. The fertilized seedlings received 250 kg/ha NH 4NO3 fertilizer. The seedling status of the unfertilized control was normalized to 100 for comparison with unfertilized treatment. Fertilization increased growth, and N content and concentration signifying N deficiency response (vector shift C in Fig. 2). Concentration of K, however, declined despite increase in growth and content indicating growth dilution (vector shift A in Fig. 2). This phenomenon is usually associated with growth dilution of non-limiting nutrients (e.g. K in this example) on addition of a limiting factor (e.g. N in this example) probably reflecting imbalanced nutrition Data from Munson and Timmer (1989).

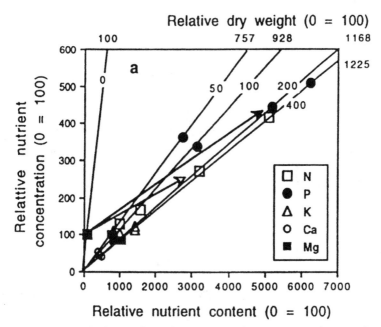

Fig. 5. Vector nomogram of relative shoot dry mass, nutrient concentration, nutrient content of mesquite seedlings grown at 0, 50, 100, 200 and 400 m g N seedling-1 fertilized with complete fertilizer. The unfertilized control was norm alized to 100 (0 = 100) for comparison with the fertilized treatments. Vector lengths indicate P as the primary and N the secondary responsive nutrients, clearly reflecting the effect of fertilization. The positive shifts in dry m ass, nutrient concentration and content after fertilizer addition signified P and N deficiency responses (Shift C in Fig. 2). Adapted from Imo and Timmer (1992).

4.2.3 Nutrient sufficiency and steady state nutrition
Nutrient sufficiency (Shift B, Fig. 2) is associated with increases in both nutrient uptake and growth, but no change in concentration indicating that the rate of nutrient uptake matched the rate of biomass accumulation, or steady state nutrition (Ingestad and Lund 1986). This response is characteristic of non-limiting nutrients that are present in sufficient amounts in the growing medium (Timmer 1991). In an experiment with mesquite seedlings (Imo and Timmer 1997) the model demonstrated the ability of exponentially based fertilization regimes to achieve steady state conditions in these seedlings (Fig. 7).

4.2.4 Luxury consumption
Luxury consumption occurs when there is no change in growth despite increased nutrient uptake, thus resulting in elevated concentration (Shift D, Fig. 2). This nutritional effect may signify nutrient loading (see for example Timmer and Munson 1991, Fig. 8).

4.2.5 Nutrient interactions
Both antagonistic dilution and nutrient toxicity are associated with reduced growth and nutrient content, often involving interaction of various factors. Antagonistic dilution (Shift

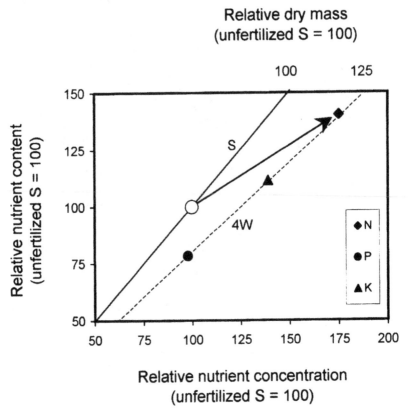

Fig. 6. Vector nutrient diagnosis of relative biomass and nutrient status of unfertilized maize grown in the *Leucaena* alleys (4W) and the treeless sole crop (S). The vector indicates a primary response of the maize crop to N deficiency (i.e. the largest vector is for N) in this treatment, presumably because of improved N availability.

Fin Fig. 2) occurs when a decline in nutrient concentration is associated with reduced growth and nutrient uptake. For example, Teng and Timmer (1990a, b) found antagonistic dilution (or induced deficiency) of Zn and Cu after fertilizing red pine seedlings with N and P (Fig. 9). Severe competition for a limiting nutrient may also cause antagonistic dilution as was found in *Pinus radiata* trees competing for nutrients with pasture in New Zealand (Mead and Mansur 1993). Nutrient toxicity (or excess uptake), on the other hand, is associated with reduced growth and nutrient uptake but elevated nutrient concentration (Shift E, Fig. 2), and occurs when growth declines m ore than corresponding reduction in nutrient uptake. This response, also referred to as concentration effect, often results from factors that stunt plant growth such as nutrient toxicity (Jarrell and Beverly 1981). For example in Fig. 9, P fertilization at high rates not only induced Zn and Cu deficiencies, but also resulted in P toxicity (Teng and Timmer 1990a, b).

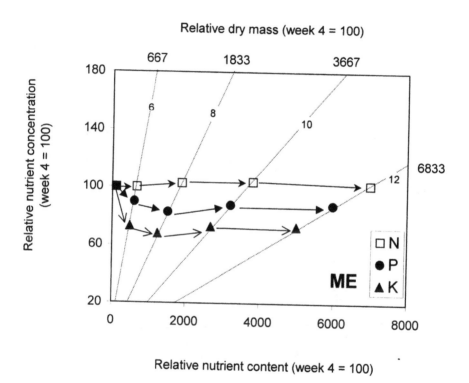

Fig. 7. Relative changes occurring at 2-week intervals in dry m ass, and nutrient concentration and content of mesquite seedlings cultured under modified exponential (M E). The seedling status at week 4 was normalized to 100 for comparison with subsequent time intervals. Vectors reflect progressions in time from week 4 to 12. The near-horizontal vector (shift B in Fig. 2) associated with the modified exponential regime indicates that growth and nutrient uptake rates were equal, exemplifying nutrient sufficiency at steady-state nutrition (Ingestad and Lund 1986). Adapted from Imo and Timmer (1997).

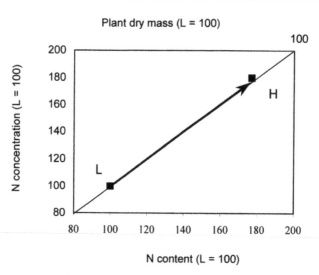

Fig. 8. Relative difference in dry m ass, and N concentration and content of seedlings raised under a low (L) and high (H) N fertilization regime. L levels were normalized to 100. The higher N status of the fertilized treatment without dry mass increase reflects luxury consumption of N (shift D in Fig. 2) that characterizes nutrient loading. Adapted from Timmer and Munson (1991).

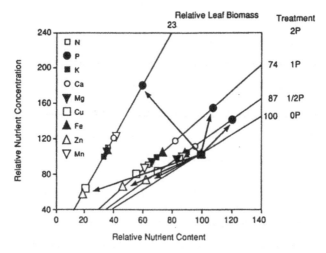

Fig. 9. Relative responses in nutrient concentration, nutrient content and biomass in hybrid poplar fertilized at various levels of P (0, 288, 576 and 1152 kg ha-1, denoted as 0P, 1/2P, 1P and 2P, respectively). Status of the reference treatment (0P) was normalized to 100 to allow comparison on a com m on base. The downward-pointing vectors (Shift F in Fig. 2) indicate that addition of P induced Zn and Cu deficiency. The upward-pointing vectors (shift E if Fig. 2) indicate excess uptake of P at higher P dose levels presumably because of stunted growth (i.e. concentration effect). Adapted from Teng and Timmer (1990a).

4.2.6 Nutrient allocation patterns

Vector diagnosis can also be used to study nutrient allocation patterns in plants, as was demonstrated by Imo and Timmer (1992b) with mesquite seedlings under differing fertilization regimes (Fig. 10). In this trial, leaf N status was markedly higher than in roots and stem (downward-pointing vectors in Fig. 8) presumably because of the higher physiological importance of the leaves. Apparently, stem biomass increased while N content and concentration declined in stems (Fig. 10), indicating nutrient depletion (Shift G, Fig. 2) probably because of retranslocation to the leaves. Malik and Timmer (1998) also used this diagnostic approach to study nutrient retranslocation in nutrient loaded and non-loaded black spruce seedlings planted on competitive boreal mixedwood forest sites. Thus, vector diagnosis approach can be used to study effects of different management regimes on both plant nutrient allocation and retranslocation processes.

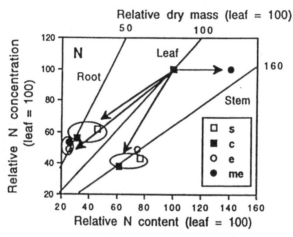

Fig. 10. Relative dry m ass and N com position of components of mesquite seedlings cultured under single dose (s), constant top dressing (c), pure exponential (e), and modified exponential (me) fertilization schedules at final harvest. Leaf status for each treatment was normalized to 100. Vectors reflect relative change in dry m ass and nutrient com position of the stem or root when compared to the leaf, and indicate N concentration and content were higher for leaf than both the stem (except 'm e' treatment) and roots. Notice that both N content and concentration in the stem declined despite increase in growth (except 'me' fertility regime), indicating N depletion from the stem (Shift G, Fig. 2) that is usually associated with nutrient retranslocation (adapted from Imo and Timmer, 1992b).

5. Practical applications: characterization of soil fertility targets

As indicated at the start of this chapter, poor diagnosis of soil fertility and crop response to soil fertility changes has been identified a major cause of poor soil fertility management in many cropping systems. Although mineral nutrition is a critical aspect of crop production and quality, precise diagnosis of soil and plant nutrient status has received little attention tropical agriculture. Current trends reflect increased interest to use fertilizers in cropping systems in order to improve the nutritional quality of field crops, but recommended guidelines are relatively unavailable for quantifying and characterizing fertility targets in these systems.

Timmer (1997) proposed a conceptual fertilizer dose response model that can be used to quantify and characterize fertility targets in cropping systems, which has since been re-configured to indicate how plant growth and nutrient status will increase with increasing fertilization, and distinguishes nutrient deficiency, sufficiency, luxury consumption and toxicity responses in plants (Fig. 11; Salifu and Timmer 2003). Traditionally based on biomass or yield parameters alone, this model has been re-configured to include nutrient uptake and nutrient concentration as well in order to improve its diagnostic capacity, thus allowing precise diagnosis of crop responses to soil fertility regimes (Salifu and Timmer 2003, Fig. 11). The application of this model been validated across a broad spectrum of soil N fertility ranging from nutrient deficiency to toxicity in conifer production systems (Salifu and Timmer, 2003; Salifu and Jacobs, 2006). Although this model has yet to be tested under multi-element interaction scenarios and various cropping systems and environmental conditions, the theoretical foundations as elaborated in Section 3 above makes the model promising for general applications.

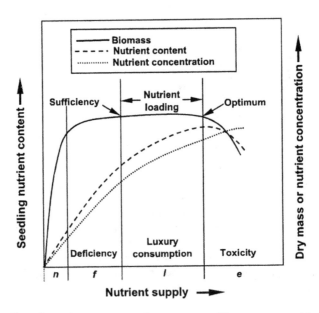

Fig. 11. Plant growth and nutrient status conform to a curvilinearpattern with increased fertilization, but artitioned here into phases to distinguish nutrient deficiency, sufficiency, luxury uptake and toxicity. Fertilizer (f) supplements native fertility (n) to avert nutrient deficiency to maximize growth at sufficiency. Extra high fertilization or nutrient loading (l) induces luxury uptake in excess of growth demand, which are stored as reserves for later utilization. Excess fertilization (e) may induce toxicity signified by diminished plant growth and N content at increasing tissue N concentration (adapted from Salifu and Timmer 2003a; Salifu and Jacobs, 2006).

Principally, this model can help quantify and define target rates (n, f, l and e, Fig. 11) for production of field crops (e.g. Cheaïb *et al* 2005). The model shows that fertilizer (f) is usually added to supplement native fertility (n) in order to avert nutrient deficiency to maximize growth at sufficiency. Any extra higher fertilization induces luxury uptake in excess of growth demand and nutrients are stored as reserves (i.e. nutrient loading, [l]) for later utilization. Excess fertilization (e) may induce toxicity, often indicated by decreased plant growth and N content but elevated tissue nutrient concentration. Such higher internal nutrient reserves acquired during nutrient loading have been shown to correlate well with improved field performance of tree seedlings (Salifu and Timmer 2003b; Malik and Timmer 1998). This simple model has been used to adapt the concept of steady-state nutrition to soil-based seedling culture by developing fertilizer delivery models which effectively induce steady state nutrition (Marney *et al*. 2010)

6. Conclusions

Diagnosis of nutritional status in cropping systems is complex given the many biochemical, physiological, ecological, socio-cultural and economic factors that determine the productivity of the target systems. The often used visual and mathematical models may not be adequate to prescribe and recommend processes for and visual methods are unlikely to confirm nutritional status of any cropping system. Vector nutrient diagnosis is an insightful tool for elucidating plant growth and nutritional responses to different cultural treatments such as fertilization and irrigation. The method also allows detection and isolation of possible nutritional effects associated with growth responses, namely: dilution and concentration effects, nutrient imbalances and interactions, and nutrient allocation patterns and retranslocation.

Originally conceived by Timmer and Stone (1978), vector nutrient diagnosis has been used widely to diagnose nutrient limitations (e.g. Joslin and Wolfe 1994; Moran and Moran 1998; Labrecqueet *et al*. 1998), explain silvicultural responses (e.g. MacDonald et al. 1998), and to assess nutrient supply from added mulch to crops in agroforestry (e.g. Anthofer *et al*. 1997; Yobterik *et al*. 1994). Some authors have also used the technique in a modified graphical form at by plotting concentration on the y-axis and growth on the x-axis, following the same diagnostic interpretations shown in Fig. 2 (e.g. Binkley *et al*. 1995; Valentine and Allen 1990). The technique has been reviewed extensively (Haase and Rose 1995; Timmer 1991), and is also described as a standard tool for soil fertility evaluation and nutrient diagnosis in several text books (for example Binkley 1986; Black 1993; Fageria *et al*. 1991; Kimmins 1996; Pritchett and Fisher 1987; Weetman and Wells 1990). The general conclusion from these reviews is that the technique is relatively simple, reliable, comprehensive, flexible, and practical in application as compared to other diagnostic techniques.

7. Acknowledgments

I am grateful to Dr. Balozi Kirongo for his critical review of this paper. Funding support from the Moi University Annual Research Grant (ARG) is greatly acknowledged.

8. References

Anthofer, J., Hanson, J. and Jutzi, S.C. (1997). Plant nutrient supply from none agroforestry tree species to wheat (*Triticum aestivum*) analysed by vector diagnosis. *Journal of Agronomy Crop Science*, Vol. 179, pp. 75-82.

Binkley, D., Smith, F. W. and Son, Y. 1995. Nutrient supply and declines in leaf area and production in lodgepole pine. *Canadian Journal of Forest Research* Vol. 25, pp 621-628.

Cheaïb A., Mollier A., Thunot, S., Lambort C., Pellerin S. and Loustau D. (2005). Interactive effects of phosphorus and light availability on early growth of maritime pine seedlings. *Annals of Forest Science,* Vol. 62, pp. 575-583.

Coleman, J.S. McConnaughay, K.D.M. and Ackerly, D.D. (1994). Interpreting phenotypic variations in plants. *Tree.* Vol. 9, pp. 187 – 191.

Farhoomand, M.B. and Peterson, L.A. (1968). Concentration and content. *Agronomy Journal,* Vol. 60, pp. 708 – 709.

Gholz, H.L. (1978). Assessing stress in *Rhododendron macrophylllum* through an analysis of leaf physical and chemical characteristics. *Canadian Journal of Botany,* Vol.56, pp. 546 – 556.

Haase, D. L. and Rose P. (1995). Vector analysis and its use for interpreting plant nutrient shifts in response to silvicultural treatments. Forest Science, Vol. 41, pp. 54-66.

Herms D.A. and Mattson W.J. (1992). The dilemma in plants: to grow or to defend? *Quarterly Review of Biology,* Vol. 67, pp. 283 – 335.

Imo, M. (1999). *Vector competition analysis: a model for evaluating interspecific plant growth and nutrient interactions in cropping systems.* PhD thesis, University of Toronto, Canada, 164 p.

Imo, M. and Timmer, V.R. (1999). Vector competition analysis of a *Leuceana*-Maize alley cropping system in western Kenya. *Forest Ecology and Management,* Vol. 126, pp. 255-268.

Imo, M. and Timmer, V.R. (1997). Vector diagnosis of nutrient dynamics in mesquite seedlings. *Forest Science,* Vol. 43, pp. 268 – 273.

Imo, M. and Timmer, V.R. (1992a). Nitrogen uptake of mesquite seedlings at conventional and exponential fertilization schedules. *Soil Science Society of America Journal,* Vol. 56, pp. 927- 934.

Imo, M. and Tim m er, V.R. (1992b). Growth, nutrient allocation and water relations of mesquite (*Prosopis chilensis*) seedlings at differing fertilization schedules. *Forest Ecology and Management,* Vol. 55, pp. 279-294.

Ingestad, T. and Lund A.B. (1986). Theory and techniques of steady state mineral nutrition and growth in plants. *Scandinavian Journal of Forest Research.* 1:439 – 453.

Jackson, D.A. and Somers K.M. (1991). The spectre of spurious correlations. *Oecologia,* Vol. 86, pp. 147 – 151.

Jarrell W.M. and Berverly R.B. (1981). The dilution effect in plant nutrition studies. *Advances in Agronomy,* Vol. 34, pp.197–2224.

Joslin, J. D. and Wolfe, M . N. (1994). Foliar deficiencies of mature southern Appalachian red spruce determined from fertilizer trials. *Soil Science Society of America Journal,* Vol. 58, pp. 1572-1579.

Kimmins, J.P. (1996). *Forest Ecology: A Foundation for Sustainable Management.* Second Edition, ISBN 0-02-364071-5, Prentice-Hall, London, pp. 71-128.

Koricheva, J. (1999). Interpreting phenotypic variation in plant allelochemistry: problems with the use of concentrations. *Ocelagaia,* Vol. 119, pp. 467-473.

Körner C.H., Pelaez-Reidle S. and Van Bel A.J.E. (1995). CO_2 responsiveness of plants: a possible link to phloem loading. *Plant Cell Environment,* Vol. 18, pp. 595 – 600.

Labrecque, M., Teodorescu, T. I. And Daigle, S. (1998). Early performance and nutrition of two willow species in short-rotation intensive culture fertilized with wastewater sludge and impact on the soil characteristics. *Canadian Journal of Forest Research*, Vol. 28, pp. 1621-1635.

MacDonald, S. E., Schm idt, M . G. and Rothwell, R. L. (1998). Impacts of mechanical site preparation on foliar nutrients of planted white spruce seedlings on mixed-wood boreal forest sites in Alberta. *Forest Ecology and Management,*. Vol. 110, pp. 35-48

Malik, V. and Timmer, V. R. (1998). Biomass partitioning and nitrogen retranslocation in black spruce seedlings on competitive mixedwood sites: a bioassay study. *Canadian Journal of Forest Research*, Vol. 28, pp. 206-215.

Marney, E.I., Adjei, O.E., Issaka R.N. nad Timmer V.R. (2010). A strategy for tree-perennial crop productivity: nursery phase nutrient additions in cocoa-shade agroforestry systems. *Agroforestry Systems*, Vol. 81, pp.147-155.

Mead, D. J. and Mansur, I. (1993). Vector analysis of foliage data to study competition for nutrient and moisture: an agroforestry example. *New Zealand Journal of Forest Science*, Vol. 23, No. 1, pp. 27-39.

Moran, J. A. and Moran, A. J. (1998). Foliar reflectance and vector analysis reveal nutrient stress in prey-deprived pitcher plants (*Nepenthes rafflesiana*). *International Journal of Plant Science*, Vol. 159, pp. 996-1001.

Munson, A. L. and Timmer, V. R. (1995). Soil nitrogen dynamics and nutrition of pine following silvicultural treatments in boreal and Great Lakes-St. Lawrence plantations. Forest Ecology and Management, Vol. 76, pp. 169 - 179.

Sokal, R. R. and Rohlf F.J. (1995). Biometry: the principles and practice of statistics in biological research, 3rd ed. Freeman: San Francisco. 887 pp. ISBN: 0-7167-2411-1.

Salifu K.F. and Timmer V.R. (2003a). Optimizing nitrogen loading in *Picea mariana* seedlings during nursery culture. *Canadian Journal of Forest Research*, Vol. 33, pp. 1287–1294.

Salifu K.F. and Timmer V.R. (2003b). Nitrogen retranslocation response of young *Picea mariana* to nitrogen-15 supply. *Soil Science Society of America Journal*, Vol. 67, pp. 905-913.

Salifu K. F. and Jacobs F.D. (2006). Characterizing fertility targets and multi-element interactionsin nursery culture of *Quercus rubra* seedlings. *Annals of Forest Science*, Vol. 63, pp. 231-237.

Smith, R.B. Waring H.R. and Pary D.A. (1981). Interpreting foliar analyses from Douglas-fir as weight per unit of leaf area. *Canadian Journal of Forest Research*, Vol.11, pp. 593 – 603.

Raubenheimer, D. and Simpson, S.J. (1992). Analysis of covariance: an alternative to nutritional indices. *Entomology and Experimental Applications,* Vol. 62, pp. 221 - 406

Tamm, C.O. (1964). Determination of nutrient requirements of forest stands. *International Review of Forestry Research* Vol. 1, pp. 115 – 170.

Teng, Y. and Timmer, V. R. (1990a). Phosphorus-induced micro-nutrient disorders in hybrid poplar. I. Preliminary diagnosis. *Plant Soil*, Vol. 126, pp. 19-29.

Teng, Y. and Timmer, V. R. (1990b). Phosphorus-induced micro-nutrient disorders in hybrid poplar. III. Prevention and correction in nursery culture. *Plant Soil*, Vol. 126. Pp. 41-51.

Timmer, V. R. (1997). Exponential nutrient loading: a new fertilization technique to improve seedling performance on competitive sites. *New Forest*, Vol. 13, pp. 279-299.

Timmer, V.R. and Stone, E.L. (1978). Comparative foliar analysis of young balsam fir fertilized with nitrogen, phosphorus, potassium and lime. *Soil Science Society and American Journal*, Vol. 42, pp. 125 – 130.

Timmer, V.R. (1991). Interpretation of seedling analysis and visual symptoms. In *Mineral Nutrition of Conifer Seedlings*, Van den Drisessche, R. (Ed), pp 113-134, ISBN 0-8493-591-6, CRC Press, Boca Raton, Florida, USA.

Timmer, V. R. and Munson, A. D. (1991). Site-specific growth and nutrient uptake of planted *Picea mariana* in the Ontario Clay Belt. IV. Nitrogen loading response. *Canadian Journal of Forest Research*, Vol. 21, pp. 1058-1065.

Valdencantos A., Cortina J.V. and Vallejo R. (2006). Nutrient status and field performance of tree seedlings planted in Mediterranean degraded areas. *Annals of Forest Science*, Vol. 63, pp. 249-256.

Valentine, D. W. and Allen, H. L. (1990). Foliar responses to fertilization identify nutrient limitations in loblolly pine. *Canadian Journal of Forest Research*, Vol. 20, pp. 144-151.

Weetman, G.F. (1989). Graphical vector analysis technique for testing stand nutritional status. In *Research Strategies for Long-Term Site Productivity*, Dyck W.J. and Mees C.A. (eds), Forest Research Institute, New Zealand, pp 93 – 109.

Weetman, G. F. and Wells, C. G. (1990). Plant analysis as an aid in fertilizing forests. *In* Soil Testing and Plant Analysis, Westerm an, R. L., Baird, J. V., Christensen, N. W., Fixen, P. E., and Whitney, D. A. (Eds.). , Soil Science Society America, Madison, Wisconsin, USA. pp. 659-690.

Yobterik, A. C., Timmer, V.R. and Gordon, A. M. (1994). Screening agroforestry tree mulches for corn growth: a combined soil test, pot trial and plant analysis approach. *Agroforestry Systems*, Vol. 25, pp. 153-166.

Part 4

Tillage and Crop Production

14

Energy Use Pattern in Millet Production in Semi-Arid Zone of Nigeria

Mohammed Shu'aibu Abubakar
Department of Agricultural Engineering Bayero University, Kano
Nigeria

1. Introduction

Agriculture is an important economic sector in Nigeria, although the country depends heavily on the oil industry for its budgetary revenues. Approximately 70 percent of the population engages in agricultural production at a subsistence level. Even though, the agriculture related activities holdings are generally small scale. Agriculture provided 41 percent of Nigeria's total gross domestic product (GDP) in 1999. This percentage represented a normal decrease of 24.7 percent from its contribution of 65.7 percent to the GDP in 1957. The decrease will continue because of the fact that when economic development occurs, the relative size of the agricultural sector usually decreases (Abdullahi et al., 2006). Nigeria's wide range of climate variations allows it to produce a variety of food crops. The staple food crops include cassava, yams, sweet potatoes, coco-yams, corn, cow-peas, beans, millet, rice, wheat, sorghum, and a variety of fruits and vegetables. Efficient use of energy is one of the principal requirements of sustainable agriculture. Energy use in agriculture has been increasing in response to increasing population, limited availability of arable land, and a desire for higher standards of living. Therefore, energy is one of the most valuable inputs in agricultural production. It is invested in various forms such as mechanical (farm machines, human power, and animal draft), chemical fertilizer (pesticides and herbicides) and electrical. The amount of energy used in agricultural production, processing and distribution needs to be adequate in order to feed the rising population and to meet other social and economic goals (Stout, 1990). Because of the subsistence nature of the millet production in the study area, most of farmers mainly produce the crop using only manual energy. Very few farmers use tractors for tillage during the land preparation stage. Apart from this single mechanical energy use, all other farm operations are executed using manual energy or animal traction. This trend of limited mechanisation is common to other crops grown in the country. Therefore, less energy input has being the case crop production like millet. Because of the lack of data on energy expenditure and benefits associated with energy analysis in the production of millet. Also information on comparative use of different energies is also lacking. And most of the producers do not have enough knowledge on the most efficient energy inputs. Consequently, it is neither possible to identify viable energy inputs and options in the production process nor plan for their conservation. Under these situations, an input–output energy analysis provides planners and policy makers an opportunity to evaluate economic interactions of energy use. This information is required in order to make deductions on the efficiencies of the energies and suggestions on which energy sources or their combinations need to be used and at what levels. Also this would serve as a data bank for any related study.

2. Energy input-output analysis in crop production

Some studies on energy use and evaluation methods elsewhere were reported. Bridges and Smith (1979) developed a method for determining the total energy input for agricultural practices. The categories of energy considered were those of manufacture, transport and repairs (MTR), fuel and labour. Fluck (1985) also in his study developed two models to quantify energy sequestered in repairs and maintenance of agricultural machinery as compared with the energy input in new machinery. Energy use analysis from the literature have shown that different authors who used different methods for evaluating human energy reported several values of the energy content for manual labour. Hence, there is no universally accepted energy value of manual labour. However, for countries where agriculture is dominated by human energy, it is reasonable to adopt the value obtained by Norman (1978). Sustainable direct energy is required to perform various tasks related to crop production processes such as for land preparation, irrigation, harvest, post harvest processing, transportation of agricultural inputs and outputs. In other word, high level of direct energy such as fuel and electricity are needed to be used at farm for crop production (Alam et al., 2005; Hoeppner et al., 2006; Khambalkar et al., 2005; Kizilaslan, 2009). Unlike direct energy which is directly consumed at the farm, indirect energy is not directly consumed at the farm rather are the energy used in the manufacture, packaging and transport of fertilizers, seeds, machinery production and pesticides (Ozkan et al., 2004). The energy input for the crop production differs to a large extent from area to area and also depending on the level of mechanization. In modern crop production is characterized by the high input of fossil energy (fuel and electricity) which is consumed as direct energy and as indirect energy (fertilizers, pesticides, machinery, etc.). In some low-input farming systems, example in large areas of Africa, the energy input on arable land is lower than 1GJ ha^{-1}, whereas in some modern high-input farming systems in west Europe, it can exceed 30GJ ha^{-1} (Pimentel, 2009; Reed et al., 1986). In the past decade, with increase in energy inputs in agriculture, an equivalent increase in crop yields occurred. Other studies have suggested that the energy use efficiency of our traditional cropping systems have been sharply going downward in recent years due to energy inputs increasing faster than energy output as a result of the growing dependency on inorganic fertilizers and fossil fuels (Hatirli et al., 2006; Jekayinfa & Bamgboye, 2007; Khambalkar et al., 2005). If the increase in the energy use in the agricultural industry continues, the only chance of producers to increase total output will be using more input as there is no chance to expand the size of arable lands. Under these circumstances, an input-output analysis provides planners and policy-makers an opportunity to evaluate economic interactions of energy use.

3. Millet production

Pearl millet *(Pennisetum glaucum (L) R. Br.)* is a cereal grain with good drought tolerance and hardiness widely grown in the hot and dry climates areas of arid and semi-arid regions of Africa and southern Asia. It is one of the four most important cereals crop (millets, sorghum, maize and rice) normally grown where rain fall is not sufficient (200-600 mm) for corn and sorghum. In 1995, the global production of millet exceeds 10 million tons per year in a total estimated area of 15 and 14 million hectares in Africa and Asia respectively. Millet production increased from 26 million tonnes in 1981 to 31 million tones in 1990 in Asia, Africa and the former USSR. The major millets producers' nations in 1990 were India (15%), China (10%), Nigeria (65%) and the former USSR (10%) (FAO, 1996). Amongst different species of millet Worldwide, four are cultivated in Africa with Pearl millet-*Pennisetum glaucum (L) R. Br.*, Finger millet-*Eleusine coracana L. Gaertn.*, Teff millet-*Eragrostis teff (Zucc)* and Fonio millet-*Digitaria exilis (acha)* accounting for 76%, 21%, 1.8% and 0.8% of the total production respectively (Andrews & Kumar, 1992). In most countries of Africa and Asia, millets production is primarily for human consumption as staple food (78%) with other uses of less than 20%. In other countries like Mexico, Australia, Canada and the United State of America, pearl millet is grown as a forage crop for livestock production. Future trends indicate that millet crop production will increase globally because of the increase in number of millet consumers. However, the production of millets is still at subsistence level by smaller scale farmers (0.5-5 hectare farm size) in most part of the Africa. Furthermore, millet crops remain the key sources for food security and energy for about 250 million people in sub-Saharan Africa. Meanwhile, the millets crop production areas in this region of sub-Saharan Africa coincide with where most of the poor people live. This coincidence has a significant effect on these poor people to their socio-economic, food/shelter, health and environment. In Nigeria, like any of the sub-Saharan African countries, millet is produced in rid and semi-arid drought-prone northern part by the low income earners farmers. Sufficient energy is needed in the right form and at the right time for adequate crop production. One way to optimize energy consumption in agriculture is to determine the efficiency of methods and techniques used. With the current increase in world population, energy consumption needs effective planning. That is, the input elements need to be identified in order to prescribe the most efficient methods for controlling them

4. Study area

A study on the pattern of energy use in millet production was conducted in the eight local government areas of Jigawa States semi-arid zone of Nigeria (Figure 1). It is situated in north-western part of the country between Latitudes 11.00oN to 13.00oN and Longitudes 8.00oE to 10.15oE. The study areas have 3-4 months rainfalls duration followed by a long dry session. The annual precipitation is between 400 and 600 mm, which vary from year to year. The main livelihood of the people is agriculture of which millet is the most important crop for consumption. Over eighty percent of the population is engaged in subsistence farming and animal husbandry.

A stratified random sampling technique was used to select the millet farmers in the study area and were classified into three groups (I-III) based on their farm sizes as small (1 ha or less), medium (2-4 ha) and large farms (5 ha and more). Sixty (60) farmers were interviewed in each of the groups. A total of 180 sample data were collected. The data for energy input resources used in all the selected farms during millet production from land preparation up to transportation to market or house were collected using structural questionnaire and oral interviews in the production years 2006 and 2007.

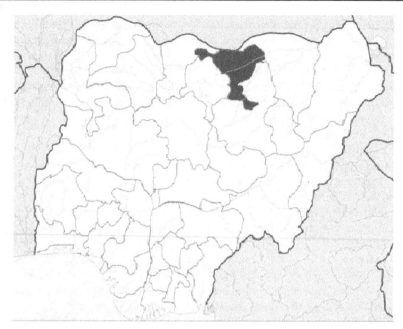

Fig. 1. Map of the study area

5. Energy use pattern in millet production

Substantial numbers of research studies have been conducted on energy use in agriculture (Abubakar & Ahmad, 2010; Ahmad, 1994; Canakci & Akinci, 2006; Hatirli et al., 2006; Hoeppner et al., 2006; Kizilaslan, 2009). Energy use in agricultural production has become more intensive due to the use of fossil fuel chemical fertilizers, pesticides, machinery and electricity to provide considerable increases in food production. However, more intensive energy use has brought some important human health and environment problems so efficient use of inputs has become important in terms of sustainable agricultural production. However, millet has been paid relatively little attention. Furthermore, this study considered the effect of farm size on energy use and input costs. According to Pimentel (1992), energy consumption per unit area in agriculture is directly related to the development of the technology in farming and the level of production. The amount of energy used in agricultural production, processing and distribution is prerequisites for improved agricultural production. Fluck and Baird (1980) hypothesized that the highest partial energy productivity is achieved at the point of minimum mechanization energy inputs and increasing mechanization energy increase crop yield at a decreasing rate.

5.1 Evaluation of energy use

The input energy consumption for different farm field operations in producing millet was classified on the basis of source and use as direct and indirect energy and then renewable and non-renewable energy. The direct energies such as human or animal power, diesel and electricity, are the energy which are released directly from power sources in millet production while indirect energy are those which are dissipated during various conversion processes like energy consumed indirectly in manufacturing, repair and transport, storage, distribution and related activities and also energies embodied in seeds, farmyard manure, pesticides and fertilizers. Non-renewable energy includes petrol, diesel, electricity, chemicals, fertilizers and renewable energy consists of human and animal power (Pimentel, 1992; Singh et al., 2002, 2003; Singh et al., 2007).

5.1.1 Direct energy inputs

The direct energy inputs per hectare during millet production include manual (human) labour, draft animal and fuel (diesel) and were computed using the equations adopted by Bockari-Gevao et al., (2005) in equations below as follows;

5.1.1.1 Energy input from manual labour

The rate of labour use in the millet production process was determined for each operation. The labour energy input (MJ ha^{-1}) at every stage of the production process was estimated by the following equation 2;

$$ME_{labour} = \frac{NLabour \; X \; Time}{Area} \; X \; FLabour \tag{1}$$

where:
ME_{labour} = Manual labour energy, MJ ha^{-1}
NLabour = Number of working labourers
Time = Operating time, h
Area = Operating area, ha
FLabour = Labour energy factor, MJ h^{-1}

5.1.1.2 Energy input from draft animal

Singh et al.,(1997) reported that pair of bullocks have power equivalent of 746 W (1.0 hp). Therefore Energy input from draft animal was evaluated as follow;

$$E_{DA} = \frac{W}{Area} \tag{2}$$

where:
E_{DA} = Draft animal energy input, MJ ha^{-1}
W = Power equivalent for pair of bullocks, MJ
Area = Operation area, ha

5.1.1.3 Energy input from fuel

The specific energy use from fuel (fossil) was evaluated by quantifying the amount of diesel consumed during each millet production process

$$S_{DFE} = AFU \; X \; PEU \; X \; NP \; X \; h \tag{3}$$

where:
S_{DFE} = Specific direct energy use (fuel) for a field operation, MJ ha^{-1}
AFU = Average quantity of fossil fuel (diesel or petrol) use per working hour, L h^{-1}
PEU = Specific energy value per unit litre of fuel, MJ L^{-1}
NP = Number of pass for applications in the considered field operation

h = Specific working hours per pass, h ha^{-1}

5.1.2 Indirect energy inputs

The indirect energy inputs per hectare during millet production include machinery, seed, fertilizer and pesticide and were computed using the equations adopted by Bockari-Gevao et al., (2005) in equations below as follows;

5.1.2.1 Energy input from machinery

The indirect energy contribution of machinery for each field operation was determined by the following equation 4 below;

$$S_{INDME} = \frac{TMW \times CMED}{SV} \times NP \times h \tag{4}$$

where:
S_{INDME} = Specific indirect energy for machinery use for a field operation, MJ ha^{-1}
TMW = Total weight of the specific machine, kg
CMED = Cumulative energy demand for machinery, MJ kg^{-1}
SV = Salvage value life of machinery, h
NP and h as defined above.

5.1.2.2 Energy input from chemical energy

The indirect chemical energy per unit area for other production inputs such as fertilizer, pesticides and farmyard manure was expressed as in equation 5 below;

$$INDE_{SFP} = Rate \times ENFMaterial \tag{5}$$

where:
$INDE_{SFP}$ = Indirect energy input such as for seed, fertilizer or pesticides, MJ ha^{-1}
Rate = Application rate of input, kg ha^{-1}
ENFMaterial = Energy factor of material used, MJ kg^{-1}

5.1.2.3 Energy input from biological energy

Mainly seeds and hormone were included as biological energy inputs. Existing data on hormones was used. The energy equivalent value of 14.00 MJ kg^{-1} was used for seed (millet) input and an assumed equivalent value higher than energy (seed) input by 1 MJ kg^{-1} of crop (millet) production output was also used (Singh & Mittal, 1992; Singh et al., 1997).

5.1.3 Total energy inputs

The energy input intensity (e_I) was determined from the summation of all the energies input (direct and indirect) and dividing by the effective area of millet production as given by the following equation 6 below;

$$e_I = \frac{E}{A} \tag{6}$$

where:
e_I = Energy input intensity, MJ ha^{-1}
E = Total energy consumption, MJ
A = The effective production area, ha

5.1.4 Total energy output

The energy output intensity (e_O) was derived by multiplying the production intensity (s) by the energy coefficient of seed (B_s) as represented in equation 7;

$$e_O = s \times B_s \tag{7}$$

where:
e_0 = Energy output intensity, MJ ha^{-1}
s = Production intensity, kg ha
B_s = Energy coefficient of seed (millet), MJ kg^{-1}

5.1.5 Energy use ratio

The overall energy use ratio (OEUR) was then determined as the ratio of the energy output intensity to the energy input intensity (Equation 8). It is assumed that, if the OEUR is greater than 1, then the production system is gaining energy, otherwise it is losing energy.

$$OEUR = \frac{e_O}{e_I} \tag{8}$$

where:
OEUR = Overall energy use ratio, dimensionless
e_0 = Energy output intensity, MJ ha^{-1}
e_I = Energy input intensity, MJ ha^{-1}
Energy equivalent value of 109 MJ kg^{-1} was used to represent the embodied energy in a piece of equipment as reported by Pimentel (1992). He further reported that the average energy value of 109 MJ kg^{-1} of weight of machinery includes 62.8 MJ kg^{-1} for steel production; 8.4 MJ kg^{-1} for the fabrication of parts and assembly; and 37.7 MJ kg^{-1} for repairs and maintenance. All practices requiring fossil fuel were evaluated with diesel and petrol as the energy sources. The energy associated with fuel use was 47.8 MJ L^{-1} and 46.3 MJL^{-1} for diesel and petrol fuels, respectively (Pimentel, 1992). This includes estimates for engine oil, grease, manufacture and transportation to the farm as reported by Bridges & Smith (1979). The human energy required to perform any operation or practice is based on the number of labourers required to perform the operation and the field capacity of the machine. In this study, the labour input in terms of manual energy was evaluated at 1.96 MJh^{-1} (Norman, 1978; Pimentel, 1992). Chemical fertilizers, farmyard manure and pesticides are main sources for chemical energy inputs.). The total chemical fertilizer input was calculated using energy equivalent values were assumed to be 78.1, 17.4 and 13.7 MJ/kg for nitrogen (N), phosphorus (P_2O_5) and potassium (K_2O) respectively (Mudahar & Hignett, 1987). These are the energy requirements for producing and transporting commercial fertilizers. The average energy inputs for the production of the active ingredients of herbicides, insecticides and fungicides were assumed to be 255, 185 and 97 MJkg^{-1}, respectively (Black, 1971; Hatirli et al., 2006). An average energy coefficient (B_s) of 14.7 MJ kg^{-1} for millet seeds was used (Abubakar & Ahmad, 2010)

6. Analysis and discussions of energy input-output in millet production

Energy analysis was performed based on field operations in millet production such as land clearing, tillage, planting, weeding, farmyard manure/fertilizer application, pesticides application, harvesting and threshing. Operational energy used in form of the direct (fuel and human labour or animal power) and indirect (machinery, farmyard manure, fertilizer, pesticide, and seed) energy sources involved in the production process were computed. The analysis of the data collected with respect of the millet reduction in the study area was reported. The major issue of concern is that farmers use more energy to increase output but they do not have enough knowledge on most efficient energy inputs to use. Thus, an input–output energy analysis provides farmers and policy makers an opportunity to evaluate economic intersection of energy use. Direct and indirect types of energy are required for agricultural production. Energy input-output relation analysis is usually used to evaluate the efficiency and environmental impacts of the production systems. On the other hand, the energy use ratios in agricultural production are closely related with production techniques, quantity of input, yield level and environmental factors. It was also reported that large farms used energy in the best possible way to achieve maximum yield than the small size farm (Sarkar, 1997; Shearer et al., 1981; Sims et al., 2006).

Field operation	Energy resource input (MJ)	Energy resource Input for different farmer groups (MJ ha⁻¹)		
		Group I	Group II	Group III
Land Clearing	ME_{labour}	130	220	70
	E_{DA}	75	30	Nil
	S_{DFE}	Nil	35	65
	S_{INDME}	Nil	65	145
Tillage	ME_{labour}	400	65	40
	E_{DA}	320	180	Nil
	S_{DFE}	Nil	820	1600
	S_{INDME}	Nil	550	700
Planting	ME_{labour}	450	145	135
	E_{DA}	70	120	Nil
	S_{DFE}	Nil	320	600
	S_{INDME}	Nil	150	250
Weeding	ME_{labour}	750	435	115
	E_{DA}	35	25	220
	S_{DFE}	Nil	120	180
	S_{INDME}	Nil	240	425
Farmyard manure/Fertilizers application	ME_{labour}	350	65	30
	E_{DA}	180	110	Nil
	S_{DFE}	Nil	125	150
	S_{INDME}	Nil	250	340
Pesticides application	ME_{labour}	320	115	45
	E_{DA}	110	Nil	Nil
	S_{DFE}	Nil	45	120
	S_{INDME}	Nil	120	180
Harvesting and Threshing	ME_{labour}	540	350	150
	E_{DA}	215	90	Nil
	S_{DFE}	Nil	75	145
	S_{INDME}	Nil	250	415

Table 1. Mean values of energy resource input for various field operations for different farmer groups

6.1 Energy use pattern

Table 1 showed the computed values of energy resource input for various field operations for different farmer groups. The study revealed that the least amount of energy input was during land clearing for the entire three farmer group (Figure 1a-c). Actually the amount of energy needed for this operation is generally low because all the farmlands have been previously cultivated. There was nothing much to be done apart from burning and collecting dry plant residues and grasses. Tillage and weeding operational activities consumed the highest energy input values for the three groups of the farmers. This could be due to the highly intensive and excessive energy use during soil breaking by tillage implements and weeding was mostly repeated manually since fewer chemicals were used by the farmers in controlling weeds. This findings is in agreement with the result reported by Nuray, (2009); Umar, (2003); Leach, (1975) and Lockeretz et al., (1978). Group I farmers consumed 20% of the energy used on weeding operation, 19% on harvesting and threshing activities with 5% energy used on land clearing. Energy used by Group II and III farmers include 33% for tillage, 17% for weeding and 7% for land clearing; and 38% for tillage, 16% for planting and 5% for land clearing respectively. Results suggest that for the group with big size farm ($\geq$ 5 hectares), tillage operation consumes the highest energy whereas for group with farm size ($\leq$ 1 hectare), weeding engulfs more energy. This is similar with research reported by Shahin et al., (2008); Pimentel & Pimentel, 1996 (1996) and Walsh et al., (1998), whose agreed that the energy consumption depends on farm size and level of production activities.

Duncan multiple range test (DMRT) for the mean comparison of the resource input for various field operations for different farmer groups (Table 2). Result shows that all means are statistically different at 95% confidence level. A linear relationship between the total energy input and output from the multiple linear regression analysis conducted for the various farmer groups with value of $R^2 = 0.97$. This indicates that millet crop yield is directly dependent on the energy resource input.

Means with the same letter are not statistically different at the 5% level of significance Figure 2a and b depicts the average total energy inputs from based on the energy sources during the millet production from the various farmer groups. Manual and fuel energy were the main contributor of the direct energy for farmers in group I and Group III respectively (Figure 2a). From the indirect energy sources fertilizer had the highest contribution followed by pesticides and machinery energy sources (Figure 2b).

Farmer groups	Field operational energy consumption (MJ [ha-1])						
	Land Clearing	Tillage	Planting	Weeding	Farmyard manure/Fertilizers application	Pesticides application	Harvesting and Threshing
Group I	205[a]	720[b]	520[c]	785[d]	530[e]	430[f]	755[g]
Group II	350[a]	1615[b]	735[c]	820[d]	550[e]	280[f]	515[g]
Group III	280[a]	2340[b]	985[c]	940[d]	490[e]	345[f]	710[g]

Table 2. Duncan multiple range tests for mean comparison of energy resource input for various field operations for different farmer groups

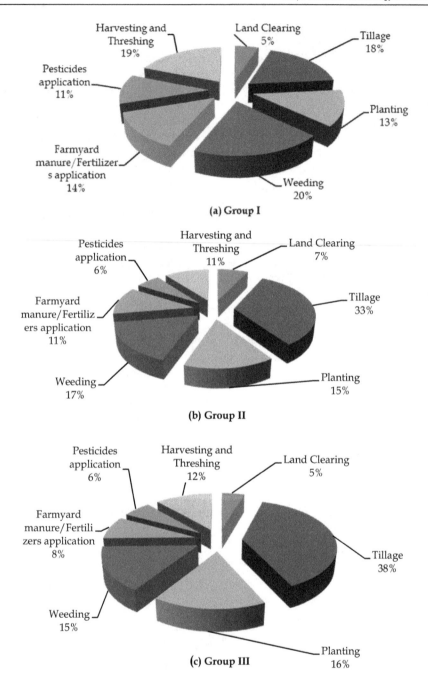

Fig. 1. a, b and C. Percentage mean total of energy input per field operation for various farmers group (a) Group I, (b) Group II and (c) Group III

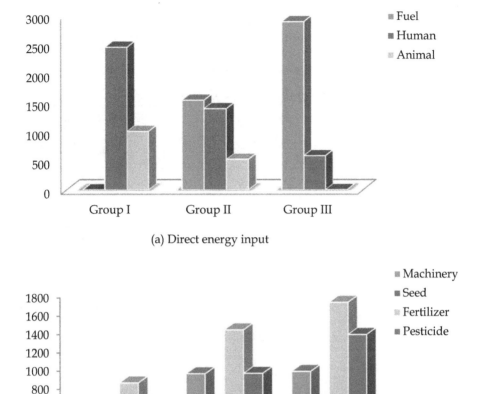

(a) Direct energy input

(b) Indirect energy input

Fig. 2. a. and b. Average total energy inputs based on sources (a) direct energy input, (b) indirect energy input for various farmer groups

6.2 Overall energy use ratio

The overall energy use ratio (OEUR) was determined from the ratio of total energy output to the total energy input (Table 3). The farmer group with OEUR value greater than 1 indicates the millet production system is gaining energy or else that it is losing energy. Result demonstrates that group II farmers have shown efficient use of energy resources. This could be because they have been using all the sources of the energy during their millet production processes. The lowest overall energy use ratio value obtained was 0.8 for group I farmers.

This shows losing of energy use or low efficient level of energy usage in the course of millet production. The reason is obvious since the farmers in this group use manual labour from human and animal in producing the millet which was laborious and time consuming, a scenario similar to the finding of Haque et al., (2000); Pimentel, (2009); Mohammadi & Omid, 2010) (2010) and Tolga et al., (2009) who conducted and reported similar work for different types of crops in different parts of the world.

Energy input and output in all field operations	Farmer groups		
	Group I	Group II	Group III
Total energy input (e_I) (MJ ha^{-1})	3945	4845	6090
Total energy output (e_O) (MJ ha^{-1})	3156	12597	12789
Overall energy use ratio (OEUR)	0.8	2.6	2.1

Table 3. Total energy inputs, energy outputs and energy use ratio

7. Conclusions

The study reported the pattern of energy use for millet production with selected farmers. Production energy indicators were evaluated using data collected from a structural questionnaire and published literatures. Result revealed the major energy sources were manual labour, animal draft and fuel energy for the direct energy and also farmyard manure, pesticides, machinery and seed for the indirect energy. Soil tillage and weeding operations were the production activities that consumed most of the energy intense operation whereas land clearing operation requires the least energy input for the entire farmer groups. However, it was observed that the cost of energy use per unit area decreased with increase of farm size. This would serve as a key guide for small size millet producers in the study area during policy making, planning and action taken as well as the government or any other stakeholder of millet production around the globe.

8. References

Abubakar, M. S. & Ahmad, D. (2010). Pattern of energy consumption in millet production for selected farms in Jigawa, Nigeria. *Australian Journal of Basic and Applied Sciences,* 4(4), 665-672.

Ahmad, B. (1994). Energetics of major crops in mixed cropping system. *Agricultural Mechanization for Africa, Latin America and Asia 25*(3), 52-54.

Alam, M. S., Alam, M. R. & Islam, K. K. (2005). Energy flow in agriculture: Bangladesh. *American Journal of Environmental Science, 1*(3), 213-220.

Andrews, D. J. & Kumar, K. A. (1992). Pearl millet for food, feed and forage. *Advance Agronomy, 48*(89-139).

Black, J. N. (1971). Energy relations in crop production a preliminary survey. *Annals of Applied Biology, 67*(2), 272-278.

Bockari-Gevao, S. M., Wan Ishak, W. I., Azmin, Y. & Chan, C. W. (2005). Analysis of energy consumption in lowland rice-based cropping system of Malaysia. *Songklanakarin Journal of Science and Technology, 27*(4), 819-826.

Bridges, T. C. & Smith, E. M. (1979). A method for determining the total energy input for agricultural practices. *Transactions of the ASAE,* 781-784.

Canakci, M. & Akinci, I. (2006). Energy use pattern analyses of greenhouse vegetable production. *Energy, 31*(8-9), 1243-1256.

FAO. (1996). *Production yearbook.* Rome, Italy: Food and Agriculture Organisation.

Fluck, R. C. (1985). Energy sequestered in repairs and maintenance of agricultural machinery. *Transactions of the ASAE, 28*(3), 738-744.

Haque, M. A., Umar, B. & Kawuyo, U. A. (2000). A preliminary survey on the use of animal power in agricultural operations in Adamawa state, Nigeria. *Outlook on agriculture, 29*(2), 123-127.

Hatirli, S. A., Ozkan, B. & Fert, C. (2006). Energy inputs and crop yield relationship in greenhouse tomato production. *Renewable Energy, 31*(4), 427-438.

Hoeppner, J. W., Entz, M. H., McConkey, B. G., Zentner, R. P. & Nagy, C. N. (2006). Energy use and efficiency in two Canadian organic and conventional crop production systems. *Renewable Agriculture and Food Systems, 21*(01), 60-67.

Jekayinfa, S. O. & Bamgboye, A. I. (2007). Development of equations for estimating energy requirements in palm-kernel oil processing operations. *Journal of Food Engineering, 79*(4), 322-329.

Khambalkar, V., Pohare, J., Katkhede, S., Bunde, D. & Dahatonde, S. (2005). Energy and economic evaluation of farm operations in crop production. *Journal of Agricultural Science, 2*(4), 15-22.

Kizilaslan, H. (2009). Input-output energy analysis of cherries production in Tokat Province of Turkey. *Applied Energy, 86*(7-8), 1354-1358.

Leach, G. (1975). Energy and food production. *Food Policy, 1*(1), 62-73.

Lockeretz, W., Shearer, G., Klepper, R. & Sweeney, S. (1978). Field crop production on organic farms in the Midwest. *Journal of Soil and Water Conservation, 33*(3), 130-134.

Mohammadi, A. & Omid, M. (2010). Economical analysis and relation between energy inputs and yield of greenhouse cucumber production in Iran. *Applied Energy, 87,* 191-196.

Mudahar, M. S. & Hignett, T. P. (1987). *Energy requirements, technology and resources in fertilizer sector In Plant nutrition and pest control: Energy in world agriculture* (Vol. 2): Elsevier.

Norman, M. J. T. (1978). Energy inputs and outputs of subsistence cropping systems in the tropics. *Agro-ecosystems, 4*(3), 355-366.

Nuray, K. (2009). Energy use and inputs-outputs energy analysis for apple production in Turkey. *Journal of Food , Agriculture and Environment, 7*(2), 419-423.

Ozkan, B., Akcaoz, H. & Fert, C. (2004). Energy input-output analysis in Turkish agriculture. *Renewable Energy, 29,* 39-51.

Pimentel, D. (1992). *Energy inputs in production agriculture.* Amsterdam: Elsevier.

Pimentel, D. (2009). Energy inputs in food crop production in developing and developed nations. *Energies, 2*(1), 1-24.

Pimentel, D. & Pimentel, M. (1996). Energy use in grain and legume production. *Food, Energy and Society,* 107–130.

Reed, W., Geng, S. & Hills, F. J. (1986). Energy input and output analysis of four field crops in California1. *Journal of Agronomy and Crop Science, 157*(2), 99-104.

Sarkar, A. (1997). Energy-use patterns in sub-tropical rice-wheat cropping under short term application of crop residue and fertilizer. *Agriculture, Ecosystems & Environment, 61*(1), 59-67.

Shahin, S., Jafari, A., Mobli, H., Rafiee, S. & Karimi, M. (2008). Effect of farm size on energy ratio for wheat production: A case study from Ardabil Province of Iran.

Shearer, G., Kohl, D. H., Wanner, D., Kuepper, G., Sweeney, S. & Lockeretz, W. (1981). Crop production costs and returns on Midwestern organic farms: 1977 and 1978. *American Journal of Agricultural Economics, 63*(2), 264-269.

Sims, R. E. H., Hastings, A., Schlamadinger, B., Taylor, G. & Smith, P. (2006). Energy crops: current status and future prospects. *Global Change Biology, 12*(11), 2054-2076.

Singh, H., Mishra, D. & Nahar, N. M. (2002). Energy use pattern in production agriculture of a typical village in arid zone of India - Part I. *Energy Conversion and Management, 43*(16), 2275-2286.

Singh, H., Mishra, D. & Nahar, N. M. (2003). Energy use pattern in production agriculture of a typical village in arid zone of India - Part II. *Energy Conversion and Management, 44*(7), 1053-1067.

Singh, H., Singh, A. K., Kushawaha, H. L. & Singh, A. (2007). Energy consumption pattern of wheat production in India. *Energy 32*(8), 1848-1854.

Singh, S. & Mittal, J. P. (1992). *Energy in production agriculture.* New Delhi, India: Mittal Publications.

Singh, S., Verma, S. R. & Mittal, J. P. (1997). Energy requirements for production of major crops in India. *Agricultural Mechanization for Africa, Latin America and Asia, 28*(4), 13-17.

Stout, B. A. (1990). *Handbook of energy for world agriculture.* London: Elsevier Applied Science.

Tolga, T., Bahattin, C. & Vardar, A. (2009). An analysis of energy use and input costs for wheat production in Turkey. *Journal of Food , Agriculture and Environment, 7*(2), 352-356.

Umar, B. (2003). Comparison of manual and manual-cum-mechanical energy uses in groundnut production in a semi-arid environment. *Agricultural Engineering International: CIGR Journal.*

Walsh, M. E., De La Torre Ugarte, D., Slinsky, S., Graham, R. L., Shapouri, H. & Ray, D. (1998). Economic analysis of energy crop production in the US--location, quantities, price and impacts on traditional agricultural crops. *Bioenergy 98: Expanding Bioenergy Partnerships, 2,* 1302-1310.

15

Tillage Effects on Soil Health and Crop Productivity: A Review

Peeyush Sharma and Vikas Abrol
Dryland Research Sub Station, Rakh Dhiansar,
SKUAST- Jammu 181 1 33
India

1. Introduction

The greatest challenge to the world in the years to come is to provide food to burgeoning population, which would likely to rise 8,909 million in 2050. The scenario would be more terrible, when we visualize per capita availability of arable land (Fig 1). The growth rate in agriculture has been the major detriment in world food production. It has been declining since past three decades.

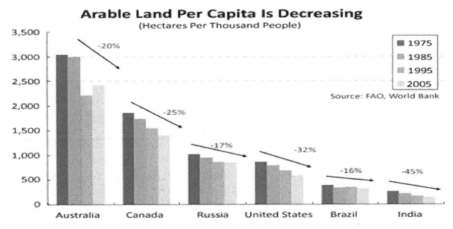

Fig. 1. Decline in arable land per capita in several countries over thirty-year period between 1975 and 2005.

The cultivation of agricultural soils has until recently predominantly been achieved by inverting the soil using tools such as the plough. Soil tillage is one of the basic and important components of agricultural production technology. Various forms of tillage are practised throughout the world, ranging from the use of simple stick or jab to the sophisticated para-plough. The practices developed, with whatever equipment used, can be broadly classified into no tillage, minimum tillage, conservation tillage and conventional tillage. Energy plays a key role in the various tillage systems. Soil tillage is defined as physical, chemical or biological soil manipulation to optimize conditions for germination, seedling establishment and crop growth (Lal, 1979a, 1983). Ahn and Hintze (1990), however, defined it as any physical loosening of the soil carried out in a range of cultivation operations, either by hand or mechanized. For any given location, the choice of a tillage practice will depend on one or more of the following factors (Lal 1980; Unger 1984): i) Soil factors includes Relief, Erodibility, Erosivity, Rooting depth, Texture and structure, Organic-matter content & Mineralogy ; ii) Climatic factors includes Rainfall amount and distribution, Water balance, Length of growing season, Temperature (ambient and soil), Length of rainless period; iii) Crop factors includes Growing duration, Rooting characteristics, Water requirements, Seed ; Socio-economic factors includes Farm size, Availability of a power source, Family structure and composition, Labour situation. Tillage is a labour-intensive activity in low-resource agriculture practised by small land-holders, and a capital and energy-intensive activity in large-scale mechanized farming (Lal 1991). Continual soil inversion can in some situations lead to a degradation of soil structure leading to a compacted soil composed of fine particles with low levels of soil organic matter (SOM). Such soils are more prone to soil loss through water and wind erosion eventually resulting in desertification, as experienced in USA in the 1930s (Biswas, 1984). This process can directly and indirectly cause a wide range of environmental problems. The conventional soil management practices resulted in losses of soil, water and nutrients in the field, and degraded the soil with low organic matter content and a fragile physical structure, which in turn led to low crop yields and low water and fertilizer use efficiency (Wang et al. 2007). Therefore, scientists and policy makers put emphasis on conservation tillage systems. Compared to conventional tillage, there are several benefits from conservation tillage such as economic benefits to labor, cost and time saved, erosion protection, soil and water conservation, and increases of soil fertility (Uri et al. 1998, Wang and Gao 2004)

Conservation tillage (reduced tillage) can lead to important improvements in the water storage in the soil profile (Pelegrín, 1990; Moreno *et al.*, 1997, 2001). Tillage operations generally loosens the soil, decreases soil bulk density and penetration resistance by increasing soil macroporosity. Under these conditions, improvements were also obtained in crop development and yield, especially in very dry years (Pelegrín et al., 1990; Murillo et al., 1998, 2001). Mahboubi et al (1993) in a 28 years long term experiment found that no-tillage resulted in higher saturated hydraulic conductivity compared with conventional tillage on a silt loam soil in Ohio. Whereas, Chang and Landwell (1989) did not observe any changes in saturated hydraulic conductivity after 20 years of tillage in a clay loam soil in Alberta. Saturated hydraulic conductivity of silt clay loam soil was higher when subject to 10 years of tillage than no-tillage in Indiana (Heard et al., 1988). They attributed the higher hydraulic conductivity of tilled soil to greater number of voids and abundance soil macropores caused by the tillage implementation. Studies comparing no-tillage with conventional tillage systems have given different results for soil bulk density. (Osunbitan et al., 2005) found that soil bulk density was greater in no-till in the 5 to 10 cm soil depth however Logsdon et al., (1999) reported no differences in bulk density between tillage systems. The ambiguous nature of these research findings call for additional studies of the effect of long-term tillage on soil properties under various tillage practices in order to optimize productivity and maintain sustainability of soils.

Since tillage strongly influences the soil health, it is important to apply that type of technology that will make it feasible to sustain soil properties at a level suitable for normal growth of agriculture crop. Appropriate tillage practices are those that avoid the degradation of soil properties but maintain crop yields as well as ecosystem stability (Lal 1981b, c, 1984b, 1985a; Greenland 1981). The best management practices usually involve the least amount of tillage necessary to grow the desired crop. This not only involves a substantial saving in energy costs, but also ensures that a resource base, namely the soil, is maintained to produce on a sustainable basis.

2. Tillage effects on crop yield

The effect of tillage systems on crop yield is not uniform with all crop species, in the same manner as various soils may react differently to the same tillage practice. Murillo et al., (2004) compared the traditional tillage, TT (the soil was ploughed by mouldboard, to a 30 cm depth, after burning the straw of the preceding crop) and conservation tillage, CT (the residues of the previous crop were left on the soil surface, as mulch, and a minimum vertical tillage (chiseling, 25 cm depth) and disc harrowing (5 cm depth) were carried out. Results revealed that crops yield was higher in CT (Table 1)

Crop	Treatment	Thousand kernel weight (g)	Yield (kg ha^{-1})
Sunflower	CT	54.5	>2,000
	TT	56.0	>2,000
Wheat	CT	47.3	3,094
	TT	46.6	2,517

Table 1. Effect of tillage on crop yield

Results presented by Nicou and Charreau (1985) showed the effect of tillage on yields of various crops in the West African semi-arid tropics (Table 2). Cotton showed the smallest yield increase with tillage within the range of crops tested. Tillage effects in semi-arid zones are closely linked to moisture conservation and hence the management of crop residues. Several authors (Unger et al.1991; Larson 1979; Brown et al. 1989; Thomas et al. 1990, Sharma et al. 2009) emphasize the link between crop residue management and tillage and recognize them as the two practices with major impact on soil conservation in the semi-arid zones. Residue retention in a cropping system in Burkina Faso significantly increased the yield of cowpeas as shown in Table 3 (IITA/SAFGRAD 1985).

Crop	Number of annual results	Yield (kg ha^{-1})		Yield increase (%)
		control	with tillage	
Millet	38	1558	1894	22
Sorghum	86	1691	2118	25
Maize	31	1893	2791	50
Rice	20	1164	2367	103
Cotton	28	1322	1550	17
Groundnut	46	1259	1556	24

Table 2. Effect of tillage on crop yields in the West African semi-arid tropics

Preceding crop	Residue management system[1]	Date of flowering[2]	Date of maturity[2]	Yield (kg ha[-1])
Maize	Residues removed	48.7	71.2	436
Crotalaria	Residues retained	46.6	69.2	918
Maize	Residues retained	45.7	68.5	921
LSD (0.05)		1.6	1.0	175

[1]No tillage in all treatments [2] Number of days after planting

Table 3. Effect of cropping sequence and residue management on cowpea reproductive physiology and grain yield in the Sudan Savannah of Burkina Faso

It is evident from the extensively published data on tillage that crop yields under conventional tillage are superior to those under conservation tillage. However, several other studies show contradictory results. In both cases the economics of the tillage input are not considered, namely energy and labour costs as well as capital investment in equipment. Underwood et al. 1984; Frengley 1983 and Stonehouse 1991 observed conservation tillage superior and a more cost effective farming practice than conventional tillage on some soils and under certain climatic conditions. Although conservation tillage is being widely adopted, there is strong evidence that soils prone to surface crusting and sealing would benefit from conventional tillage once every 2 or 3 years. Rao et al. (1986) found that conventional tillage is superior to no tillage, reduced tillage or mulching with a number of crops - sun hemp (*Crotalaria juncea*), barley (*Hordeum vulgare*), mustard (*Brassica juncea*) and chickpea (*Cicer arietinum*) grown in the dry season. Nicou (1977) and Charreau (1972; 1977) showed that soil inversion and deep ploughing increases plant-available water and crop yields as compare to the no tillage in West African semi-arid regions. Similar data showing greater responses to tillage than no tillage or greatly reduced tillage were reported by Karaca et al. (1988), Prihar and Jalota (1988) and Willcocks (1988) on a variety of soils.

Mulch management	Tillage Treatments									
	Grain Yield of Maize (kgha[-1])					Grain Yield of Wheat (kgha[-1])				
	CT	MT	NT	RB	Mean	CT	MT	NT	RB	Mean
No Mulch (NM)	1370	1365	1246	1255	1308	1080	1063	930	1025	1024
Straw Mulch (StM)	2020	1990	1776	1896	1920	1410	1430	1210	1335	1346
Polythene Mulch (PM)	2183	2137	1930	2007	2065	1505	1510	1360	1450	1456
Soil Mulch (SM)	1890	1860	1730	1851	1832	1320	1360	1110	1265	1263
Mean	1865	1837	1670	1752		1328.7	1340	1152	1268	
CD (P=0.05)	M=150 S=180 M at S=160 S at M=253					M=145 S=193 M at S=301 S at M= NS				

Where Conventional tillage (CT), Minimum Tillage (MT), No Tillage (NT), Raised Bed (RB); M= Tillage Treatments, S= mulching Treatments, M at S= Interaction of tillage on same level of Mulch, S at M= Interaction of Mulch on same level of tillage (Sharma et al. 2011)

Table 4. Effect of tillage and mulching on grain yield of Maize and Wheat in semi arid tropics, India (Average of three years)

Sharma et al. (2011) reported that the greatest maize yield of 1865 kgha⁻¹ was achieved with conventional tillage (CT) system while not significantly lower yield was achieved with minimum tillage (MT) system (1837 kgha⁻¹). However, higher wheat yield was recorded in MT as compare to the CT in maize –wheat rotation (Table 4).

3. Tillage effects on soil properties

3.1 Tillage effects on soil degradation

Soil erosion has conventionally been perceived as one of the main causes of land degradation and the main reason for declining yields in tropical regions. Intensive or inappropriate tillage practices have been a major contributor to land degradation. The last four decades has seen a major increase in intensive agriculture in the bid to feed the world population more efficiently than ever before. In many countries, particularly the more developed countries, this intensification of agriculture has led to the use of more and heavier machinery, deforestation and landuse changes in favour of cultivation. This has led to several problems including loss of organic matter, soil compaction and damage to soil physical properties. Soil tillage breaks down aggregates, decomposes soil organic matter, pulverizes the soil, breaks pore continuity and forms hard pans which restrict water and air movement and root growth. On the soil surface, the powdered soil is more prone to sealing, crusting and erosion. Improving soil physical fertility involves reducing soil tillage to a minimum and increasing soil organic matter (Fig. 2).

Fig. 2. Physical degradation of a soil as a result of intensive tillage

Tillage-induced soil erosion in developing countries can entail soil losses exceeding 150 t/ha⁻¹. annually and soil erosion, accelerated by wind and water, is responsible for 40 percent of land degradation world-wide. Several more recent studies have shown that no-tillage systems with crop residue mulch can increase nutrient use efficiency (Lal 1979a, b, c; Hulugalle et al. 1985). The no-till system seems to have a broad application in humid and sub-humid regions, for which 4-6 tons ha⁻¹ of residue mulch appears optimal (Lal 1975; Aina et al. 1991). The beneficial effect of conservation tillage systems on soil loss and runoff have been demonstrated in studies conducted by ICRISAT (1988) and Mensah bonus and Obeng (1979) (Table 5 & 6).

Treatment	Sorghum grain yield[1]	Runoff[2] (mm)	Soil loss[2] (t ha⁻¹)
10 cm deep traditional ploughing	2.52	128	1.66
15 cm non inverted primary tillage	2.83	102	1.62
15 cm deep mouldboard ploughing	2.76	106	1.70
25 cm deep mouldboard ploughing	3.22	85	1.41
S.E	+0.07	+4.9	+0.279

[1]Average values of four years (1983, 1984, 1986 and 1987) [2] Average values of 1986 and 1987

Table 5. Effect of different tillage treatments on sorghum grain yield, runoff and soil loss under Luvisols (ICRISAT Centre 1983-1987)

Treatment	Soil loss (t ha⁻¹yr⁻¹)		Runoff (%)	
	Kwadaso	Ejura	Kwadaso	Ejura
Bare fallow	313.0	18.3	49.8	36.4
No-tillage	1.96	9.2	3.4	0.52
Mulching	0.42	1.9	1.4	0.33
Ridging (across slope)	2.72	4.5	1.9	1.30
Minimum tillage	4.90	3.8	1.7	1.10
Traditional mixed cropping	33.6	2.5	13.2	5.10

Table 6. Effects of tillage systems on soil loss and runoff in Ghana (1976)

3.2 Tillage effects on water content

Tillage effects differ from one agro-ecological zone to the other. In semi-arid regions moisture conservation is one of the key factors to consider. Nicou and Chopart (1979) showed that tillage and residue management increased soil profile water content. The soil was mechanically tilled to a depth of 20-30 cm (Table 7).

Tillage system	Profile water content (mm)
No till, residues burnt	49.4
Ploughing, residues incorporated	95.8
Ploughing, residues incorporated followed by addition of external mulch	103.7

Table 7.Effect of tillage system on profile water content to a depth of 1 m at 2 weeks after planting

Sharma et al. (2011) showed that the no tillage retained the highest moisture followed by minimum tillage, raised bed and conventional tillage in inceptisols under semi arid regions of India (Fig 3). Tillage treatments influenced the water intake and infiltration rate (IR)

increased in the order of NT > MT > RB > CT and in mulching treatment the order was PM > StM > SM > NM. The maximum mean value of IR (182.4 mm/day) was obtained in case of no tillage and polythene mulch combination and minimum (122.4 mm/day) was recorded in CT and no mulch combination (Fig 4).

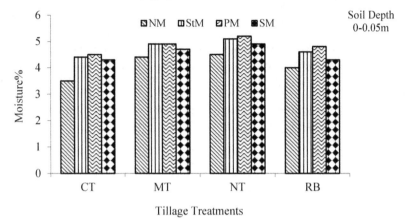

Fig. 3. Effect of tillage & water management practices on soil water content at harvesting of maize (3 years average), were, CT=Conv. Till., MT=Min. Till., NT=No Till., RB=Raised bed, NM=No Mulch, StM=Straw mulch, PM=Polythene Mulch and SM=Soil Mulch

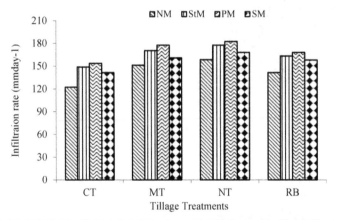

CD (P=0.05), M=9.6, S=5.52, M at S=14.4, Sat=NS, where, M=Till. Treats.; S= Mulch Treats

Fig. 4. Several researchers also show the importance of tillage on soil moisture (Lal 1977; Klute 1982; Norwood et al. 1990). Tillage enhances soil water storage by increasing soil surface roughness and controlling weeds during a fallow. This stored water may improve subsequent crop production by supplementing growing season precipitation (Unger and Baumhardt, 1999). Several studies shown that deep tillage has immense potential for water storage and better crop production. Schillinger (2001) and Lampurlanes et al. (2002) observed no difference in water storage efficiency of reduced tillage in comparison with other tillage systems.

3.3 Tillage effects on porosity

Soil porosity characteristics are closely related to soil physical behavior, root penetration and water movement (Pagliai and Vignozzi 2002, Sasal et al. 2006) and differ among tillage systems (Benjamin 1993). Lal et al. (1980) revealed that straw returning could increase the total porosity of soil while minimal and no tillage would decrease the soil porosity for aeration, but increase the capillary porosity; as a result, it enhances the water capacity of soil along with poor aeration of soil (Wang et al.1994, Glab and Kulig 2008). However, Borresen (1999) found that the effects of tillage and straw treatments on the total porosity and porosity size distribution were not significant. Allen et al. (1997) indicated that minimal tillage could increase the quantity of big porosity. Tangyuan, et al. (2009) showed that the soil total porosity of 0–10 soil layer was mostly affected; conventional tillage can increase the capillary porosity of soil and the porosities were C > H > S (Figure 5) but the non-capillary porosity of (S) was the highest. Returning of straw can increase the porosity of soil.

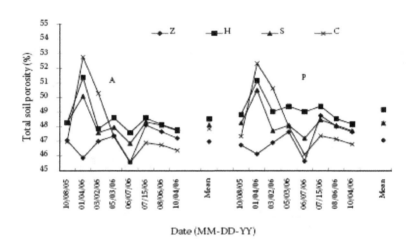

Where, Conventional tillage (C), Zero-tillage (Z), Harrow-tillage (H) and Subsoil-tillage (S),

Straw absent (A) or Straw present (P)

Fig. 5. Tillage and residue management effect on soil porosity

The increase in plant available water capacity of the soil under different tillage treatments was found to decrease with an increase in the level of compaction. Because compaction results in the breaking down of larger soil particle aggregates to smaller ones, it is difficult for water to drain out of the soils because of the greater force of adhesion between the micropores and soil water. For the same tillage treatment, the effect of increasing the axle load upon a soil is to decrease the total porosity and to increase the percentage of smaller pores as some of the originally larger pores have been squeezed into smaller ones by compaction (Hamdeh, 2004) (Fig 6).

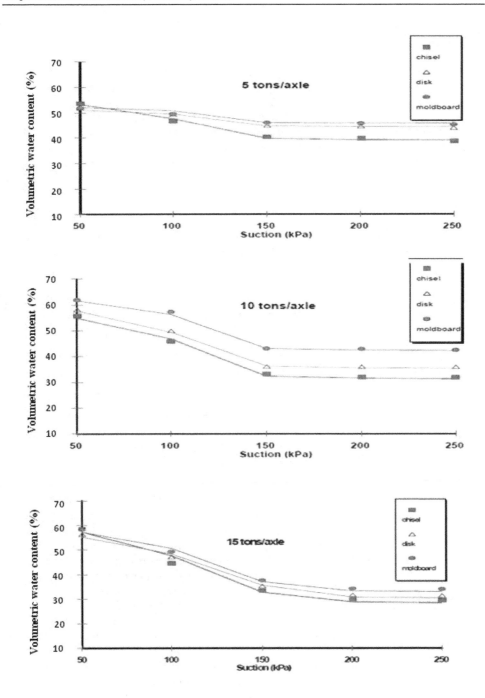

Fig. 6. Water retention curves for different axle load levels and different tillage system

3.4 Tillage effects on bulk density

The two of the most commonly measured soil physical properties affecting hydraulic conductivity are the soil bulk density and effective porosity as these two properties are also fundamental to soil compaction and related agricultural management issues (Strudley et al. 2008). The studies comparing no-tillage with conventional tillage systems have given different results for soil bulk density. Several studies showed that soil bulk density was greater in no-till in the 5 to 10 cm soil depth (Osunbitan et al. 2005). No differences in bulk density were found between tillage systems (Logsdon et al. 1999). However Tripathi et al. (2005) found increase in bulk density with conventional tillage in a silty loam soil. Moreover, there are few studies that have examined changes in soil physical properties in response to long term tillage and frequency management (> 20 yr) in the northern Great Plains. Rashidi and Keshavarzpour (2008) observed that the highest soil bulk density of 1.52 g cm^{-3} was obtained for the NT treatment and lowest (1.41 g cm^{-3}) for the CT treatment (Table 8). The highest soil penetration resistance of 1250 kPa was obtained for the NT treatment and lowest (560 kPa) for the CT treatment (Table 8). The highest soil moisture content of 19.6% was obtained for the CT treatment and lowest (16.8%) for the NT treatment .

Treatments	Soil bulk density (gcm^{-3})	Soil penetration resistance (kPa)	Soil moisture content (%)
CT	1.41 c	560 c	19.6 a
RT	1.47 b	815 b	18.4 b
MT	1.50 ab	1105 a	17.1 c
NT	1.52 a	1250 a	16.8 c

CT=Conv. Till., RT=Reduced till., MT=Minimum Till., NT=No till.

Table 8. Effect of different tillage treatments on soil physical properties (mean of 2006 and 2007). Means followed by the same letter in the same column are not significantly different at the 1% level

Hamdeh, 2004 reported that, the vehicle significantly increased soil dry density to a depth of 40 cm for all treatments at 10 cm depth. The MB treatment caused the maximum percentage increase of dry bulk density at all depths. This indicated the significant effects of axle load on soil physical properties. The percentage difference for each treatment was less at the 10-20 cm depth than at the 0-10 cm depth. These results reflect a more compact soil layer at the 0-10 cm depth than at the 10-20 cm depth. The averages of percentage increase of dry density at the 0-20 cm depth show that the MB treatment had the highest effect while the CS treatment had the lowest effect. These results suggested that tyre traffic followed by tillage might have a significant affect on the resulting soil physical properties. There is no significant difference (P < 0.1) between the ML and the CB at the 20-30 cm depth. At 20-40 cm depth MB treatment had the greatest percentage increase of dry bulk density while the CS treatment had the lowest percentage increase of dry bulk density. Results demonstrate that the axle load is crucial factor for the depth of subsoil compaction. An increase in axle wheel loads resulted in greater soil compaction due to increased in both shear and vertical soil stresses.

Treatment	Water use (cm)	Grain yield (kg ha⁻¹)	Water use efficiency (kg ha⁻¹cm⁻¹)	Plant population at harvest (per plot)[1]
Conventional tillage	32.15	3106a	96.61a	180a
Plough	29.64	2923a	98.62a	179a
Zero tillage	30.44	2639b	86.70b	160b
Manual	29.19	2692b	92.22ab	188a
Conventional tillage	48.19	5240a	108.74a	203a
Plough	47.64	5067a	106.36a	198ab
Zero tilage	49.20	4612b	96.08b	194ab
Manual	48.00	4612b	96.08b	194ab
Conventional tillage	49.60	5533a	111.55a	207a
Plough	50.01	4998b	99.94b	205a
Zero tillage	50.14	5949c	118.65c	203a
Manual	49.01	4303d	87.69d	199a
Conventional tillage	49.54	5259a	106.16a	200a
Plough	48.92	5174a	105.16a	206a
Zero tillage	49.69	5887b	118.47b	198a
Manual	49.01	4103c	83.72c	204a
Conventional tillage	49.62	5384a	108.50	202a
Plough	49.94	5238a	104.80a	199a
Zero tillage	49.21	5678b	115.38b	205b
Manual	48.64	3713c	76.34c	197a

[1]Means followed by the same letter in the same column are not significantly different at the 5% level.

Table 9. Effect of tillage practices on water use, maize yield and water-use efficiency (early season) (Osuji 1984)

3.5 Tillage effects on water use efficiency

Nigeria, Osuji (1984) observed that water-use efficiency and maize grain yields were significantly higher under zero tillage than under other tillage treatments (Table 9). Lal (1985c) showed that soil physical properties and chemical fertility were substantially worse in ploughed watersheds after six years of continuous mechanized farming and twelve crops of maize, while the decline in the soil properties was decidedly less in the no-tillage watershed. The lower maize yields of the ploughed watershed are related to erosion, compaction, fall in organic matter content and fall in pH. After 10 years of continuous comparative no-tillage and conventional tillage trails in Southwest Nigeria, Opara-Nadi and Lal (1986) observed that total porosity, moisture retention, saturated and unsaturated hydraulic conductivity, and the maximum water-storage capacity increased under no-tillage with mulch.

3.6 Tillage effect on environment

CT Tillage may affect the production of nitrous oxide through its effect on soil structural quality and water content (Ball et al., 1999). CT can prevent nutrient loss (Jordan et al. 2000) (Table 10).

Measurements	Plough	Non- inversion tillage	Benefit compared to ploughing
Runoff (L ha^{-1})	213328	110275	48 % reduction
Sediment loss (kg ha $^{-1}$)	2045	649	68 % reduction
Total P loss (kg P ha $^{-1}$)	2.2	0.4	81 % reduction
Available P loss (kg P ha^{-1})	3x 10^{-2}	8 x 10^{-3}	73 % reduction
TON (mg Ns^{-1})	1.28	0.08	94 % reduction
Soluble phosphate (ug Ps^{-1})	0.72	0.16	78 % reduction
Isoproturon	0.011ugs^{-1}	Not detected	100 % reduction

Table 10. Effect of soil tillage on soil erosion and diffuse pollution

Comparison of herbicide and nutrient emissions from 1991 to 1993 on a silty clay loam soil. Plots 12 m wide were established and sown with winter oats in 1991 followed by winter wheat and winter beans. De-nitrification in anaerobic soil and nitrification in aerobic soil produce nitrous oxide, with the former being more important. As soil structure improves, the potential for creating anaerobic conditions and nitrous oxide emissions is reduced (Arah et al., 1991). Intensive soil cultivations break-down SOM producing CO_2 thereby lowering the total C sequestration held within the soil. Building SOM the adoption of CT, especially if combined with the return of crop residues, can substantially reduce CO_2 emissions (West and Marland, 2002). In the UK, where CT was used soil C was 8% higher compared to conventional tillage, equivalent to 285g SOM m^{-2}. In the Netherlands SOM was 0.5% higher using an integrated approach over 19 years, although this increase was also achieved because of higher inputs of organic matter (Kooistra et al., 1989). Murillo et al. (2004) in a long term experimentation, observed that in CT (0-10 cm depth) organic matter values have been reached close to the minimum content of 2% (1.1% organic C, Table 11) considered necessary for most agricultural practices carried out in European Occidental soils (Bullock, 1997). These are indeed moderate values, and would not justify the implementation of conservation tillage systems (aimed at achieving high surface organic matter content).

Soil depth (cm)	Treatment	Year		
		2001 (November)	2002 (January)	2002 (March)
0-5	CT	9.8*	9.3*	11.1*
	TT	8.1	8.1	8.6
5-10	CT	9.5*	9.6	10.2*
	TT	8.1	8.5	8.3
10-25	CT	6.5	5.9	8.5
	TT	6.7	6.4	7.6
25-40	CT	-	4.4	6.9
	TT	-	5.0	6.1

* Significant differences between treatments per year per depth

Table 11. Mean values of organic carbon in the soil treated by conservation tillage and traditional tillage for the years 2001 sunflower and 2002 (wheat)

After 12 years of integrated farming incorporating CT, the SOM content was 25% higher at 0–5 cm and overall from 0 to 30 cm, 20 % higher (El Titi, 1991). Similar increases in SOM in the upper surface layers were also found in a number of studies conducted throughout Scandinavia (Rasmussen, 1999, Paustian et al., 2000). With CT, there is a risk that SOM may be reduced below this surface layer, but no evidence for this was found in Sweden (Stenberg et al., 2000).The significant build up in SOC is well documented in long term experiments with conservation tillage.

4. Strategies for mitigating challenges

Conservation agriculture (CA) is a concept for resource-saving agricultural crop production that strives to achieve acceptable profits together with high and sustained production levels while concurrently conserving the environment. Interventions such as mechanical soil tillage are reduced to an absolute minimum and the use of external inputs such as agrochemicals and nutrients of mineral or organic origin are applied at an optimum level and in a way and quantity that does not interfere with or disrupt the biological processes. One of the soil conservation techniques developed in USA is known as 'conservation tillage'(CT), this involves soil management practices that minimise the disruption of the soil's structure, composition and natural biodiversity, thereby minimising erosion and degradation, but also water contamination (Anonymous, 2001).

5. Principles of conservation agriculture

Conservation agriculture systems utilize soils for the production of crops with the aim of reducing excessive mixing of the soil and maintaining crop residues on the soil surface in order to minimize damage to the environment. This is done with objective to:

- Provide and maintain an optimum environment of the root-zone to maximum possible depth.
- Avoid physical or chemical damage to roots that disrupts their effective functioning.
- Ensure that water enters the soil so that (a) plants never or for the shortest time possible, suffer water stress that will limit the expression of their potential growth; and so that (b) residual water passes down to groundwater and stream flow, not over the surface as runoff.
- Favour beneficial biological activity in the soil

CT is now commonplace in areas where rainfall causes soil erosion or where preservation of soil moisture because of low rainfall is the objective. World-wide, CT is practised on 45 million ha, most of which is in North and South America (FAO, 2001) but is increasingly being used in other semi-arid (Lal, 2000a) and tropical regions of the world (Lal, 2000b). In USA, during the 1980s, it was recognized that substantial environmental benefits could be generated through soil conservation and to take advantage of this policy goals were changed. These were successful in reducing soil erosion; however, the social costs of erosion are still substantial, estimated at $37.6 billion annually (Lal, 2001). World-wide erosion-caused soil degradation was estimated to reduce food productivity by 18 million Mg at the 1996 level of production (Lal, 2000b). Because of the increasing population and rising standards of living, it is essential to develop those agricultural practices that maximize agricultural production while also enhancing ecosystem services. Eco-efficiency is related to both "ecology" and "economy," and denotes both efficient and sustainable use of resources in farm production and land management (Wilkins, 2008). Experience has shown that conservation agriculture systems achieve yield levels as high as comparable conventional agricultural systems but with less fluctuations due, for example, to natural disasters such as drought, storms, floods and landslides. Conservation agriculture therefore contributes to food security and reduces risks for the communities (health, conditions of living, water supply), and also reduces costs for the State (less road and waterway maintenance).

6. Conclusion

Soils are one of the world's most precious commodities. Continuing soil degradation is threatening food security and the livelihood of millions of farm households throughout the world. Soil types and their various reactions to tillage are of paramount importance in determining the superiority of one practice over the other. Socio-economic considerations, however, should always be taken into account in decision making for the adoption of one practice over another. Soil health refers to the soil's capacity to perform its three principal functions e.g. economic productivity, environment regulation, and aesthetic and cultural values. There is a need to develop precise objective and quantitative indices of assessing these attributes of the soil. Training of professional staff must include developing their capacities in interdisciplinary collaboration and interpersonal relations. Research programmes and activities need to do more to address the real-life problems of farmers, and to include farmers in the design and implementation of programmes relevant to their needs. Research methodologies should be standardized and information dissemination should be an indispensable component of any common tillage network programme to be developed.

7. References

Ahn, P.M. and Hintze, B. 1990. No tillage, minimum tillage, and their influence on soil properties. In: *Organic-matter Management and Tillage in Humid and Sub-humid Africa.* pp. 341-349. IBSRAM Proceedings No.10. Bangkok: IBSRAM.

Allen M., Lachnicht S.L., McCartney D., Parmelee R.W. 1997. Characteristics of macroporosity in a reduced tillage agroecosystem with manipulated earthworm populations: implications for infiltration and nutrient transport. *Soil Biology and Biochemistry* 29: 493-498.

Anonymous, 2001. Conservation Agriculture in Europe. http://www.ecaf.org.uk.

Arah, J.R.M., Smith, K.A., Crichton, I.J., Li, H.S., 1991. Nitrous oxide production and denitrification in Scottish arable soils. *J. Soil Sci.* 42: 351-367

Aulakh, M.S., Rennie, D.A., Paul, E.A. 1984. Gaseous N losses from soils under zero-till as compared to conventional-till management systems. *J. Environ. Qual.* 13 : 130-136.

Ball, B.C., Scott, S., Parker, J.P. 1999. Field N_2O, CO_2 and CH_4 fluxes in relation to tillage, compaction and soil quality in Scotland. *Soil Till. Res.* 53: 29-39.

Benjamin J.G. 1993. Tillage effects on near-surface soil hydraulic properties. *Soil and Tillage Research* 26: 277-288.

Biswas, M.R., 1984. Agricultural production and the environment:a review. Environ. Conserv. 11, 253-259.

BØrresen T. 1999. The effect of straw management and reduced tillage on soil properties and crop yields of spring-sown cereals on two loam soils in Norway. *Soil and Tillage Research* 51: 91-102.

Brown, H.J., Cruse, R.M. and Colbin, T.S. 1989. Tillage system effects on crop growth and production costs for a corn-soybean rotation. *J. Production Agriculture* 2:273-279.

Bullock, P. 1997. Sustainable development of soils in western Europe–an overview. Proc. L Aniversario de la Sociedad Española de la Ciencia del Suelo, Madrid. pp. 109-123.

Chang C., Lindwall C.W.1989. Effect of long term minimum tillage practices on some physical properties of Chernozemic clay loam. *Can. J. Soil Sci.* 69:433-449.

Charreau, C. 1972. Problèmes posés par l'utilisation agricole des sols tropicaux par les cultures annuelles. *Agronomie Tropicale* 27:901-929.

Charreau, C. 1977. Controversial points in dryland farming practices in semi-arid West Africa. In: *Symposium on Rainfed Agricultural Semi-arid Regions.* G.H. Cannell (ed.). 313-360. Univ. of California, Riverside.

El Titi, A., 1991. The Lautenbach Project 1978–1989: integrated wheat production on a commercial arable farm, south-west Germany. In: Firbank, L.G., Carter, N., Potts, G.R. (Eds.), The Ecology Of Temperate Cereal Fields. Blackwell Scientific Publishers, Oxford, pp. 399–412.

Food and Agriculture Organization of the United Nations, 2001a *The Economics of Conservation Agriculture,* FAO Land and Water Development Division, Rome

Frengley, G.A.G. 1983. Economic benefits of conservation tillage in New Zealand. 1. Mixed crop and livestock farms. *Soil and Tillage Research* 3: 347-356.

Gill, S. M., M. S. Akhtar, and Z. Saeed. 2000. Soil water use and bulk density as affected by tillage and fertilizer in rainfed wheat production system. Pak. J. Biol. Sci. 3: 1223-1226.

Głab T., Kulig B. 2008. Effect of mulch and tillage system on soil porosity under wheat (*Triticum aestivum*).*Soil and Tillage Research 99*: 169–178.

Greenland D.J. 1981. Soil management and soil degradation. *J. Soil Science* 32: 301-322.

*Hamdeh- Abum N. H.*2004. The *effect of tillage treatments on soil water holding capacity and on soil physical condition In:* 13th International Soil Conservation Organisation Conference – Brisbane, July 2004 *Conserving Soil and Water for Society: Sharing Solutions*

Heard J.R., Kladivko E.J. Manning J.V. 1988. Soil macroporosity, hydraulic conductivity and air permeability of silt soils under long-term conservation tillage in Indiana. *Soil Till. Res.* 11: 1-18.

ICRISAT. 1988. International Crops Research Institute for the Semi-arid Tropics. *Annual Report 1987.* ICRISAT, Patancheru, India.

IITA/SAFGRAD. 1985. (International Institute of Tropical Agriculture/Semi-Arid Food Grains Research and Development Project). *Annual Report for 1984.* IITA/SAFGRAD, Ouagadougou, Burkina Faso.

Jordan, V.W., Leake, A.R., Ogilvy, S.E., 2000. Agronomic and environmental implications of soil management practices in integrated farming systems. Aspects Appl. Biol. 62, 61–66.

Karaca, M., Guler, M., Durutan, N., Pala, M. and Unver, I. 1988. Effect of fallow tillage systems on wheat yields in Central Anatolia, Turkey. In: *Challenges in Dryland Agriculture.* P.W. Unger et al. (eds.) pp. 131-133. Proc. Int. Confer. on Dryland Agriculture, Amarillo/Bushland, Texas, August 1988.

Klute, A. 1982. Tillage effects on hydraulic properties of soil. A review. In: *Predicting Tillage Effects on Soil Physical Properties and Processes.* P.W. Unger and Van Doren, D.M. (eds.) ASA Special Publication No.44:29-43.

Kooistra, M.J., Lebbink, G., Brussaard, L., 1989. The Dutch programme on soil ecology of arable farming systems 2: Geogenesis, agricultural history, field site characteristics and present farming systems at Lovinkhoeve experimental farm. Agric. Ecosys. Environ. 27, 463–469.

Lal   R. 1981c. Soil management in the tropics. In: *Characterisation of Soils of the Tropics: Classification and Management.* D.J. Greenland (ed.). Oxford University Press, UK.

Lal   R. 1984b. Soil erosion from tropical arable lands and its control. *Advances in Agronomy* 37:183-248.

Lal   R. 1985a. No-till in the lowland humid tropics. In: *The Rising Hope of our Land Conference* (Georgia, 1985). pp. 235-241.

Lal   R. 1981b. Soil conditions and tillage methods in the tropics. *Proc. WARSS/WSS Symposium on No-tillage and Crop Production in the Tropics* (Liberia 1981).

Lal, R. 1977. Importance of tillage systems in soil water management in the tropics. In: *Soil Tillage and Crop Production.* R. Lal (ed.) pp. 25-32. IITA, Ibadan, Nigeria.

Lal, R. 1979a. Importance of tillage systems in soil and water management in the tropics. In: *Soil Tillage and Crop Production.* R. Lal (ed.). pp. 25-32. IITA Proc. Ser. 2.

Lal, R. 1980. Crop residue management in relation to tillage techniques for soil and water conservation. In: Organic recycling in Africa 74-79. *Soils Bulletin 43.* FAO, Rome.

Lal, R. 1983. No-till farming: Soil and water conservation and management in the humid and sub-humid tropics. *IITA Monograph No. 2,* Ibadan, Nigeria.

Lal, R. 1985c. Mechanized tillage systems effects on properties of a tropical Alfisol in watersheds cropped to maize. *Soil and Tillage Research* 6:149-161.

Lal, R. 1991. Tillage and agricultural sustainability. *Soil and Tillage Research* 20: 133-146.

Lal, R. 2000b. Physical management of soils of the tropics: priorities for the 21st century. *Soil Sci.* 165: 191–207.

Lal, R. 2001. Managing world soils for food security and environmental quality. *Adv. Agron.* 74:155–192.

Lampurlanes, J., P. Angas, and C. Martinez. 2002. Tillage effects on water storage during fallow, and on barley root growth and yield in two contrasting soils of the semi-arid Segarra region in Spain. *Soil Till. Res.* 65: 207-220.

Larson, W.E. 1979. Crop residues: Energy production or control? In: *Effects of Tillage and Crop Residue Removal on Erosion, Runoff and Plant Nutrients.* pp. 4-6. Soil Conservation Society of America Special Publication No. 25.

Logsdon S.D., Kasper T.C., Camberdella C.A. 1999. Depth incremental soil properties under no-till or chisel management. *Soil Sci. Soc. Am. J.* 63:197-200.

Logsdon, S.D., Kasper, T.C., Camberdella. C.A.,1999. Depth incremental soil properties under no-till or chisel management. *Soil Sci. Soc. Am. J.* 63:197-200.

Mahboubi, A.A., Lal R. Fausey N.R.1 993.Twenty-eight years of tillage effects on two soils in Ohio. *Soil Sci. Soc. Am. J.* 57:506-512.

Mensah-Bonsu, and Obeng, H.G. 1979. Effects of cultural practices on soil erosion and maize production in the semi-deciduous rainforest-savanna transitional zones of Ghana. In: *Soil Physical Properties and Crop Production in the Humid Tropics.* Greenland, D.J. and Lal, R. (eds.). John Wiley, Chichester. pp. 509-519.

Moreno F., Murillo J.M.,Pelegrin F.,Fernandez J.E. 2001. Conservation and traditional tillage in years with lower and higher precipitation than the average (south-west Spain). In: Conservation Agriculture, a Worldwide Challenge (García-Torres L., Benites J., Martínez- Vilela A., ed.) ECAF, FAO, Cordoba, Spain, pp. 591-595.

Moreno F.,Pelegrin F, Fernandez J.MurilloJ.M, 1997. Soil physical properties, water depletion and crop development under traditional and conservation tillage in southern Spain. *Soil Till Res.* 41: 25-42.

Murillo J. M., Moreno F., Girón I. F. and Oblitas M. I. 2004.Conservation tillage: long term effect on soil and crops under rainfed conditions in south-west Spain (Western Andalusia). Spanish Journal of Agricultural Research 2 (1): 35-43

Murillo J.M., Moreno F., Pelegrin F. 2001. Respuesta del trigo y girasol al laboreo tradicional y de conservación bajo condiciones de secano (Andalucía Occidental). *Invest Agr: Prod Prot Veg* 16: 395-406.

Murillo J.M., Moreno F., Pelegrin F., Fernandez J.E. 1998. Responses of sunflower to traditional and conservation tillage under rainfed conditions in southern Spain. *Soil Till Res.* 49: 233-241.

Nicou, R. 1977. *Le travail du sol dans les terres exondées du Sénégal: motivations, contraintes.* Centre National de Recherche Agronomique (CNRA), Bambey, Senegal. Mimeo. 50 pp.

Nicou, R. and Charreau, C. 1985. Soil tillage and water conservation in semi-arid West Africa. In: *Appropriate Technologies for Farmers in Semi-arid West Africa.* H. Ohm and J.G. Nagy (eds.). pp. 9-32. Purdue University Press, West Lafayette.

Nicou, R. and Chopart, J.L. 1979. Water management methods for sandy soils of Senegal. In: *Soil Tillage and Crop Production.* R. Lal (ed.). pp. 248-257. IITA, Ibadan, Nigeria. Proc. Ser. No.2.

Norwood, C.A., Schlegel, A.J., Morishita, D.W. and Gwin, R.E. 1990. Cropping system and tillage effects on available soil water and yield of grain sorghum and winter wheat. *J. Production Agriculture* 3:356-362.

Opara-Nadi, O.A. and Lal, R. 1986. Effects of tillage methods on physical and hydrological properties of a tropical Alfisol. *Zeitschrift fur Pflanzenernahrung und Bodenkunde* 149:235-243.

Osuji, G.E. 1984. Water storage, water use and maize yield for tillage systems on a tropical Alfisol in Nigeria. *Soil and Tillage Research* 4:339-348.

Osunbitan J.A. ,Oyedele D.J., Adekalu K.O. 2005. Tillage effects on bulk density, hydraulic conductivity and strength of a loamy sand soil in southwestern Nigeria. *Soil Till. Res.* 82:57-64.

Pagliai M., Vignozzi N. 2002. Soil pore system as an indicator of soil quality. *35*: 69–80.

Paustian, K., Six, J., Elliott, E.T., Hunt, H.W., 2000. Management options for reducing CO2 emissions from agricultural soils. *Biogeochemistry* 48: 147–163.

Pelegrin F.,Moreno F.,Martin–Aranda J.,Camps M. 1990. The influence of tillage methods on soil physical properties and water balance for a typical crop rotation in SW Spain. *Soil Till Res.* 16: 345-358.

Prihar, S.S. and Jalota, S.K. 1988. Role of shallow tillage in soil water management. In: *Challenges in Dryland Agriculture.* P.W. Unger *et al.* (eds.) pp. 128-130. Proc. Int. Confer. on Dryland Agriculture, Amarillo/Bushland, Texas, August 1988.

Rao, P., Agrawal, S.K. and Bishnoc, O.P. 1986. Yield variations in winter crops under different soil tillage and moisture conservation practices. *Indian Journal of Ecology* 13:244-249.

Rashidi, M. and F. Keshavarzpour, 2008. Effect of different tillage methods on soil physical properties and crop yield of Melon (Cucumis melo).American-Eurasian *J. Agric. & Enviorn. Sci.* 3(1):43-48

Rockwood, W.G. and Lal, R. 1974. Mulch tillage: a technique for soil and water conservation in the tropics. *Span* 17:77-79.

Sasal M.C., Andriulo A.E., Taboada M.A.2006. Soil porosity characteristics and water movement under *Advances in Geoecology,*zero tillage in silty soils in Argentinian Pampas. *Soil and Tillage Research* 87: 9–18.

Schillinger, W. F. 2001. Minimum and delayed conservation tillage for wheat-fallow farming. *Soil Sci. Soc. Am. J.* 65:1203-1209.

Sharma, Peeyush ; Abrol, Vikas; Maruthi Sankar, G. R.; and Singh, Brinder.2009. Influence of tillage practices and mulching options on productivity, economics and soil physical properties of maize (*zea maize*) - wheat (*Triticum aestivum*) system. *Indian Journal of Agriculture Scinece* 79 (11):865-870.

Sharma, Peeyush; Abrol, Vikas and Sharma, R. K. 2011. Impact of tillage and mulch management on economics, energy requirement and crop performance in maize-wheat rotation in rainfed subhumid inceptisols, India *Europ. J. Agronomy* 34: 46–51

Stenberg, M., Stenberg, B., Rydberg, T., 2000. Effects of reduced tillage and liming on microbial activity and soil properties in a weakly-structured soil. Agric. *Ecosys. Environ.* 14: 135-145.

Stonehouse, D.P. 1991. The economics of tillage for large-scale mechanized farms. *Soil and Tillage Research,* 20:333-351.

Strudley MW, Green TR, Ascough II JC 2008. Tillage effects on soil hydraulic properties in space and time. *Soil Tillage Research* 99: 4-48.

Tangyuan, N., Bin H., Nianyuan, J., Shenzhong, T., Zengjia, L. 2009. Effects of conservation tillage on soil porosity in maize-wheat cropping system Plant Soil Envir on. *55* (8): 327-333

Tangyuan, N., H. Bin, J. Nianyuan, T. Shenzhong L. Zengjia1.2009. Effects of conservation tillage on soil porosity in maize-wheat cropping system *Plant Soil Envi on. 55,*(8): 327-333 327

Thomas, G.A., Standley, J., Webb, A.A., Blight, G.W. and Hunter, H.M. 1990. Tillage and crop residue management affect vertisol properties and grain sorghum growth over seven years in the semi-arid sub-tropics. I. Crop residue and soil water during fallow periods. *Soil and Tillage Research* 17:181-197.

Tripathi, R.P.; Sharma, Peeyush and Singh, S.2005. Tilth index: an approach for optimizing tillage in rice – wheat system. *Soil & tillage Research.* 80:125-137.

Underwood, R.L., Unger, P.W., Wiese, A.F. and Allen, R.R. 1984. Conservation tillage for water conservation. *Conservation Tillage No. 110.* Great Plains Agric. Council Pub. 227-234.

Unger PW, Baumhardt RL. 1999. Factors related to dryland grain sorghum yield increases. *Agron. J.* 91: 870-875.

Unger, P.W. 1984b. Tillage and residue effects on wheat, sorghum, and sunflower grown in rotation. *Soil Sci. Soc. Am. J.* 48:885-891.

Unger, P.W., Stewart, B.A., Parr, J.F. and Singh, R.P. 1991. Crop residue management and tillage methods for conserving soil and water in semi-arid regions. *Soil and Tillage Researc,* 20:219-240.

Uri, N.D., Atwood, J.D., Sanabria, J., 1998. The environmental benefits and costs of conservation tillage. Sci. Total Environ. 216, 13–32.

Wang D.W., Wen H.D. 1994. Effect of protective tillage on soil pore space status and character of micro morphological structure. Journal of Agricultural University of Hebei, 17: 1-6. (In Chinese)

Wang X.B., Cai D.X., Perdok U.D., Hoogmoed W.B., Oenema O. 2007: Development in conservation

Wang Z.C., Gao H.W. (2004): Conservation Tillage and Sustainable Farming. Agricultural Science and Technology Press, Beijing.

West, T.O., Marland, G., 2002. A synthesis of carbon sequestration, carbon emissions, and net carbon flux in agriculture: comparing tillage practices in the United States. *Agric. Ecosyst. Environ.* 91: 217–232.

Willcocks, T.J. 1988. Tillage systems for sustainable rainfed crop production in semi-arid regions of developing countries. In: *Challenges in Dryland Agriculture.* P.W. Unger *et al.* (eds.) pp. 131-133. Proc. Int. Confer. on Dryland Agriculture, Amarillo/Bushland, Texas, August 1988.

Permissions

The contributors of this book come from diverse backgrounds, making this book a truly international effort. This book will bring forth new frontiers with its revolutionizing research information and detailed analysis of the nascent developments around the world.

We would like to thank Dr. Peeyush Sharma and Dr. Vikas Abrol, for lending their expertise to make the book truly unique. They have played a crucial role in the development of this book. Without their invaluable contribution this book wouldn't have been possible. They have made vital efforts to compile up to date information on the varied aspects of this subject to make this book a valuable addition to the collection of many professionals and students.

This book was conceptualized with the vision of imparting up-to-date information and advanced data in this field. To ensure the same, a matchless editorial board was set up. Every individual on the board went through rigorous rounds of assessment to prove their worth. After which they invested a large part of their time researching and compiling the most relevant data for our readers. Conferences and sessions were held from time to time between the editorial board and the contributing authors to present the data in the most comprehensible form. The editorial team has worked tirelessly to provide valuable and valid information to help people across the globe.

Every chapter published in this book has been scrutinized by our experts. Their significance has been extensively debated. The topics covered herein carry significant findings which will fuel the growth of the discipline. They may even be implemented as practical applications or may be referred to as a beginning point for another development. Chapters in this book were first published by InTech; hereby published with permission under the Creative Commons Attribution License or equivalent.

The editorial board has been involved in producing this book since its inception. They have spent rigorous hours researching and exploring the diverse topics which have resulted in the successful publishing of this book. They have passed on their knowledge of decades through this book. To expedite this challenging task, the publisher supported the team at every step. A small team of assistant editors was also appointed to further simplify the editing procedure and attain best results for the readers.

Our editorial team has been hand-picked from every corner of the world. Their multi-ethnicity adds dynamic inputs to the discussions which result in innovative outcomes. These outcomes are then further discussed with the researchers and contributors who give their valuable feedback and opinion regarding the same. The feedback is then collaborated with the researches and they are edited in a comprehensive manner to aid the understanding of the subject.

Apart from the editorial board, the designing team has also invested a significant amount of their time in understanding the subject and creating the most relevant covers. They scrutinized every image to scout for the most suitable representation of the subject and create an appropriate cover for the book.

The publishing team has been involved in this book since its early stages. They were actively engaged in every process, be it collecting the data, connecting with the contributors or procuring relevant information. The team has been an ardent support to the editorial, designing and production team. Their endless efforts to recruit the best for this project, has resulted in the accomplishment of this book. They are a veteran in the field of academics and their pool of knowledge is as vast as their experience in printing. Their expertise and guidance has proved useful at every step. Their uncompromising quality standards have made this book an exceptional effort. Their encouragement from time to time has been an inspiration for everyone.

The publisher and the editorial board hope that this book will prove to be a valuable piece of knowledge for researchers, students, practitioners and scholars across the globe.

List of Contributors

Simone Graeff, Johanna Link, Jochen Binder and Wilhelm Claupein
University Hohenheim, Crop Science (340a), Germany

Genesis T. Yengoh
Department of Earth and Ecosystem Sciences Division of Physical Geography and Ecosystem Analysis Lund University Sölvegatan 12, SE-223 62 Lund, Sweden

Sara Brogaard and Lennart Olsson
Lund University Centre for Sustainability Studies Geocentrum 1, Sölvegatan 10, Lund, Sweden

Ali Reza Kiani
Department of Agricultural Engineering, Agricultural and Natural Resources Research Centre of Golestan Province, Gorgan, Iran

Fariborz Abbasi
Agricultural Engineering Research Institute, AERI, Karaj, Iran

J. Mzezewa
Department of Soil Science, University of Venda, South Africa

E.T. Gwata
Department of Plant Production, University of Venda, South Africa

Paxie W. Chirwa
University of Pretoria, Faculty of Natural & Agricultural Sciences, South Africa

Ann F. Quinion
45 Hanby Ave Westerville, USA

Deivanai Subramanian
Department of Biotechnology, Faculty of Applied Sciences, AIMST University, Malaysia

M. M. Buri, R. N. Issaka and J. K. Senayah
CSIR - Soil Research Institute, Academy Post Office, Kwadaso- Kumasi, Ghana

H. Fujii
Japan International Research Center for Agricultural Sciences (JIRCAS), Japan

T. Wakatsuki
Faculty of Agriculture, Kinki University, Nara, Japan

Mohamed S. Alhammadi
Research & Development Division, Abu Dhabi Food Control Authority, UAE

Shyam S. Kurup
Faculty of Food and Agriculture, United Arab Emirates University, United Arab Emirates

Melkamu Jate
Research Centre Hanninghof, Yara International, Hanninghof, Duelmen, Germany

Pilar Mazuela and Elizabeth Bastias
Universidad de Tarapacá, Chile

Miguel Urrestarazu
Universidad de Almería, Spain

Susana B. Rosas, Nicolás A. Pastor, Lorena B. Guiñazú, Javier A. Andrés, Evelin Carlier, Verónica Vogt, Jorge Bergesse and Marisa Rovera
Laboratorio de Interacción Microorganismo–Planta. Facultad de Ciencias Exactas, Físico-Químicas y Naturales, Universidad Nacional de Río Cuarto, Campus Universitario. Ruta 36, Km 601, (5800) Río Cuarto, Córdoba, Argentina

O.M. Adesope and V.C. Ugwuja
Department of Agricultural Economics and Extension, University of Port Harcourt, Nigeria

E.C. Matthews-Njoku
Department of Agricultural Extension, Federal University of Technology, Owerri, Nigeria

N.S. Oguzor
Federal College of Education (Technical), Omoku, Nigeria

Moses Imo
Chepkoilel University College, Moi University, Kenya

Mohammed Shu'aibu Abubakar
Department of Agricultural Engineering Bayero University, Kano, Nigeria

Peeyush Sharma and Vikas Abrol
Dryland Research Sub Station, Rakh Dhiansar, SKUAST- Jammu 181 1 33, India

Printed in the USA
CPSIA information can be obtained
at www.ICGtesting.com
JSHW011454221024
72173JS00005B/1076